Life Cycle Assessment
A Metric for the Circular Economy

Life Cycle Assessment
A Metric for the Circular Economy

Edited by

Aiduan Borrion
University College London, UK
Email: a.borrion@ucl.ac.uk

Mairi J. Black
University of Surrey, UK
Email: m.black@surrey.ac.uk

and

Onesmus Mwabonje
Imperial College London, UK
Email: o.mwabonje@imperial.ac.uk

ROYAL SOCIETY
OF CHEMISTRY

Print ISBN: 978-1-78801-445-8
PDF ISBN: 978-1-78801-620-9
EPUB ISBN: 978-1-83916-136-0

A catalogue record for this book is available from the British Library

The Royal Society of Chemistry is a charity, registered in England and Wales, Number 207890, and a company incorporated in England by Royal Charter (Registered No. RC000524), registered office: Burlington House, Piccadilly, London W1J 0BA, UK, Telephone: +44 (0) 20 7437 8656.

Visit our website at www.rsc.org/books

Printed in the United Kingdom by CPI Group (UK) Ltd, Croydon, CR0 4YY, UK

Foreword

N. D. MORTIMER

Email: n.d.mortimer@btinternet.com

The scene is set for understanding the potential role of life cycle assessment as a metric for circular economy in Chapter 1 of this book, which introduces the theories and tools for assessing the impact of human activities on resources and the environment. The increasing magnitude of these impacts is traced to the growth of the human population and the continued industrialisation of human society. Ongoing responses, largely based on economic, social and environmental sustainability, are summarised, with those concluding the need for circular economy as a means of extending resources and reducing environmental damage, being highlighted. The importance of quantifying progress with the transition from linear to circular economy is clearly established. The relevance of life cycle assessment as a suitable tool for such quantification is explored through its historical development and its links with theories concerning planetary carrying capacity and boundaries, industrial ecology and cleaner production. Whilst demonstrating such relevance, notable methodological challenges are raised regarding the consistent and transparent application of full life cycle assessment, which covers all resource and environmental impacts, to circular economy.

The foundations for appreciating the fundamental aspects of circular economy are laid in Chapter 2. The negative impacts of existing linear economy are contrasted with the potential benefits of circular economy. However, despite these well-known attractions, the current slow progress with implementing circular economy is illustrated. Although the many definitions of circular economy are acknowledged, the most helpful are recognised as those that apply practical perspectives to maximising utility, whilst creating value from the cycling of resources. Key principles are identified,

Life Cycle Assessment: A Metric for the Circular Economy
Edited by Aiduan Borrion, Mairi J. Black and Onesmus Mwabonje

Published by the Royal Society of Chemistry, www.rsc.org

which entail adopting cradle-to-cradle approaches in the provision of products and services, taking into account loops in what are referred to as technical and biological materials. Important aspects of circular economy are summarised for the specific industries addressed by case studies in subsequent chapters. Tools for assessing circular economy are introduced, including materials flow analysis, environmental extended input-output analysis and life cycle assessment. It is noted that, in order to provide adequate evaluation of circular economy, such tools must often be used in combination with careful consideration of their contexts, systems boundaries, metrics, indicators, documented data sources and methodological limitations. Three principles are described for business models, which are intended to untap opportunities for value creation from circular economy. Stages for implementing circular economy in businesses are summarised and examples of policy intervention in the European Union and China are provided. The chapter concludes by outlining some of the conditions that will promote and challenge the widespread adoption of circular economy.

An essential introduction to the application of life cycle assessment in the implementation of circular economy is given in Chapter 3. To those who do not work with or encounter life cycle assessment on a regular basis, this technique can sometimes seem to be an impenetrable specialism. This may be largely due to the use of its own dedicated terminology. It is, of course, important to understand the specific terms that are used in life cycle assessment to appreciate the meaning of its results. However, it is helpful to appreciate that life cycle assessment is based on regarding any individual activity or collection of activities as a system in which there are flows of inputs and outputs. Once such systems thinking is embraced and understood, much of the apparent complexities of life cycle assessment should become more accessible. Amongst the four distinct stages of any life cycle assessment exercise, it can be argued that the most crucial is the first. This is because, as well as identifying the system under investigation, this first stage should involve specifying, in a clear, coherent and comprehensive manner, the purpose of the investigation which, in turn, determines all aspects of the methods of calculation and analysis subsequently used. In doing so, this ensures that the actual results answer the particular question being asked. Subsequent stages consist of producing reliable estimates of the inputs and outputs of the system, translating inputs and outputs into indicators of impact, and, finally, interpreting results for communication and eventual use in decision-making. An extensive summary of the variety of international and national standards and guidelines for the life cycle assessment of products and organisations is provided. Possible extension of life cycle assessment from resource and environmental considerations to economic and social concerns is outlined. Practical challenges in applying life cycle assessment to circular economy are addressed, especially in relation to potential trade-offs between environmental impacts and materials circularity.

Both concrete and steel are very important components in the construction of buildings and infrastructure. They have contrasting features, as explained in Chapter 4, which provides a life cycle assessment case study of two designs for an office building in France: one constructed mainly in concrete (with reinforced steel) and one mainly in steel (with concrete foundations and floors). By unit weight, the manufacture of concrete generally has lower impacts, particularly in terms of fossil fuel consumption and greenhouse gas emissions, than the production of steel. However, in the different designs, a lower weight of steel is required relative to the weight of concrete. The combination of these two factors is shown to give a slight advantage for the steel design over the concrete design. As well as the construction phase, the use phase and end-of-life phase, including the potential for materials recovery and recycling, of these designs are evaluated in the case study. Both designs have the same ongoing energy demands and satisfy the same environmental regulations, and, as is typical currently, the use phase dominates the magnitude of impacts in the case study. The role of Environmental Certification of Buildings is considered, especially in relation to reducing operational energy demands. Future widespread deployment of passive, near zero and net zero energy building designs is expected to reduce the importance of the use phase, thereby elevating the relevance of the construction phase in the life cycle assessment of buildings. Hence, Environmental Product Declarations, which incorporate information from life cycle assessment of specific building components, should become increasingly important as a means to reduce the impacts of the construction phase. However, the limitations of current approaches that are unable to accommodate other concerns, including social value, are emphasised.

The construction industry will play an important part in establishing the infrastructure necessary for the transition to the net zero carbon economy. Hence, it is necessary to determine the magnitude of possible reductions in the greenhouse gas emissions associated with the main materials, such as cement and steel, consumed by this industry. The use of life cycle assessment in evaluating greenhouse gas emissions savings in the production of cement, aggregates and concrete is explored, using case studies from Hong Kong, in Chapter 5. Significant potential for reducing greenhouse gas emissions, along with savings in non-renewable energy consumption, are reported by adopting the principles of circular economy in the relevant production processes. This involves considering alternatives to clinker based on limestone, sand and iron ore, in the form of fly ash and waste glass powder, alternative fuel to coal, in the form of co-fired pellets made from waste wood, alternatives to natural aggregates, in the form of construction and demolition wastes and waste glass sand, and alternative supplementary cementitious materials, in the form of fly ash, granulated blast furnace slag and silica fume. The importance of avoiding landfilling, as a particularly local concern, is emphasised and taken into account in these life cycle assessments. It is recognised that realisation of estimated benefits by the construction industry is only developing slowly due to the lack of necessary methodologies, guidelines and indicators.

Chapter 6 gives a closer look at the application of life cycle assessment to the use of waste glass powder as a supplementary cementitious material in concrete. This is achieved by means of a case study involving the provision of a pedestrian bridge deck over a 100-year period in Canada. As such, the use of waste glass powder is an example of open-loop circular economy and life cycle assessment is adopted to derive indicators of resource and environmental impacts of new concrete mixes relative to conventional concrete. In generating appropriate results for meaningful comparison, the need to specify the essential details of life cycle assessment carefully is clearly emphasised and suitably demonstrated. In particular, it is established that a whole range of environmental impacts, and not just global warming potential, must be taken into consideration to avoid possible trade-offs between different burdens. Additionally, locational and technological variations may have to be addressed. Furthermore, the choice of any co-product allocation must be relevant and properly justified. Most critically, the correct functional unit must be selected. In this specific case study, in which different concrete mixes have different properties of strength and durability, the lifespan of the bridge deck, rather than the volume of concrete or the structure of an individual bridge, becomes an important element of the functional unit. This has a fundamental influence on the results, which are tested for robustness using sensitivity and uncertainty analyses, allowing subsequent outcomes to be qualified accordingly.

Chapter 7 focuses on the application of life cycle assessment to specific polymer materials and packaging, in the form of food trays, which incorporate starch derived from biomass feedstocks. Two case studies are reported, which demonstrate the need to consider quite a large variety of factors in the life cycle assessment of new materials and products. This is mainly because polymers produced from both biomass feedstocks and petrochemicals have the potential to exhibit circular economy regarding possible recycling of biogenic and fossil carbon, respectively. It is shown that whether such potential can be achieved depends crucially on the recyclability of the packaging, which can affect the practical choice of end-of-life treatment. At present, some biopolymers cannot be recycled, unlike a number of conventional plastics derived from petrochemicals. This can tip the balance of comparative environmental benefits away from certain biopolymers and towards their established petrochemical alternatives. However, other factors are capable of swinging the balance back the other way. In the case studies examined, the particular type of biomass feedstock used to produce biopolymers and the specific technique for forming packaging trays from biopolymers are found to exert important influences on the comparative outcomes from life cycle assessment.

Both the complexity and rapidly evolving nature of the textiles industry present important difficulties for its successful transition to circular economy. These difficulties are compounded by increased pre- and post-consumer waste generation due to traditionally short product lifetimes, which are being exacerbated by the growth of fast fashion. Nevertheless, such transition is regarded as a possible solution for an industry that can be viewed as unsustainable, not only environmentally but also economically. Chapter 8 demonstrates the complexity of this industry by reviewing the wide variety of processes involved in producing textiles from natural and synthetic fibres. The general results of life cycle assessment studies are summarised to present a picture of the daunting array of resource and environmental impacts from textile production, use and end-of-life phases. A specific case study is used to demonstrate the practical work involved in determining the impacts of polyethylene terephthalate yarn production in Turkey. The role of life cycle assessment is emphasised as a means of providing the necessary information for environmental performance labels which can initiate and maintain wider application of circular economy to textiles.

Processing of a range of different biomass feedstocks in biorefineries provides the means to create many different chemicals and/or fuels. If these chemicals, in particular, can be converted into plastics, which are equivalent to those currently produced from fossil fuels, an intriguing window of opportunity is opened for establishing a circular economy in biogenic carbon, which will be needed by a world tackling global climate change seriously within the short timescale to 2050. The possibility of achieving this is investigated in Chapter 9, which adopts inventory analysis from life cycle assessment to determine the annual mass balance of biogenic carbon between global sources of relevant biomass feedstocks and their potential conversion into chemicals and fuels in biorefineries. Very promising results are reported, especially for chemicals. However, caution is also noted regarding the practicalities of realising such promise. Specifically, further development is required to design bioplastics with use and end-of-life phases similar to those of existing products derived from petrochemicals, so that loops can be closed through present waste management practices. Additionally, further research is needed to exploit all the biogenic carbon components available in biomass feedstocks and to resolve the currently unfavourable economics of future biorefineries.

There is a well-established focus on the resource and environmental impacts of packaging, which is mainly due to its routine visibility to almost all members of society. This has resulted in a long history of applying life cycle assessment to packaging and its potential circular economy. Hence, Chapter 10 is able to draw on an extensive body of literature in the form of case studies covering many types of packaging from across the world. This provides the basis to identify key issues and challenges associated with the application of life cycle assessment to the circular economy of packaging. In particular, it is apparent that there is ongoing debate about the definition of appropriate functional units and the establishment of suitable systems boundaries. Packaging may have more than one function since it is not only an outer layer to a given product but is also a potential source of recyclable material. Repeated recycling of packaging materials with possible degradation introduces the need to consider temporal as well as spatial systems boundaries especially in dynamic rather than static circular economies. Using numerous examples, it is explained that these and other matters are not simply methodological problems for life cycle assessment practitioners. This is because effective means are required to enable any comparative resource and environmental benefits from the circular economy of packaging identified by life cycle assessment to be translated into product design, to influence business organisation and to be communicated to consumers.

Valorisation, or the re-classification and transformation of previously regarded wastes into useful resources, is the main subject of Chapter 11, which focuses on agricultural crop production. In its current form, agriculture is viewed as unsustainable due to its use of non-renewable energy and minerals, and its unacceptable environmental impacts. Improvements in efficiency are needed by adopting the principles of circular bioeconomy, which involves turning former wastes into more valuable co-products and by-products. Life cycle assessment can assist this process by evaluating the options available and by identifying those which result in the lowest resource and environmental impacts and/or the greatest economic and social benefits. It is proposed that this requires an approach based on different levels of detail in life cycle assessment, which reflects the level of maturity of the technologies needed to transform agricultural wastes into suitable feedstocks. The potential of this approach is demonstrated using five case studies from America, Asia and Europe, which exhibit features of linear systems, and open- and closed-loop circular systems. General principles are drawn for applying the circular economy to food production systems. Limitations are also identified with the application of life cycle assessment which is currently unable to capture the restoration and regeneration of natural systems.

Some of the complexities in evaluating the impacts of activities on a global scale illustrated in Chapter 12, which considers livestock production. The main challenges are identified as food security and climate change with a growing global population increasing the demand for meat production. The context of livestock production and its impacts are illustrated by case studies based on beef production in Brazil and dairy production in Kenya. The many different sources of greenhouse gas emissions that have to be taken into account for both local and global systems modelling are outlined. These sources include enteric emissions from the animals themselves, emissions from their manure, net carbon dioxide emissions or absorption from soil and biomass caused by land use changes, and greenhouse gas emissions associated with the use of fossil fuels in the processes involved in producing and delivering food for eventual consumption. Since competition for land for different uses is important, such modelling must accommodate not only livestock production but also alternatives such as crop and bioenergy production. Additionally, the behaviour and preferences of consumers must be addressed. Variations in these and other factors result in different outcomes, as exemplified by estimated global greenhouse gas emissions from livestock production up to 2050. In particular, these estimates are shown to be influenced most strongly by levels of daily meat consumption by individuals, the choice of type of meat and livestock yields.

Bioenergy, as a potentially renewable form of energy based on the use of biomass as a feedstock, can be regarded as an example of the circular economy applied to the recycling of biogenic carbon. In the case of fuels made from such feedstocks, this is because the carbon dioxide released during their combustion will have been absorbed during the original growth of the biomass. However, this does not necessarily mean that, in practice, bioenergy is carbon neutral. There are many possible sources of greenhouse gases that have to be taken into account when considering the production and use of biofuels. Life cycle assessment is ideally suited to evaluating the relevance and magnitude of these sources in the essentially closed loop circular biogenic economy of bioenergy systems. The synergy of the methodology of life cycle assessment and the principles of circular economy to maximise resource efficiency and minimise environmental impacts is demonstrated in Chapter 13, which presents five case studies of biofuel production and use. Overall, these case studies illustrate the importance of determining significant negative and positive factors on the calculation of total greenhouse gas emissions. Negative factors include greenhouse gas emissions from the fossil fuels, fertilisers, chemicals and other products used in the provision and conversion of biomass feedstocks, leakages due to the burning of agricultural wastes, direct land use change which occurs when land is converted to biomass feedstock cultivation, and indirect land use change which might arise when cultivation of a biomass feedstock displaces an existing crop to another location. Positive factors include the use of agricultural residues as biomass feedstock and the co-production of other outputs from the biomass conversion process, which substitute for products previously made from and/or with fossil fuels.

The necessarily rapid transition from current energy systems based on fossil fuels to those depending on renewable energy technologies and their supporting storage systems will have a significant effect on the demand for relevant materials. In addressing this important consideration, Chapter 14 shows clearly that transforming energy systems from fossil fuels to renewables will not so much alleviate concerns over depletion, but will rather switch attention from resources of fossil fuels to the availability of metals. This is achieved by estimating the metal requirements up to 2050 of five energy scenarios based on assumptions about the mix of different renewable energy technologies and battery storage options. Two methods are used to derive depletion scores in terms of life cycle impact characterisation. This enables critical metals to be identified and possible options for avoiding problems to be outlined. Such options include the possible use of alternative storage methods, the direct substitution of critical metals, and the potential for metal recovery and recycling by introducing circular economy to the life cycle of renewable energy systems. Additionally, it is noted that competition for other uses of critical metals is a relevant factor that has to be taken into account.

Beyond the details of specific case studies and their particular conclusions there is a wider debate about the concept of circular economy and the application of life cycle assessment. This necessary debate, which challenges any possible nebulous thinking over the claimed benefits of circular economy and their apparently precise quantification by life cycle assessment, is thought-provokingly addressed in Chapter 15. Areas of fundamental problems are identified with reference to strong *versus* weak sustainability, relative *versus* absolute sustainability, incorporation or separation of humans and the environment, and negative and positive interpretations of sustainability. Unless properly resolved, some of the underlying issues that are revealed have the potential to undermine the practical realisation of any actual environmental, economic and social benefits from circular economy. Such resolution recognises that, at this particularly critical time, the most appropriate concepts and tools must be adopted and adapted to support serious decision-making. It is argued that life cycle assessment, as a tool to assist implementation of the concept of circular economy, needs to evaluate absolute, rather than relative, sustainability with regard to unbreachable planetary boundaries as part of its impact assessment. Additionally, economic aspects need to be incorporated into life cycle assessment to demonstrate whether circular economy has economic benefits and, if not, what alternative economic systems are required to ensure they can be realised. Ideally, an encompassing relationship is envisaged between the inspiring vision of circular economy and the thorough testing of its environmental and economic outcomes by life cycle assessment.

Preface

By the end of the 20th century, concerns about the environmental impact of the ever-increasing global population were already reaching worrying proportions. In the first 20 years of the 21st century, these concerns have only increased with the threat of climate change, loss of habitats and biodiversity, and water scarcity. Other existing and persistent threats include those to our terrestrial, marine and aerial environments, as the result of the specific impacts we see from our increased and unsustainable demand on natural resources and ecosystems, and our management of emissions and waste. Reassessing the way in which we have been utilising resources, understanding the impacts of increased consumerism, and poor management of goods and materials after use, are critical in helping us to re-dress the balance and reduce our impacts for future generations. It is indeed worrying, given the body of evidence that already exists, that we may already be past the tipping point for some of the environmental impacts.

Circular Economy thinking is a conceptual way of managing resources in non-linear, more efficient manner, and there is a need to better understand how the re-use, re-cycling and re-purpose model will impact on the techno- and biospheres proposed in the Circular Economy model. Life Cycle Assessment (LCA) has been in development as a methodology since the late 1960s, to quantify environmental impacts. As the LCA methodology matures, it brings into existence an increasing body of data and methodological approaches across several impactful activities, materials and products. However, this provides a daunting array of approaches to the LCA methodology and despite the development of well-set rules, commencing an LCA study can present an enormous challenge to newcomers and established practitioners alike.

The editors of this book, Dr Aiduan Borrion (University College London), Dr Mairi J. Black (University of Surrey/University College London) and

Life Cycle Assessment: A Metric for the Circular Economy
Edited by Aiduan Borrion, Mairi J. Black and Onesmus Mwabonje

Published by the Royal Society of Chemistry, www.rsc.org

Dr Onesmus Mwabonje (Imperial College London), have worked collaboratively for a number of years and believe that LCA is better informed by such collaboration and stakeholder engagement, to verify data, rationalise the approaches taken and methodologies employed in a transparent manner. In fact, LCA requires that we draw on and share the expertise and knowledge of a number of disciplines and perspectives. This book has brought together a number of international, multidisciplinary experts in LCA, to address a range of products, processes and services, with the aim of relating the use of LCA as a metric for Circular Economy. The diligent and insightful chapters provided by all the contributors to this book are key to providing a practical view on approaches taken in LCA. It is hoped that the book will provide a good reference point and offer guidance in addressing some of the key industries that are so impactful to the natural environment. The editors of the book would like to thank all the chapter authors and contributors for their expert insights.

At the time of writing this preface, the world as we know it has faced a new threat that has impacted populations across the globe. The Covid-19 pandemic has necessitated significant shifts in societal structures and behaviours, as a means of managing the impacts on the human population, but still tragic losses of life have already occurred. We don't know what the future will hold and, for many, the disruptive force of the pandemic has led us to – and will necessitate – new systems of management in healthcare; as well as in the workplace and in transport systems, food and industrial resources supply chains, and in the social activities we may all have taken for granted. To this end, it is hoped that the approaches taken to a global recovery will take a 'grow back better' approach, which will genuinely allow us to address the environmental concerns of 2020, both pre- and post-pandemic.

Aiduan Borrion, Mairi J. Black and Onesmus Mwabonje

Contents

Life Cycle Assessment: A Metric for the Circular Economy
Edited by Aiduan Borrion, Mairi J. Black and Onesmus Mwabonje

Published by the Royal Society of Chemistry, www.rsc.org

CHAPTER 1

Theories and Tools for the Assessment of Environmental Impacts of Human Activities

M. J. BLACK,[a,b] A. L. BORRION*[b] AND O. MWABONJE[c]

[a] Centre for Environment and Sustainability, University of Surrey, Guildford GU2 7XH, UK; [b] Department of Civil, Environmental and Geomatic Engineering, University College London, London WC1E 6BT, UK; [c] Centre for Environmental Policy, Imperial College London, London SW7 2AZ, UK
*Email: a.borrion@ucl.ac.uk

1.1 Introduction

The rapid expansion of the global population in the last three centuries corresponds to our advancements in technology, improved medicines and healthcare, and the provision of goods and services.[1,2] Technological advancements began during the industrial revolutions in the 18th and 19th centuries. The first industrial revolution saw mechanisation and factory production driven by steam (coal). This was followed by the second industrial revolution, which utilised liquid fossil fuels and electricity to increase productivity in heavy and light industries.

Since then, we have seen the advancement of electronics and IT in the third industrial revolution, which has led to the automation of production, and we now stand at the frontier of the fourth industrial revolution, with advancements in digital technologies leading the way forward in unprecedented ways that are not yet fully comprehended.[3]

Life Cycle Assessment: A Metric for the Circular Economy
Edited by Aiduan Borrion, Mairi J. Black and Onesmus Mwabonje

Published by the Royal Society of Chemistry, www.rsc.org

The increase in manufacturing capacity facilitated by our use of fossil energies was seen across a diversity of industries, from the iron and steel used in construction and transport, food production and processing, to consumer goods such as household wares, textiles and clothing. These phases of industrialisation also correspond to growing global concentrations of carbon dioxide (CO_2) and methane (CH_4) in the atmosphere, which can be measured by analysing the air trapped in polar ice over time.[4]

It has been suggested that this period marks the beginning of new geological epoch, termed 'The Anthropocene', which is governed by human activities (Figure 1.1). Whilst the term has not been formally recognised as a 'Geologic Time Scale' by the International Commission on Stratigraphy, it has become widely used to define the era in which we now live, where the impacts of human activities on the local and global environment have become a source of increasing concern to many.[5]

Finding ways in which to quantify the impacts of human activities on the environment has led to the development of several macro-level models and conceptual ways of thinking, however, the complexity of our interactions with the environment and the sheer number of developments in consumer-driven activities, has also led to thinking at a more granular level. In this chapter, we introduce some of the 'big picture' thinking about human impacts on the earth's systems and introduce the concept of the Circular Economy (Chapter 2 of this book will provide a detailed account of Circular Economy thinking). We also introduce ways in which we might understand and manage our impacts at the granular level and introduce Life Cycle Assessment (LCA) as a means of quantifying these impacts (Chapter 3 will provide a more detailed account of LCA methodology and principles). The following chapters in this book will describe case study approaches to LCA, covering a wide range of production systems which have developed to fulfil our needs for food, shelter and consumer products.

1.2 The Increasing Global Population, Potential Outcomes and Environmental Impacts

In 2017, the United Nations reported that the world population of 7.6 billion would be expected to reach 8.6 billion in 2030, 9.8 billion in 2050 and 11.2 billion in 2100.[6] The potential outcome of the increasing human population has been theorised in several different ways. The Malthusian theory[†] proposes simplistically that, as population growth occurs exponentially whilst food production increases linearly, populations will outgrow their resources unless they are 'checked' through preventative checks (avoiding contributing

[†]The Malthusian theory of population growth was presented in 1798 in 'An Essay on the Principle of Population', which was also a social commentary on the economic and political issues of its time. It is considered controversial for many reasons, such as the promotion of neo-Malthusian theory, which advocates means of population control which are considered morally unjustifiable.[7]

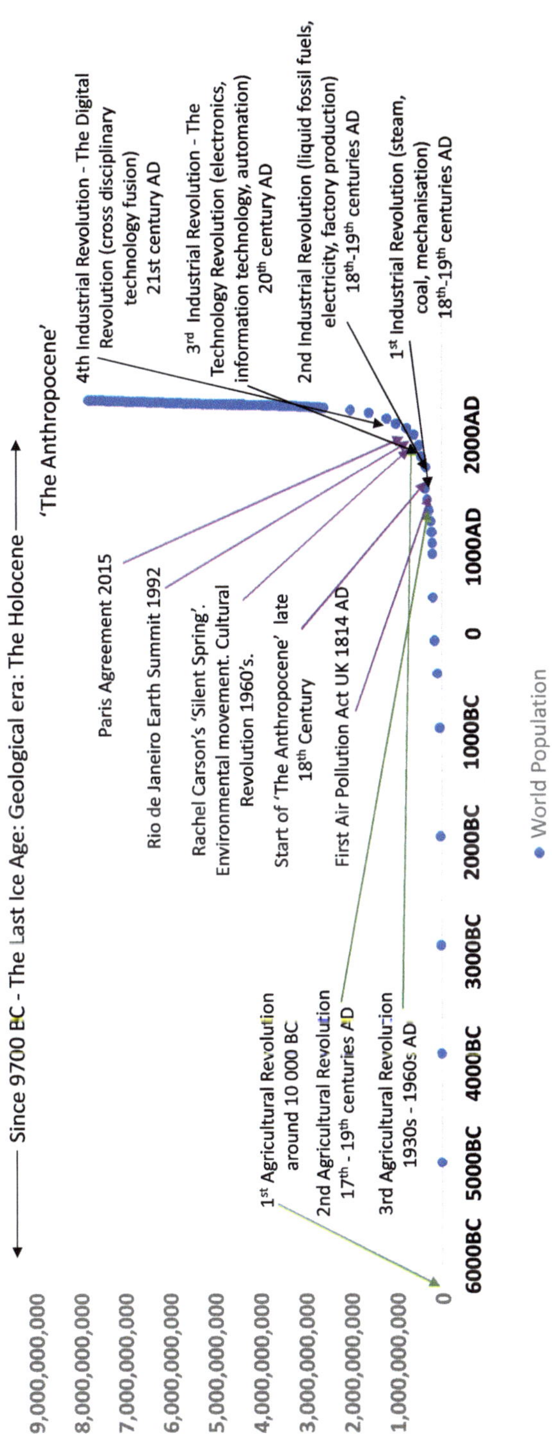

Figure 1.1 World population growth and industrial development.

to population) and positive checks [the reduction (possibly catastrophic) of populations as the result of disease, warfare, famine, and poor living and working conditions].[8] Diamond (2004), has taken a broader outlook, considering the outcome of population increase to be drastic population collapse, as the result of eight 'existing' environmental factors that are impacted by human activities. These are: deforestation and habitat destruction; soil problems (erosion, salinisation and soil fertility losses); water management problems; overhunting; overfishing; effects of introduced species on native species; human population growth and increased per capita impact of people. Four 'recent' environmental factors are: the cumulative impacts of our population growth, activities and needs, *i.e.* human-caused climate change; the build-up of toxic chemicals in the environment; energy shortages, and maximised utilisation of the earth's photosynthetic capacity (or net primary production).[9]

With global population increase, in the simplest terms, more people will require more food, water and shelter and many have feared that famine and poverty will follow. Whilst this is undoubtedly true for some countries, global per capita Gross Domestic Product (GDP)[‡] has grown 10-fold since the 1960s.[10] The relationship between population growth and GDP is highly complex and can be contradictory, and the way in which it manifests in terms of environmental impact is difficult to quantify; however, access to disposable income clearly increases access to goods and services (once basic needs have been fulfilled) and leads to increased consumerism.[11] Larger numbers of people living on the planet has led to increased consumption of resource (land, food, water, air, fossil fuels and minerals) and increased generation of waste and pollution, as the result of our consumption activities.

A report by the United Nations Environment Programme in 2019 assessed global resource use and estimated that global material extraction in 2017 was 92 billion tonnes, increasing from 27 billion tonnes in 1970,[§] *i.e.* more than 3 times the amount used in 1970.[12] The assessment covers biomass (wood, crops and crop residues for food, energy and plant-based materials; grazed biomass; wild fisheries); fossil fuels (coal, gas and oil); metals (iron, aluminium, copper and non-ferrous metals) and non-metallic minerals (sand, gravel and limestone) as well as land (crop and pastureland) and water resources (for agriculture, industries and municipal use) (Table 1.1).

As a result of our use of resources, environmental impacts such as climate change, habitat loss, loss of biodiversity and the impacts of waste and pollution have been part of our everyday conversations and concerns in the late 20th/early 21st century, with climate change at the top of the agenda.[13,¶] Many of the key environmental impacts are inter-related, *e.g.* land use change can

---

[‡]Gross Domestic Product provides an economic snapshot of a country or region, and is used to estimate the size of an economy and its growth rate.

[§]Based on a 50-year material resources database for 191 countries.

[¶]The IPPC report of 2018 suggests that catastrophic climate change will occur unless global temperatures are held below 1.5 °C by 2050.[13]

Table 1.1 Global consumption of natural resources.

Natural resource	Use in 1970	Use in 2017
Metals	2.6 billion tonnes	9.1 billion tonnes
Non-metallic minerals	9 billion tonnes	44 billion tonnes
Fossil fuels	6 billion tonnes	15 billion tonnes
Biomass	9 billion tonnes	24 billion tonnes
Water	2500 km^3 per year	3900 km^3 per year (in 2010)
Land (cropland)	15.2 million km^2 (in 2000)	15.4 million km^2 (in 2010)
Land (pastureland)	31.3 million km^2 (in 2000)	Decreased to 30.9 million km^2 (in 2010)

contribute to climate change, as the conversion of forest to cropland causes a release of carbon stored in trees and the depletion of carbon sinks, reduced absorption of CO_2 may then occur as the result of replacement activities (*e.g.* different crops or livestock systems or, more directly, urbanisation) as well as loss of biodiversity and impacts on water cycles. Climate change, intensification of farming systems and over-farming may then result in desertification, water depletion and so on.[||]

In terms of waste and pollution, it has been estimated that one third of all food produced ends up as waste, due to inefficient consumer and retailer management, or through spoiling as the result of transportation and harvesting practices.[15] According to the United Nations, the food sector accounts for 30% of the world's total energy consumption, 22% of total GHG emissions, and food *waste* equates to 1.3 billion tonnes of food.

It has been reported that solid waste generation (from broken household items, ash, food waste and packaging) is linked to urbanisation and affluence. A study published in 2013 states that in 1900, only 13% of the global population lived in urban settings (200 000 million people) and generated less than 300 000 tonnes of rubbish per day; by 2000, 49% of the global population were urban citizens (2.9 billion people), generating 3 million tonnes of solid waste per day. The same study estimates that by 2025, the amount of solid waste generated by our cities will reach 6 million tonnes per day.[16] In recent years, the drastic impact of our use and mismanagement of the disposal of plastics has been realised and has taken a significant place on the environmental agenda. It has been reported that, since 1950, 8300 million metric tonnes of virgin plastics have been produced to date** (Figure 1.2).[17] Of this, 2500 million tonnes is in use, whilst 6300 million tonnes of plastic waste has

---

[||]It is also noted here that many of these issues are linked to OECD member countries. The distribution of resource use and resulting impacts on climate change is recognised in the expectations of the country-specific Nationally Determined Contribution plans in the UN Paris Agreement for Climate Change.[14]

**Based on global data on production, use and end-of-life fate of polymer resins, synthetic fibers and plastic additives, including: thermoplastics; thermosets; polyurethanes (PUR); elastomers; coatings; sealants and focussing on the most prevalent resins and fibers: high density polyethylene (PE); low density and liner low-density PE; polypropylene (PP); polystyrene (PS); polyvinylchloride (PVC); polyethylene terephthalate (PET); polyurethane (PUR) resins; polyester; polyamide; acrylic fibers.[17]

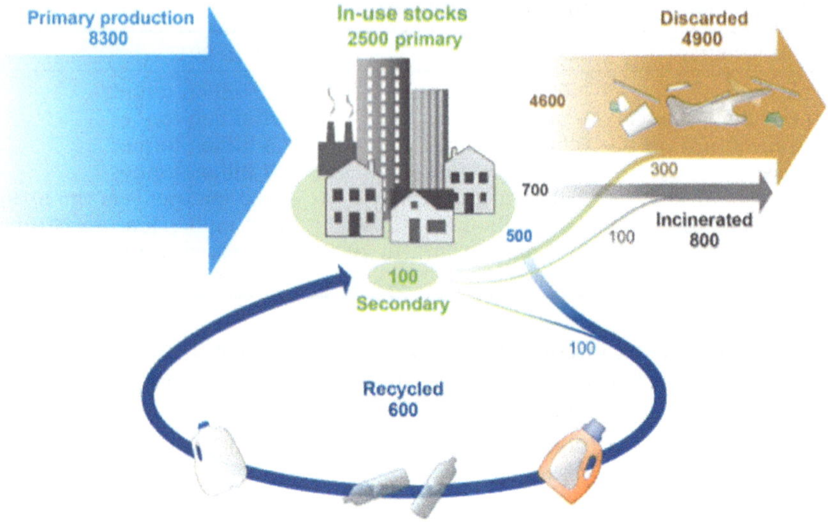

Figure 1.2 Global production, use and fate of polymer resins, synthetic fibers and additives (1950–2015 in million metric tonnes).
Reproduced from ref. 17 with permission from AAAS, Copyright 2017. The authors some rights reserved; exclusive licensee American Association for the Advancement of Science. Distributed under a Creative Commons Attribution NonCommercial License 4.0 (CC BY-NC) http://creativecommons.org/licenses/by-nc/4.0/.

been generated. Of this waste, 600 million tonnes have been recycled, 800 million tonnes have been incinerated and 4900 million tonnes have been discarded, to end up in landfill or the natural environment.[17]

Reports on the amount, and fate, of plastics that end up in the marine and terrestrial environments are of increasing concern, not only because of the direct impact on wildlife as the result of entanglement and digestion, but also due to the widespread infiltration of microplastics into the complex food webs of the land and the oceans, which can also eventually lead to human ingestion.[18,19] A framework has been developed to calculate the proportion of marine debris (solid waste) which originates from land,[20] by calculating the waste generated per capita in 192 coastal countries, the percentage of waste that is plastic and the percentage of plastic waste that is mis-managed and has the potential to enter the ocean. The report states that, of 275 million tonnes of plastic waste generated in 2010, 4.8–12.7 million tonnes has the potential to enter the marine environment.

1.3 Development and Sustainability

Although concerns about the impact of industrialisation on the environment were already being recognised as an outcome of the industrial revolution (leading to the development of the first air pollution Act in the UK in 1814),[21] it could be argued that the 20th century 'environmental movement' really began once our societies began to recover from the devastation of two world wars. Immediately after World War II, food production and manufacturing were being actively encouraged, with agriculture being supported by policies to promote self sufficiency,[††] and technological advancements, *e.g.* the intensification of farming brought about by increased mechanisation and the use of artificial fertilisers and pesticides.[22] However, by the 1960s, publications such as Rachel Carson's *Silent Spring*[23] led to the recognition that modern farming systems[‡‡] and the chemicals used to achieve higher productivity were having significant effects on the surrounding ecosystems.[24] Other social and political issues are synonymous with the environmental, conservation and ecology movements of this era, *e.g.* the anti-nuclear and anti-war movement represented by many of the youth at this time. The period of social change that took place in the 1960s has greatly influenced the next generations, so that climate change and the environment are now an integral part of global thinking. For many, climate change and pressures on the environment are the key challenges for our future generations.[25] Environmental checks and, more broadly, sustainability, are increasingly being integrated into policy and corporate thinking (with varying degrees of actual, practical uptake).

'Sustainability' is defined in the Cambridge dictionary as '1: the quality of being able to continue over a period of time and 2 (environment): the quality of causing little or no damage to the environment and therefore able to continue for a long time'.[26] The concept of 'Sustainable Development' was first defined in the Bruntland report,[27] and recognises the link and need for interaction between economic, social and environmental concerns. Sustainable Development is defined as 'development that meets the needs of the present without compromising the ability of future generations to meet their own needs'. It defines the concepts of 'needs' and 'limitations', addressing the essential needs of the world's poor as a priority, and the limitations that technology and social organisation will have on the environment's ability to meet present and future needs. As globalisation has continued, there has been increasing recognition that these issues cannot be addressed by individual countries, leading to seminal events such as the Rio de Janeiro Earth Summit[§§] in 1992.[28] The Earth Summit led to the development of Agenda 21, which was an action plan for sustainable development for the United Nations, multilateral groups and individual member states. Agenda 21 contains a section on the conservation and management of resources for development (Section II), which includes atmospheric protection; planning and management of land resources; management of fragile ecosystems; the promotion of sustainable agriculture and development; conservation of biological diversity; environmentally sound management of biotechnology; protection of the oceans; protection of the quality and supply of freshwater; management of toxic chemicals, hazardous wastes, solid waste and sewage-related issues, as well and the management of

††In the UK, the 1947 Agriculture Act led to the development of modern agricultural systems.
‡‡Agriculture is recognised as the dominant cause of current species loss.[24]
§§Formally, the United Nations Conference on Environment and Development (1992).[28]

Table 1.2 The Millennium Development Goals.

Goal 1	Eradicate Extreme Poverty and Hunger
Goal 2	Achieve Universal Primary Education
Goal 3	Promote Gender Equality and Empower Women
Goal 4	Reduce Child Mortality
Goal 5	Improve Maternal Health
Goal 6	Combat HIV/Aids, Malaria and Other Diseases
Goal 7	Ensure Environmental Sustainability
Goal 8	Develop a Global Partnership for Development

radioactive waves. Agenda 21 also addresses Social and Economic dimensions (Section I), such as co-operation to accelerate sustainable development; combating poverty; changing consumption patterns and identifies 'Major Groups and their Roles' (Section III), *i.e.* women, children and youth, indigenous people, workers and so forth.

Section IV of Agenda 21 addresses the means of implementing the challenges laid out in the Sections of the report, and led to the development of specific targets described in the Millennium Development Goals (MDGs).[29] The MDGs were eight challenges that were to be met by 2015, in order to ensure a sustainable future for the planet (Table 1.2).

Since then, the United Nations Conference of Parties (COPs) have been held annually, moving forward on such issues as climate change, with the Paris Agreement[30] and further development of the MDGs in Agenda 2030.[31] Agenda 2030 now describes seventeen challenges, called the Sustainable Development Goals (SDGs), which are intended to be guiding principles and targets towards which all countries and stakeholders will work to achieve the equitable balance across countries for the three dimensions of sustainable development (*i.e.* economic, social and environmental). The seventeen SDGs have been defined (Table 1.3) and SDG 12 is highlighted in the context of this book. SDG 12 presents just some of the challenges faced in the development of a Circular Economy, and the means to quantify the impacts of changing activities, which will be applicable and practical for the sustainability of the planet and future generations.

1.4 What is Life Cycle Assessment

Life Cycle Assessment (LCA) is a cradle-to-grave analysis tool, used to evaluate the environmental aspects and potential impacts associated with all the stages of the production, life and disposal of a product, system or service. An LCA typically includes raw material extraction, processing, manufacturing, distribution, use, reuse and disposal (*i.e.* cradle-to-grave). LCA can be utilised as a comparative tool for evaluating potential environmental impacts between different products, systems or services. It can be used to highlight opportunities to improve the environmental performance of products, systems and services, at various points in their life cycles, as well as informing decision-makers for effective strategic planning, priority

Table 1.3 Agenda 2030 Sustainable Development Goals.

Goal 1	End poverty in all forms everywhere
Goal 2	End hunger, achieve food security and improved nutrition and promote sustainable agriculture
Goal 3	Ensure heathy lives and promote well-being for all at all ages
Goal 4	Ensure inclusive and equitable quality education and promote lifelong learning opportunities for all
Goal 5	Achieve gender equality and empower all women and girls
Goal 6	Ensure availability and sustainable management of water and sanitation for all
Goal 7	Ensure access to affordable, reliable, sustainable and modern energy for all
Goal 8	Promote sustained, inclusive and sustainable economic growth, full and productive employment and decent work for all
Goal 9	Build resilient infrastructure, promote inclusive and sustainable industrialisation and foster innovation
Goal 10	Reduce inequality in and among countries
Goal 11	Make cities and human settlements inclusive, safe, resilient and sustainable
Goal 12	Ensure sustainable consumption and production patterns
12.1	Implement the 10-Year Framework of Programmes on Sustainable Consumption and Production Patterns, all countries taking action, with developed countries taking the lead, taking into account the development and capabilities of developing countries
12.2	By 2030, achieve the sustainable management and efficient use of natural resources
12.3	By 2030, halve per capita global food waste at the retail and consumer levels and reduce food losses along the production and supply chains, including post-harvest losses
12.4	By 2020, achieve the environmentally sound management of chemicals and all wastes throughout their life cycle, in accordance with agreed international frameworks, and significantly reduce their release to air, water and soil in order to minimize their adverse impacts on human health and the environment
12.5	By 2030, substantially reduce waste regeneration through prevention, reduction, recycling and re-use
12.6	Encourage companies, especially large and transnational companies, to adopt sustainable development and lifestyles in harmony with nature
12.7	Promote public procurement practices that are sustainable, in accordance with national policies and priorities
12.8	By 2030, ensure that people everywhere have the relevant information and awareness for sustainable development and lifestyles in harmony with nature
12.a	Support developing countries to strengthen their scientific and technological capacity to move towards more sustainable patterns of consumption and production
12.b	Develop and implement tools to monitor sustainable development impacts for sustainable tourism that creates jobs and promotes local cultures and products
12.c	Rationalise inefficient fossil-fuel subsidies that encourage wasteful consumption by removing market distortions, in accordance with national circumstances, including by restructuring taxation and phasing out those harmful subsidies where they exist, to reflect their environmental impacts, taking fully into account the specific needs and conditions of developing countries, and minimizing the possible adverse impacts on their development in a manner that protects the poor and the affected communities

Table 1.3 (*Continued*)

Goal 13	Take urgent action to combat climate change and its impacts[a]
Goal 14	Conserve and sustainably use the oceans, seas and marine resources for sustainable development
Goal 15	Protect, restore and promote sustainable use of terrestrial ecosystems, sustainably manage forests, combat desertification, and halt and reverse land degradation and halt biodiversity loss
Goal 16	Promote peaceful and inclusive societies for sustainable development, provide access to justice for all and build effective, accountable and inclusive institutions at all levels
Goal 17	Strengthen the means of implementation and revitalise the Global Partnership for Sustainable Development (including finance; technology; capacity building; trade; and systemic issues in policy and institutional coherence multi-stakeholder partnerships and date monitoring and accountability)

[a]Acknowledging that the United Nations Framework Convention on Climate Change is the primary international, intergovernmental forum for negotiating the global response to climate change.

setting or re-design of products or associated processes. LCA methodology can also be used as a tool to assist with the selection of appropriate indicators of environmental performance and ecolabelling marketing. The international standards ISO 14040:2006[32] and ISO 14044:2006,[33] provide the principles, framework and guidelines needed to conduct and report an LCA. The European Platform on Life Cycle Assessment has built on these standards to produce a series of comprehensive documents, collectively called the International Life Cycle Data Systems Handbooks (ILCD Handbooks)[34] which provide additional interpretation and guidance on each aspect of carrying out LCA. The ILCD Handbooks are essential references for those who seek in-depth understanding of each aspect of the processes involved in developing and carrying out LCA. Chapter 3 of this book will provide a further overview of the processes involved in LCA methodology.

1.4.1 Early Development of Life Cycle Assessment

Since the 1990s, much attention has been paid to LCA by individuals and organisations in the environmental field. However, the first attempts to look at the impacts of extended product systems can be traced back to the late 1950s and 1960s. The term 'life cycle concept' was first introduced by Novick in 1959,[35] in a report by the RAND Corporation that focused on 'Life Cycle Analysis'[¶¶] of costs associated with the development and end of life of operations for activities under the US budget. Life Cycle Analysis then became a technique used by government bodies to improve budget management, linking the total cost of ownership to functionality.[36]

[¶¶]Life Cycle Analysis is a term that has been used historically to describe the process of examining the life cycle of a product or process. It has widely been replaced by Life Cycle Assessment (LCA) or Life Cycle Inventory Assessment (LCIA).

In 1969, The Midwest Research Institute (MRI), today known as Franklin Associates Inc., developed a methodology know as Resource and Environmental Profile Analysis (REPA). In what is often cited as the first official use of LCA by a commercial organisation, the methodology was used to conduct an analysis on beverage containers for the Coca-Cola company, comparing different containers to determine which one produced the least effects on the use of natural resources and the environment.[37]

LCA was also applied in response to the oil crisis of the early 1970s, as concerns over the limitations of raw materials and energy resources sparked interest in finding ways to cumulatively account for energy use, and to project future resource supplies and use. Both the USA and British governments commissioned industries to conduct extensive energy analysis studies.[38] Although the original emphasis was on the consumption of energy resources, several studies and conceptual workshops lead to a far broader framework, considering all raw material inputs, across the life cycle (*i.e.* associated with manufacture, processing, distribution *etc.*), as well as the associated emissions.[39]

After a decline in interest in these studies in the late 1970s, the second rise of environmental awareness in the late 1980s saw attention focused again on LCA as a potential valuable environmental management tool.[38,39] Since then, a considerable amount of research has been conducted to develop the LCA methodology. Active players in the field of LCA include the Society for Environmental Toxicology and Chemistry (SETAC), the Joint Research Council (JRC) of the European Commission, the United Nations Environment Programme (UNEP) and the International Standards Organization (ISO), to name a few.

LCA has now evolved into a major support tool for decision-making in sustainable product and process design and management, as well as to monitor progress in improving environmental practices. From the beginning of LCA, with its application to making a relatively simple technical choice, *i.e.* choosing one material over another in relation to packaging, over time, there has been a shift in LCA thinking to more encompassing questions, such as the benefits of replacing fossil-based materials with biofuels and biochemicals in the fight against climate change. LCA remains the only existing logical methodology for quantifying the effects of human activities on the environment, from extraction of raw materials to production of a product, to end-of-life disposal. LCA can be applied to many products and services including agricultural systems and practices, industrial processes and waste management, among others, and can be used to assess a number of impact categories.

1.4.2 Life Cycle Assessment Environmental Impact Categories

LCA has developed into a sophisticated tool that allows the quantitative assessment of a number of environmental impacts. LCA methodology is supported by the inputs and understanding of a large number of different

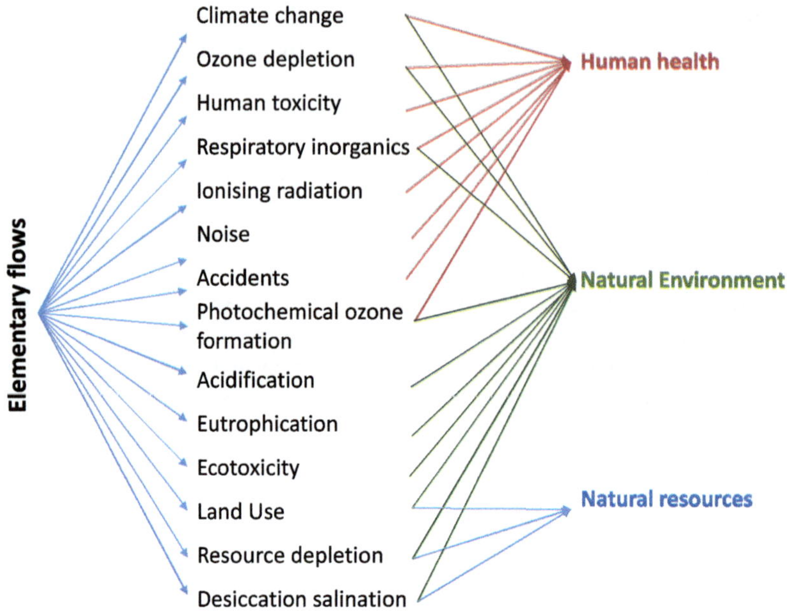

Figure 1.3 Environmental Impact Categories for LCA as defined in ILCD Handbook. Reproduced from ref. 34, https://eplca.jrc.ec.europa.eu/uploads/ILCD-Handbook-LCIA-Framework-Requirements-ONLINE-March-2010-ISBN-fin-v1.0-EN.pdf under the terms of a CC-BY 4.0 license, https://creativecommons.org/licenses/by/4.0/.

disciplines, bringing together the engineers, scientists and economists from the spectrum of industries that support our daily lives, with the natural and social scientists, who are involved in developing our understanding of all aspects of our interactions with the natural world.[‖] The stakeholder approach to LCA is essential to bettering our understanding of environmental impacts and assessment methodologies and over time a number of methodologies*** have been developed that allow the quantification of these impacts. Broadly, the environmental impact categories are similar across the methodologies, and their relevance has been assessed and reported in the ILCD Handbook.[34] The mid-point environmental impact categories are related to 'Areas of Protection', or the end-point outcome of an environmental impact, *i.e.* on human health, the natural environment and/or on natural resources (Figure 1.3). However, there is currently no harmonised, singly recognised approach and methodology for quantifying life cycle impact categories. Chapter 3 of this book will go into more detail about the protocols for carrying out LCA, and the importance of clearly defining the

‖Defined as the technosphere and biosphere in the circular economy concept.

***The ILCD Handbook currently ref. 11 LCIA methodologies, *i.e.* CML 2002; Eco-Indicator 99; EDIP97 and EDIP2003; EPS 2000; IMPACT 2002+; LIME; LUCAS; ReCiPe; Ecological Scarcity Method (Ecopoints 2006); TRACI and MEEuP.[34]

approach taken in carrying out LCA. The need to follow the protocols for carrying out LCA, and for clearly defining the methodology used in making LCA calculations, is re-iterated here.

1.5 Approaches to Understanding the Impacts of Human Activities

Understanding how far the resources of planet earth can be pushed in supporting human life is highly complex, and several macro-models and frameworks have evolved, which look to understand the natural systems of the earth, in the context of human activities. LCA is one tool that has been developed with the intention of quantifying environmental impacts, and can be considered as a metric to inform these macro-models and frameworks. The following section will provide examples of some of the conceptual ways in which we are trying to better understand the impacts of the Anthropocene period. The sections are not intended to provide a detailed review of all of these, as many are presented elsewhere in far greater detail than we can provide in one chapter, but we give an outline of some of the approaches, where LCA can be utilised as a tool in quantifying specifically defined activities within the complexity of human interactions with the environment.

1.5.1 Earth's Carrying Capacity

Earth's Carrying Capacity has been used to define 'the maximum number of individuals of population that the environment can support',[40] considering the resources available in that environment (food, water, habitat) and the level of technology deployed to utilise resources. The concept of carrying capacity has been devised as a means of conceptualising and modelling the capacity to support populations of different species in given environments and, for humans, can be viewed from social and/or environmental perspectives within local, regional or the global framework. Several studies have attempted to model and quantify the maximum human population the earth can support based on various factors, and it has been reported that estimates range from 1 billion to 1000 billion people.[41] A study based on population, biophysical resources[†††] (land) and photosynthetic capacity reports a figure of 11.4 billion people (assuming an annual per capita consumption of 1 million kcal).[41] However, the diversity of the global human population and its varying ability to grow, manage and access food, resources, and additional provisions (access to medical care and sanitation) make this a far more complex prediction. Carrying capacity of the earth can also be considered in terms of social organisation encompassing human choice, economic and cultural conditions

[†††]Building on photosynthetic capacity (*i.e.* kilograms of carbohydrate produced per hectare of land per year, assuming only radiation and temperature as the constraining factors), a selection of crop types that produce carbohydrates for edible consumption (edible calorific value) and land use patterns (based on 2005 figures).

and quality of life (leading to lower predictions of maximum populations). Further studies have considered carrying capacity in terms of human population, ecosystems and 'biocapacity', which specifically considers the impact feedback mechanisms that ecosystems can provide, such the ability to absorb waste, and the limitations of resource provision as the result of carbon sequestration rates.[42] The link between carrying capacity and LCA has been highlighted in studies which aim to link the 'normalisation' stage of LCA across environmental impact categories which can be considered globally and regionally, *e.g.* climate change, ozone depletion and terrestrial acidification.[43,44]

1.5.2 Planetary Boundaries

The planetary boundaries concept is a further iteration of ways in which carrying capacity can be modelled and quantified, to understand the absolute limitations for human activities and the resulting impacts on the environment at the global scale. Planetary boundaries thinking considers planet earth as a self regulating system, which has been relatively stable during the Holocene era, but which is now transitioning into the Anthropocene,[3] as human activities cumulatively impact on the systems which regulate the planet. Earth systems can be considered as the physical, chemical and biological processes (the biogeochemical cycles) that interact between the geosphere, the atmosphere, the hydrosphere and the biosphere, and within which we see the individual cycles that are essential to the planet's function (*e.g.* the carbon cycle; the nitrogen cycle; the water cycle and photosynthesis). The planetary boundaries concept was first defined in 2009,[45,46] and describes nine key earth systems which are seen as critical to maintaining the stable conditions that have allowed humanity to thrive and develop complex civilisations, through agricultural and societal development. The planetary boundaries systems have been considered with spatial and temporal perspectives and are summarised in Table 1.4.[36] Temporal and spatial perspective is also important when we are considering LCA, as it is

Table 1.4 Spatial and temporal consideration of key earth systems in the planetary boundaries concept.

Boundary character		
Scale of process	Processes with global-scale thresholds	Slow processes without known global-scale thresholds
Systematic processes at planetary scale	Climate change Ocean acidification	Stratospheric ozone
Aggregated processes from local/regional scale		Global P and N cycles Atmospheric aerosol Loading Freshwater use Land use change Biodiversity loss Chemical pollution

part of the consideration of scope and systems boundaries, as a key element of the process of carrying out LCA (described in Chapter 3).

Analysis of the nine planetary boundaries systems has also been presented, in the first effort to quantify them and place boundaries beyond which there is an increasing risk of impacts to our current environment before 'deleterious or even catastrophic' impacts might be seen. The planetary boundaries systems define the safe operating space of the Holocene and consider the potential challenges facing us as we move towards the Anthropocene epoch. When the concept was published in 2009, the initial analysis suggested that three of the boundaries are already beyond the safe operating space, *i.e.* biodiversity loss (extinction rate of species), climate change (increased CO_2 levels in the atmosphere) and the nitrogen cycle (N_2 removal from the atmosphere for industrial and agricultural uses). A revision of the planetary boundaries systems and framework, and further quantification of the concept was published in 2015,[47] defining core boundaries for 'safe operating space'; 'zone of uncertainty' (where there will be an increasing risk of adverse impacts); and 'dangerous level' (where there will be a high risk of serious impacts). The new assessment of planetary boundaries is shown in Figure 1.4,[47] demonstrating that we are already in the danger zone for biosphere integrity and biogeochemical flows, and in areas of uncertainty for climate change and land use systems. By defining a planetary boundary on the outer limit of the safe operating space, the

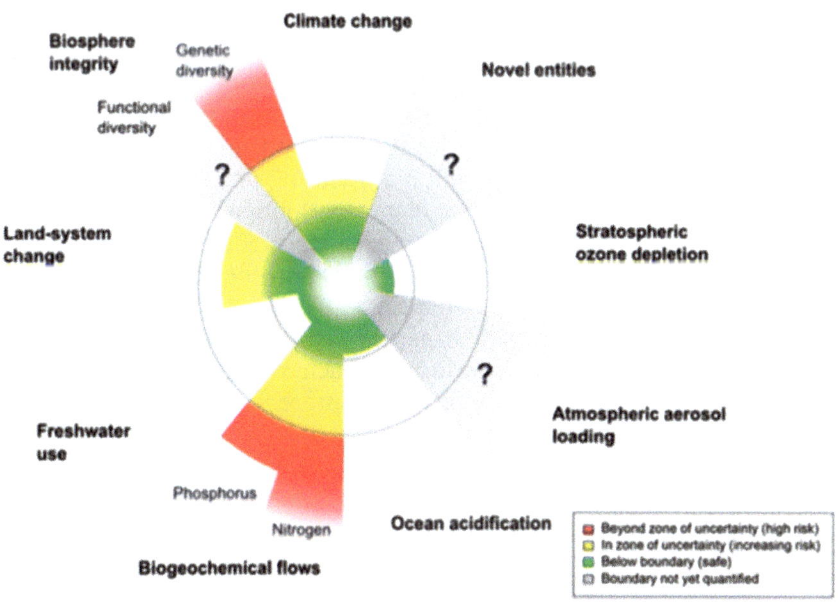

Figure 1.4 Quantification of Planetary Boundaries systems and boundaries for safe operating space to high risk of impacts.
Reproduced from ref. 47 with permission from AAAS, Copyright 2015.

intention is that actions can be taken to address the activities that are contributing to the shift in planetary boundaries systems.

In considering these macro-level concepts, we might also consider how we begin to understand the impacts human activities are having on these systems, as a result of *e.g.* growing food, providing energy and manufacturing products, a well as managing the waste generated by our societies. As seen in Table 1.4, the planetary boundaries concept takes a spatial and temporal approach, so it is possible to relate those systems to different human activities, in different locations, and consider methodologies that have been developed to assess the environmental impact at a more local level.

Environmental footprinting is a methodology that has been presented to quantify the impacts of human activities, linking to planetary boundaries thinking,[48] where both approaches benefit from input and feedback in an environmental assessment framework. In this case, the elements of footprint methodologies are considered in their context and contribution to planetary boundaries thinking, *e.g.* for blue water footprint; carbon footprint; chemical footprint; ecological footprint and grey water footprint.

It has been observed that many of the impacts of products and processes are seen and assessed at the 'local' level but accumulate at the global level (*e.g.* land-use change, P and N cycles). Building on the previous work cited, which relates LCA to carrying capacity,[43] LCA and specifically Life Cycle Impact Assessment[‡‡‡] is presented as a means of quantifying local level impacts that can be linked to the planetary boundaries framework.[49] The concept and limitations of linking the current LCA impact categories (Figure 1.3) to the nine planetary boundaries systems (Figure 1.4) are considered, and generally identify that the main differences and challenges are based on the approach taken for LCA, which considers the specific areas of protection – *i.e.* human health, the natural environment and natural resources – whilst planetary boundaries thinking considers the safe operating earth systems parameters which will maintain the pre-Anthropocene status of the planet. The development of a Planetary Boundaries-Life Cycle Impact Assessment (PB-LCIA) methodology has been reported,[50] and further iterates the links between life cycle impact categories and the control variables defined for planetary boundaries systems.

A recent framework, referred to as the Lyngby Framework,[51] has been developed as a Life Cycle Engineering approach, which aims to address and link the top-down planetary boundaries approach with ground-up assessment of products and processes, the impacts of which can be defined using LCA. The framework is shown in Figure 1.5, to provide a visual representation that conceptualises the pathways and considerations that have to be made, to link LCA to the global planetary boundaries perspective.

[‡‡‡]Life Cycle Impact Assessment will be introduced more fully in Chapter 3, but is defined here briefly as 'understanding and evaluating the magnitude and significance of the potential environmental impacts of a product system throughout the life cycle of a product' (ISO14044).

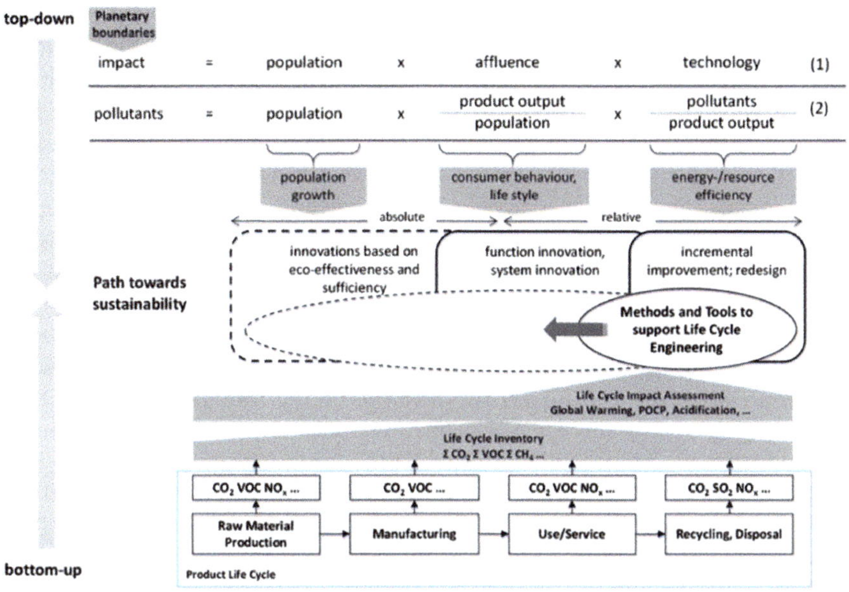

Figure 1.5 The Lyngby Framework.
Reproduced from ref. 51 with permission from Elsevier, Copyright 2017.

There are a number of other worthy approaches which have been developed to address environmental impacts and sustainability in the manufacturing of products and associated processes, *e.g.* industrial ecology; cleaner production; life cycle management; industrial symbiosis and Circular Economy. The following sections in this chapter provide a brief introduction to industrial ecology and Circular Economy, after which, this book will consider LCA as a tool in Circular Economy thinking.

1.5.3 Industrial Ecology and Cleaner Production

According to many scientific communities, human interference with global biogeochemical cycles has grown to a level that will trigger epochal changes, including climatic change and state shifts in the earth's biosphere. These changes have a substantial impact on humanity; they force humans to adapt or to be proactive and mitigate negative impacts on the environment. The spectrum of options for future action is wide. It includes technology development and deployment, economic instruments including taxes and subsidies, regulation and standards and changes in consumer behaviour and lifestyle.[52]

Industrial ecology, over the past decade, has successfully focused attention on improving the resource efficiency of systems of production. Reusing and recycling end-of-life products, redesigning products, processes and supply chains for improved efficiency, all offer clear environmental benefits to industries. However, addressing consumption in modern industrial society is deemed vital in reducing its impact to the environment.[53]

Lifset and Graedel (2002) have defined industrial ecology as the study of the flows of materials and energy within industrial and consumer society, the associated effects of these flows on the environment, alongside its influence to the economic, political, regulatory and social factors on the flow, use and transformation of resources.[54] Industrial ecology emphasises the critical need for a systems perspective in environmental analysis and decision making. The goal is to avoid narrow, partial analyses that can overlook important variables that can lead to unintended consequences. There are ample testimonies to the analytical approach to quantify emissions and resource loss on a life-cycle basis, dating from the 1970s.[55,56] Nevertheless, LCA's rapid growth and its close relationship with industrial ecology began around 1990, and its initial codification came in a 1993 handbook,[57] which set out an agenda for industries to develop, taking account of both of these approaches.[58,59] The life cycle approach within industrial ecology examines the environmental impacts of products, processes, facilities or services, from resource extraction through manufacture to consumption and finally to waste management. The use of LCA and the attention to the details in its approach, demonstrates that the cradle-to-grave perspective has huge benefits, when applied in managerial and policy decision-making, as well as in a research context. This includes product chain analysis, integrated product policy, greening of the supply chain and extended producer responsibility.[54] The adoption of LCA as an industrial ecology tool has become increasingly widespread, both in industry, civic society and government.[57] No doubt, LCA will continue to provide a vital perspective on industrial product and process design activities within the field of industrial ecology.

There are clear conceptual resonances between industrial ecology and cleaner production. Both are motivated by concerns about the increasing environmental impacts of industrial economic systems. The concepts of cleaner production can be described as the continuous application of an integrated, preventative environmental strategy to both processes and products, to reduce risks to humans and the environment. On the other hand, industrial ecology is simply described as an integrated system, in which the consumption of energy and materials is optimised, and the effluents of one process serve as the raw material for another process.[60] Adaptation and mitigation will both lead to a transformation of the biophysical basis of our society, which includes agriculture, industry, infrastructure, building stocks and vehicle fleets, and consumer products; and of the way we build, maintain, and operate this basis. The coming transformation is the continuation of a historic sequence of socio-metabolic transitions of mankind; first, from the hunter/gatherer to the agrarian, and later from the agrarian to the industrialised. This transformation represents a special global challenge, however, because it is likely to happen under environmental conditions that are significantly different from those that enabled the previous two transitions.[52]

The general principles of prospective modelling describe the current development status of two prospective model types: extended dynamic material flow analysis and Technology-hybridised Environmental-Economic Model with Integrated Scenarios.[52] These models combine the high level of technological detail known from LCA and material flow analysis (MFA) with the comprehensiveness of dynamic stock models and input/output analysis (I/O). These models are capable of building future scenarios with a time horizon until 2050 and beyond. Pauliuk and Hertwich[52] outline future applications and options for model development and discuss the relation between prospective industrial ecology models and the related concept of consequential LCA (cLCA). The prospective models for industrial ecology can answer questions that were previously in the exclusive domain of integrated assessment models (IAMs). It is evident that IAMs have a more comprehensive scope than the prospective industrial ecology models, yet they often do not obey central industrial ecology principles such as the life cycle approach and mass balance consistency. Integrating core industrial ecology principles into IAMs increases the scientific quality and policy relevance of the scenarios of society's future metabolism generated by IAMs, while placing industrial ecology concepts more prominently at the same time.[52]

The concept of industrial ecology continues to be the most plausible model for the industrial-environmental nexus of the future. In addition, major corporations that are environmental leaders are, in effect, already putting industrial ecology into practice. Its component elements are evident in their policies and practices.[57] It has to be emphasised that industry in general is constantly moving into an era of new values concerning the environment, in which corporate environmentalism appears to be essential for profitability and business survival. The speed with which a corporation understands and addresses these changing norms and values will define a large part of its competitive edge in the future. The benefit offered by industrial ecology is that it provides a coherent framework for shaping and testing strategic thinking about the entire spectrum of environmental issues confronting industry. Executives and policymakers who take steps to absorb and appreciate this mode of thinking will continue to find themselves and their organisations at a very real advantage in the world of the future.

1.5.4 Circular Economy

Since the beginning of the third industrial revolution, linear thinking, based on the logic of "take, make and dispose", has dominated in our economy. Although it has led to growth and prosperity in many parts of the world, it is a contributing factor to our current sustainability problems, because the linear model implies using resources in an unsustainable way and producing large quantities of waste. Indeed, global resource use has almost tripled in the last 40 years and, if the current consumption trend continues, it is estimated that global resource use in 2060 could reach 190 billion tonnes, with an associated 43% increase in GHG emissions from 2015 levels.[12]

The Circular Economy aims to decouple economic activities from the consumption of finite resources and to design waste out of the system.

In contrast to the linear economy, the Circular Economy proposes a system where flows of materials are circular, and energy comes from sustainable sources.[61,62] The concept highlights the value opportunities associated with products, components and material re-use, repair, remanufacturing and ultimately recycling and recovery. It proposes a whole value chain and cradle-to-cradle scope in the way products, components and materials are utilised in the economy. By circulating materials for longer and keeping products and materials at their highest utility and value, the system is expected to contribute to reduced environmental impacts linked to resource and energy use, while generating opportunities for job creation and business development.[61] The idea of circular use of resources is by no means new. Its foundations lie in the fields of planetary boundary, industrial ecology, cradle-to-cradle, environmental economics and related fields of knowledge. These concepts have promoted ideas about the preservation of energy and circular flows of resources, to reduce the entropy of socio-economic metabolisms, and new ways to sustainable production and consumption.

A particular interest of the Circular Economy concept lies in its compatibility and consistency with sustainable development through the three associated pillars (economic, social and environmental). Underpinned by a transition to renewable energy sources, the circular model builds economic, natural and social capital. Promoted by various countries and organisations, including the Ellen MacArthur Foundation (EMF), Circular Economy thinking is now an item on the top of the environmental agenda. A shift towards a Circular Economy will impact different sectors of the economy and at different scales. This poses questions about how different stakeholders might effectively assess the transition and monitor progress towards the long-term goals of circularity and sustainable development.

Various methods and frameworks for assessing the progress towards circularity have been proposed as decision-supporting tools. Most of these tools provide a systemic understanding of the environmental (and sometimes social and economic) implications of adoption of circular products, processes and/or strategies. Circular Economy strategies can be applied at the macro (*i.e.* region, nation, sector), meso (*i.e.* eco-industrial parks) and micro (*i.e.* product and organisation) levels. Most attempts to define indicators for measuring the Circular Economy strategy have so far addressed the macro and meso levels, and only a limited number of indicators are available at the product level.[63,64] Life cycle approaches can offer promising possibilities at the product level decision support.[65] Studies have already been developed, which show that LCAs play a key role in the development of meaningful indicators for the Circular Economy, regarding product level assessment.[63,66]

LCA has already been used as a metric to assess the environmental sustainability of different Circular Economy strategies, *e.g.* in narrowing resource flows,[67] slowing resource flows[68] and closing resource flows.[66] LCA-based metrics can be combined with other metrics, such as the assessment of industrial symbioses, food waste management systems, the evaluation of business models, or for the identification of more sustainable supply chain partners. This offers the flexibility of the LCA approach combined with other Circular Economy metrics. The Circular Economy concept has achieved widespread attention with the concepts, such as extending life cycles, creating multiple functionalities, and closing material loops. However, there are often multiple choices, for example, remanufacturing, recycling and reuse in closing the material loops. The environmental benefits, as well as economic benefits need to be assessed to support decision-making in Circular Economy choices. However, this presents challenges in LCA methodology, in particular in choosing functional units, expanding system boundaries and taking in account avoided environmental burdens *etc.* Inevitably, there are various challenges and barriers to incorporating LCA into Circular Economy strategies, which requires a clear understanding of the LCA approaches applied in different sectors, and calls for good LCA practices in transparency and consistency.

1.6 Summary and Introduction to Life Cycle Assessment Case Studies

The process of carrying out LCA follows an established protocol, which allows the quantification of environmental impacts of products, processes and services. The range of applications of LCA to specific industries, products and processes continues to develop and on-going considerations include the collection and rationalisation of data to inform LCA studies, as well as consideration of temporal and geographical aspects of data collection and availability. The refinement of methodological approaches to quantify life cycle impacts also continues and there is a vibrant community of LCA practitioners in academia and industry who continue to move the LCA process forward.

Circular Economy thinking is conceptual way of thinking about the impacts of consumption. Taking a closed loop approach, it provides a framework for influencing behaviours and practices to minimise these impacts. As these two approaches converge, we can expect to see the development of more linked studies, between the boundaries set in the closed loop systems of the Circular Economy and the cradle-to-grave approaches of LCA methodology. A recent position paper from the UNEP Life Cycle Initiative highlights the opportunities presented by LCA as a metric for the Circular Economy but also emphasises the need for on-going activities to address current gaps in the methodology,[69] namely the need for:

- Consistent accounting for changes in stocks of resources respecting mass balance principles
- Consistent modelling of open recycling loops
- The inclusion of all relevant resources and impacts, *i.e.* a full economy-wide LCA perspective
- Transparency of assumptions, reliability of data, and critical interpretation of results and trade-offs between a globally agreed number of impact categories.

This book aims to review the principles of LCA methodology whilst extending LCA systems boundary thinking to create better metrics that link to the Circular Economy. The book is intended to provide a robust systematic approach to the Circular Economy concept, using the established methodology of LCA. The book will provide a practical guide and companion reading for those who wish to use LCA as a research tool or to inform policy, process, and product improvement. The following chapters of the book aim to provide more detailed and specific applications of LCA to different areas of industry on a case study basis, to provide examples of how to align LCA as a metric in Circular Economy thinking.

This first chapter has provided a general summary of the increasing demand for resources and increasing consumerism, and introduces the environmental impacts, sustainability as a concept and the impacts of waste, re-use and recycling of resources. It has described some of the ways in which the academic community try to rationalise our impacts in the new Anthropocene age.

Chapter 2 introduces the concept of Circular Economy thinking and present the Circular Economy map and framework, highlighting specific areas of resource insecurity and waste impacts. Chapter 3 presents technical guidance for carrying out LCA as a metric, describing the procedure required for robust LCA methodology and its application to Circular Economy thinking. The following chapters are industry-specific and detailed LCA approaches to studies for buildings and infrastructure, paper and packaging (traditional plastics and biobased options) and textiles, agriculture (crops and livestock), renewable energies and bioenergies, can be explored further by the interested reader. Finally, the last chapter provides an interesting perspective and contrast of Circular Economy as seen through the lens of the LCA methodological approach.

References

1. UNEP Global Environmental Alert Service (GEAS), One Planet, How Many People? A Review of the Earth's Carrying Capacity, 2012, Available from: www.unep.org/geas.
2. R. A. Bendis, *The Global Innovation Imperative*, http://www.innovationamerica.us/images/stories/pdf/AUTM-11-13-2012.pdf, Accessed June 7 2020.
3. The Fourth Industrial Revolution: what it means and how to respond, World Economic Forum, https://www.weforum.org/agenda/2016/01/the-fourth-industrial-revolution-what-it-means-and-how-to-respond/, Accessed July 3 2020.
4. P. J. Crutzen, Geology of Mankind in *Paul J Crutzen: A Pioneer on Atmospheric Chemistry and Climate Change in the Anthropocene*. ed. P. J. Crutzen and H. G. Brauch, Springer International Publishing, Cham, 2016, pp. 211–215, (SpringerBriefs on Pioneers in Science and Practice), https://doi.org/10.1007/978-3-319-27460-7_10.

5. J. Zalasiewicz, M. Williams, A. Smith, T. L. Barry, A. L. Coe and P. R. Bown, *et al.*, *GSA Today*, 2008, **18**(2), 4.
6. United Nations, Department of Economic and Social Affairs, Population Division, World Population Prospects: The 2017 Revision, Key Findings and Advance Tables, 2017, Report No.: Working paper No, ESA/P/WP/248.
7. E. Neumayer, *Popul. Environ.*, 2006, **27**(4), 327–336.
8. T. Malthus, An Essay on the Principle of Population, Electronic Scholarly Publishing Project 1998, 1978, http://www.esp.org/books/malthus/population/malthus.pdf.
9. J. Diamond, *Collapse: How Societies Choose to Fail or Succeed*, Viking, New York, 2004.
10. Review of the Earth's Carrying Capacity [Internet], 2012, Available from: www.unep.org/geas.
11. E. W. F. Peterson, *SAGE Open*, 2017, **7**(4), 215824401773609.
12. IRP, *Global Resources Outlook 2019: Natural Resources for the Future We Want*, United Nations Environment Programme, Kenya, https://wedocs.unep.org/bitstream/handle/20.500.11822/27517/GRO_2019.pdf?sequence=3&isAllowed=y, Accessed June 4 2020, 2019.
13. Summary for Policymakers of IPCC Special Report on Global Warming of 1.5 °C approved by governments — IPCC, https://www.ipcc.ch/2018/10/08/summary-for-policymakers-of-ipcc-special-report-on-global-warming-of-1-5c-approved-by-governments/, Accessed June 4 2020.
14. UNFCCC, Paris Agreement, FCCC/CP/2015/L.9/Rev.1 2015, http://unfccc.int/resource/docs/2015/cop21/eng/l09r01.pdf.
15. United Nations, United Nations Conference on Environment and Development: Agenda 21, Rio de Janerio, 3 to 14 June 1992, United Nations, 1993, https://sustainabledevelopment.un.org/content/documents/Agenda21.pdf, Accessed Nov 13 2018.
16. D. Hoornweg, P. Bhada-Tata and C. Kennedy, *Nat. News*, 2013, **502**(7473), 615.
17. R. Geyer, J. R. Jambeck and K. L. Law, *Sci. Adv.*, 2017, **3**(7), e1700782.
18. W. C. Li, The Occurrence, Fate, and Effects of Microplastics in the Marine Environment, in *Microplastic Contamination in Aquatic Environments*, ed. E. Y. Zeng, Elsevier, 2018, ch. 5, pp. 133–173, http://www.sciencedirect.com/science/article/pii/B9780128137475000059, Accessed June 7 2020.
19. O. Setälä, M. Lehtiniemi, R. Coppock and M. Cole, Microplastics in Marine Food Webs, in *Microplastic Contamination in Aquatic Environments*, ed. E. Y. Zeng, Elsevier, 2018, ch. 11, pp. 339–363. http://www.sciencedirect.com/science/article/pii/B9780128137475000114, Accessed June 7 2020.
20. J. R. Jambeck, R. Geyer, C. Wilcox, T. R. Siegler, M. Perryman and A. Andrady, *et al.*, *Science*, 2015, **347**(6223), 768–771.
21. P. Brimblecombe, *J. Air Pollut. Control Assoc.*, 1978, **28**(2), 115–118.
22. R. A. Robinson and W. J. Sutherland, *J. Appl. Ecol.*, 2002, **39**(1), 157–176.

23. R. Carson, *Silent Spring*, Houghton Mifflin Harcourt, 404 p.
24. WWF., M. Grooten and R. E. A. Almond, *WWF Living Planet Report – 2018: Aiming Higher*, WWF, Gland, Switzerland, https://www.wwf.org.uk/sites/default/files/2018-10/LPR2018_Full%20Report.pdf, Accessed Nov 13 2018, 2018.
25. Our Work, Earth Day, https://www.earthday.org/our-work/, Accessed July 3 2020.
26. SUSTAINABILITY | meaning in the Cambridge English Dictionary, https://dictionary.cambridge.org/dictionary/english/sustainability, Accessed June 7 2020.
27. World Commission on Environment and Development, *Our Common Future*, Oxford University Press, http://www.un-documents.net/our-common-future.pdf, Accessed Nov 13 2018, 1987.
28. United Nations, United Nations Conference on Environment and Development: Agenda 21, Rio de Janerio, 3 to 14 June 1992, United Nations, 1993, https://sustainabledevelopment.un.org/content/documents/Agenda21.pdf, Accessed Nov 13 2018.
29. J. Sachs and U. Project, Investing in development: a practical plan to achieve the Millennium Development Goals, *Overview*, 2005.
30. UNFCCC, Paris Agreement [Internet], FCCC/CP/2015/L.9/Rev.1 2015, Available from, http://unfccc.int/resource/docs/2015/cop21/eng/l09r01.pdf.
31. Rosa W., ed. Transforming Our World: The 2030 Agenda for Sustainable Development, in *A New Era in Global Health*, Springer Publishing Company, New York, NY, http://connect.springerpub.com/lookup/doi/10.1891/9780826190123.ap02, Accesed Nov 14 2018, 2017.
32. ISO 14040:2006, ISO, https://www.iso.org/cms/render/live/en/sites/isoorg/contents/data/standard/03/74/37456.html, Accessed July 3 2020.
33. ISO 14040:2006, ISO, https://www.iso.org/cms/render/live/en/sites/isoorg/contents/data/standard/03/74/37456.html, Accessed July 3 2020.
34. ILCD Handbook - European Commission Service Site, https://eplca.jrc.ec.europa.eu/ilcdHandbook.html, Accessed July 3 2020.
35. D. Novick, *The Federal Budget as an Indicator of Government Intentions and the Implications of Intentions*, Rand Corporation, Santa Monica, CA, publication P-1803, 1959.
36. G. Huppes and M. A. Curran, *Life Cycle Assessment Handbook: A Guide for Environmentally Sustainable Products*, ed. M. A. Curran, Scrivener Publishing and John Wiley & Sons Inc, Beverly, Massachusetts, USA, 2012, ch. 1, pp. 1–14.
37. H. Baumann and A. M. Tillman, *The Hitch Hikers's Guide to LCA: An Orientation in Life Cycle Assessment Methodology and Application*, Studentlitteratur, Lund, Sweden, 2004, ch. 2, pp. 43–69.
38. M. A. Curran, *Environmental Life Cycle Assessment*, McGraw-Hill, New York, USA, 1996, ch. 1, pp. 1–3.
39. J. A. Fava, R. Denison, B. Jones, M. A. Curran, B. Vigon, S. Selke and J. Barnum, *A Technical Framework for Life Cycle Assessments*, Society of

Environmental Toxicology and Chemistry (SETAC), Brussels, Belgium and Washington DC., USA, 1991.
40. F. Pulselli and L. Coscieme Earth's Carrying Capacity. in *Encyclopedia of Quality of Life and Well-Being Research*. ed. A. C. Michalos, Springer Netherlands, Dordrecht, 2014, pp. 1748–1751, https://doi.org/10.1007/978-94-007-0753-5_800, Accessed June 7 2020.
41. S. Franck, W. von Bloh, C. Müller, A. Bondeau and B. Sakschewski, *Ecol. Modell.*, 2011, **222**(12), 2019–2026.
42. C. Hui, Carrying Capacity of the Environment, in *International Encyclopedia of the Social & Behavioral Sciences*, ed. J. D. Wright, Elsevier, Oxford, 2nd edn, 2015, pp. 155–160, http://www.sciencedirect.com/science/article/pii/B978008097086891002X, Accessed June 7 2020.
43. A. Bjørn and M. Z. Hauschild, *Int. J. Life Cycle Assess.*, 2015, **20**(7), 1005–1018.
44. A. Bjørn, M. Margni, P.-O. Roy, C. Bulle and M. Z. Hauschild, *Ecol. Indic.*, 2016, **63**, 1–13.
45. J. Rockström, W. Steffen, K. Noone, Å. Persson, F. S. Chapin and E. F. Lambin, *et al.*, *Nature*, 2009, **461**(7263), 472–475.
46. J. Rockström, W. Steffen, K. Noone, Å. Persson, F. S. I. Chapin and E. Lambin, *et al.*, Planetary Boundaries: Exploring the Safe Operating Space for Humanity, *Ecol. Soc.*, 2009, **14**(2). Accessed Aug 27 2018.
47. W. Steffen, K. Richardson, J. Rockström, S. E. Cornell, I. Fetzer and E. M. Bennett, *et al.*, *Science*, 2015, **347**(6223), 1259855.
48. K. Fang, R. Heijungs and G. R. De Snoo, *Ecol. Econ.*, 2015, **114**, 218–226.
49. M. W. Ryberg, M. Owsianiak, K. Richardson and M. Z. Hauschild, Challenges in implementing a Planetary Boundaries based Life-Cycle Impact Assessment methodology, *J. Cleaner Prod.*, 2016, **139**, 450–459.
50. M. W. Ryberg, M. Owsianiak, K. Richardson and M. Z. Hauschild, *Ecol. Indic.*, 2018, **88**, 250–262.
51. S. Kara, M. Z. Hauschild and C. Herrmann, *Procedia CIRP*, 2018, **69**, 3–10.
52. S. Pauliuk and E. G. Hertwich, Prospective Models of Society's Future Metabolism: What Industrial Ecology Has to Contribute, in *Taking Stock of Industrial Ecology*, ed. R. Clift and A. Druckman, Springer International Publishing, Cham, 2016, pp. 21–43, Available from: https://doi.org/10.1007/978-3-319-20571-7_2, Accessed June 8 2020.
53. T. Jackson, *J. Ind. Ecol.*, 2005, **9**(1–2), 19–36.
54. R. Lifset and T. E. Graedel, Industrial ecology: goals and definitions, *A Handbook of Industrial Ecology*, Edward Elgar Publishing, https://www.elgaronline.com/view/1840645067.00009.xml, Accessed June 8 2020, 2002.
55. P. Chapman, *Fuels paradise: Energy Options for Britain*, Penguin books, Harmondsworth, UK, 6th edn, 1975, ch. 6.
56. I. Boustead Handbook of industrial energy analysis, *Chichester, Eng*, E. Horwood, New York, 422 p.
57. H. B. C. Tibbs, *Industrial Ecology An Environmental Agenda for Industry*, Global Business Network, Emeryville, CA, USA, 1993, pp. 3–28.

58. T. E. Graedel and R. J. Lifset Industrial Ecology's First Decade, in *Taking Stock of Industrial Ecology*, ed. R. Clift and A. Druckman, Springer International Publishing, Cham, 2016, pp. 3–20, https://doi.org/10.1007/978-3-319-20571-7_1, Accessed June 8 2020.
59. SETAC, *Guidelines for Life-Cycle Assessment: A 'Code of Practice'*, ed. F. Consoli, D. Allan, I. Boustead, J. Fava, W. Franklin, A. A. Jensen, N. de Oude, R. Parrish, R. Perriman, D. Postlethwaite, B. Quay, J. Seguin and B. Vigon, Society of Environmental Toxicology and Chemistry (SETAC), Brussels, Belgium and Pensacola, Florida, USA, 1st edn, 1993.
60. L. Baas, *Cleaner Production and Industrial Ecology; Dynamic Aspects of the Introduction and Dissemination of New Concepts in Industrial Practice*, Eburon, Delft, 2005.
61. Circular Economy Report – The Circular Economy – Towards a Circular Economy: Business Rationale for an Accelerated Transition, https://www.ellenmacarthurfoundation.org/publications/towards-a-circular-economy-business-rationale-for-an-accelerated-transition, Accessed June 8 2020.
62. J. Korhonen, C. Nuur, A. Feldmann and S. E. Birkie, *J. Cleaner Prod.*, 2018, **175**, 544–552.
63. V. Elia, M. G. Gnoni and F. Tornese, *J. Cleaner Prod.*, 2017, **142**, 2741–2751.
64. M. Linder, S. Sarasini and P. van Loon, *J. Ind. Ecol.*, 2017, **21**(3), 545–558.
65. European Environment Agency Report no 6/2017, 1–52, Circular by design - Products in the circular economy, https://www.eea.europa.eu/publications/circular-by-design, Accessed June 8 2020.
66. M. Niero and S. I. Olsen, *Resour., Conserv. Recycl.*, 2016, **114**, 18–31.
67. S. Huysman, S. Sala, L. Mancini, F. Ardente, R. A. F. Alvarenga and S. De Meester, *et al.*, *Resour., Conserv. Recycl.*, 2015, **95**, 68–76.
68. C. Bakker, F. Wang, J. Huisman and M. den Hollander, *J. Cleaner Prod.*, 2014, **69**, 10–16.
69. Must-Read: the Initiative's Latest Position Paper on Circular Economy, Life Cycle Initiative, https://www.lifecycleinitiative.org/must-read-the-initiatives-latest-position-paper-on-circular-economy/, Accessed July 8 2020.

CHAPTER 2

The Circular Economy: Concept, Tools and Implementation

TERESA DOMENECH*[a,c] AND JULIA A. STEGEMANN[b,c]

[a] University College London Institute for Sustainable Resources, Central House, 14 Upper Woburn Place, London WC1H 0NN, United Kingdom; [b] University College London, Civil, Environmental and Geomatic Engineering, Chadwick Building, Gower Street, London WC1E 6BT, United Kingdom; [c] University College London, The Circular Economy Lab (CircEL), Gower Street, London WC1E 6BT, United Kingdom
*Email: t.domenech@ucl.ac.uk

2.1 The Circular Economy as a Solution to Impacts of the Linear Economy

The global economy is supported by the extraction of resources and their transformation into forms that fulfil human needs and desires, with a generic anthropogenic (man-made) material life cycle depicted in Figure 2.1. The currently dominant linear materials life cycle follows the horizontal arrows, proceeding from left to right across the diagram, from "cradle to grave". Raw materials are produced from natural resources, manufactured into products, which are used, and then disposed of at product end-of-use. They may reside for a period (ranging from days to years) as stocks in each stage. In the developed world, end-of-use materials are usually collected and treated before disposal; they may be recycled between all stages, as indicated

Life Cycle Assessment: A Metric for the Circular Economy
Edited by Aiduan Borrion, Mairi J. Black and Onesmus Mwabonje

Published by the Royal Society of Chemistry, www.rsc.org

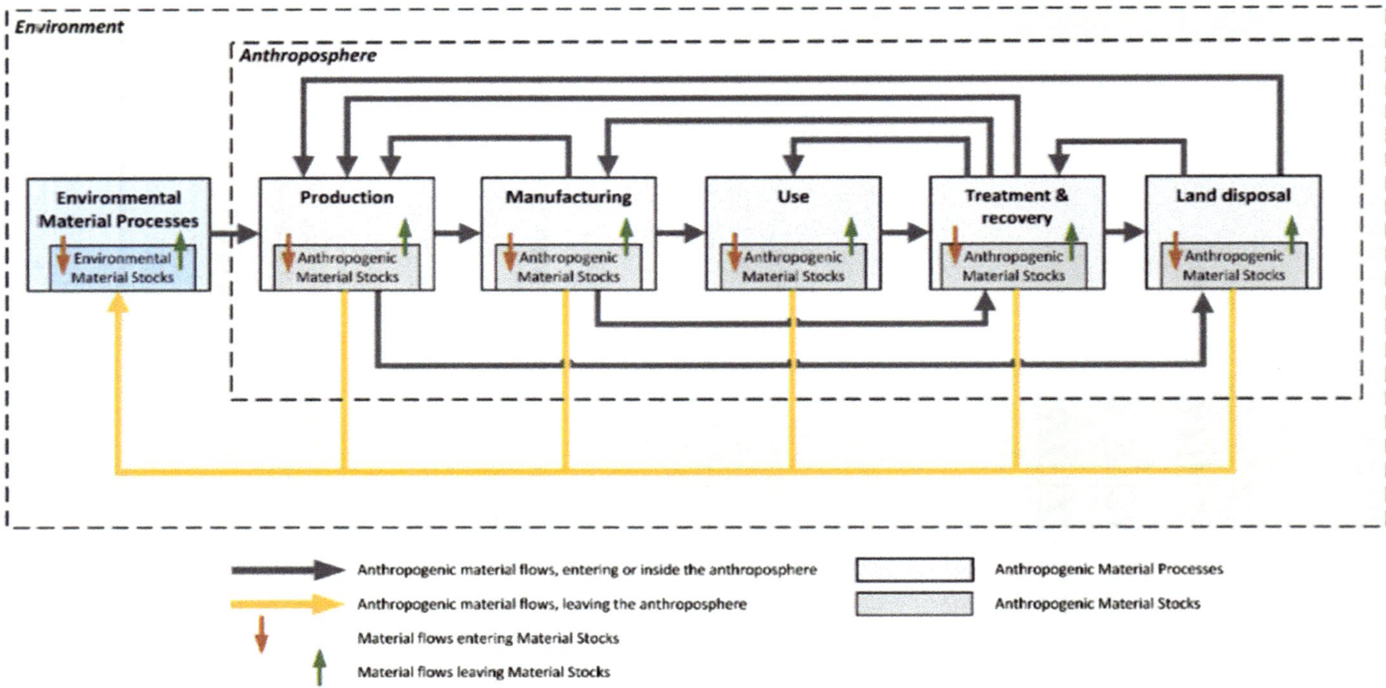

Figure 2.1 The anthropogenic material life cycle for resource use by humans.[1]
Reproduced from ref. 1, Specifications for the application of the United Nations Framework Classification for Resources to Anthropogenic Resources, by Heiberg, Sigurd, Heuss-Aßbichler, Soraya, Hilton, Julian, Horváth, Zoltán, Kral, Ulrich, Krook, Joakim *et al.*, Copyright 2018 United Nations. Reprinted with the permission of the United Nations.

by the recursive arrows returning from right to left in Figure 2.1, but for most materials this practice is still the exception rather than the rule.

Global trends show an exponential increase in resource extraction over time.[1,2] In 1970, global resource extraction was estimated at around 27.1 billion tonnes, while in 2017 it had more than tripled to 92.1 billion tonnes.[3] These data do not include mobilised materials that do not enter the value chain; for example, the annual generation of mining wastes is estimated at 350 billion tonnes.[4] Therefore, the overall environmental and social impacts associated with the full resource life cycle (Figure 2.1), are significantly larger than indicated by commodity extraction statistics.

In the last decades, increased international commodity trade has also substantially influenced how resource extraction and consumption are geographically distributed across nations.[5] Divergences in the level and composition of resource extraction and consumption reveal profound economic differences between nations. Metal ore uses have grown steadily – in terms of both quantity and variety of composition – as well as other non-renewable materials. These include non-metallic minerals, shifting the share of biomass and renewable materials in the composition of the demand.[3] Our exponentially growing reliance upon depleting resources that cannot be regenerated has become a global problem. Geographically, there has been a substantial shift in extraction from developed nations, largely Europe and North America, to Central and South-East Asia. However, a large proportion of that extraction is still embedded in products consumed or accumulated by developed nations. Global material demand, per capita, has grown from 7.4 tonnes in 1970, to around 12.2 tonnes in 2017. However, high-income countries consume, on average, 27 tonnes of materials per capita, which is 60% more than upper-middle-income countries, and as much as thirteen times the consumption of low-income countries.[3]

While most of the environmental impacts of material consumption once concentrated on upstream activities in this linear life cycle (*i.e.*, production, including mining and refining, and manufacturing), changes in demand and discarding behaviour are now exacerbating impacts associated with the use and end-of-use (waste management) phases of products. Landfills are responsible for about 5% of global greenhouse gas (GHG) emissions,[6] mainly CH_4 and CO_2, as well as leachate that contaminates groundwater with a host of environmental pollutants, including nutrients, metals, and persistent organic pollutants. A recent concern is the proliferation of environmental plastics (see Section 2.3.3), with observable impacts on organisms and ecosystems, and unknown impacts on human health. A prediction that the mass of plastics in the oceans could exceed that of fish by 2050[7] highlights the limitations of the throw-away culture in which the linear life cycle (Figure 2.1) sits.

Resource consumption is also associated with greenhouse gas (GHG) emissions, as production and manufacturing are highly energy intensive and have been mostly powered by fossil fuels. Beyond climate change implications, it also results in a range of other environmental impacts, including on human health (*e.g.*, toxicity, carcinogenicity), the natural environment (*e.g.*, ecotoxicity, acidification, eutrophication, erosion), and depletion of natural resources (*e.g.*, land and water, as well as specific materials). For example, aquatic toxicity associated with copper refining has increased faster than production;[3] and although a degree of decoupling of water use from population growth has been experienced since the 1990s, global water use nearly tripled between 1970 and 2017.[3] This has been accompanied by increased human land use for croplands and cities. Thus, another problem with the linear economy is that the energy use, GHG emissions, and other impacts embodied in our materials are lost at product end-of-use. In addition to other measures to reduce material demand, transition to a sustainable economy requires transformation of processes along the full product life cycle to enable material retention and recovery.[8] Such a Circular Economy requires full activation of the recovery flows represented by the recursive arrows in Figure 2.1, both within and between individual product life cycles, and along the whole supply chain.

If current consumption trends persist, it is estimated that global resource use in 2060 will reach 190 billion tonnes, with an associated 43% increase in GHG emissions from 2015 levels.[3] A number of modelling approaches have tried to estimate the potential for management of future environmental pressures under a Circular Economy. One such contribution is a Material Economics[9] report that estimates the contribution of circular economy approaches to mitigate climate change. The report concludes that circular economy approaches (such as product life extension and recirculation through reuse, repair and remanufacturing, and material recycling (Figure 2.1), *etc.*), could, by 2050, contribute to the reduction of CO_2 emissions from heavy industries, including steel, plastics, aluminium and cement, by 56%. The measures investigated included harvesting of materials embedded in current stocks and promoting secondary markets.

Current levels of circularity, measured as the share of secondary materials in total processed materials, are still low. Haas *et al.*[10] estimated global circularity at only 6% in 2005. Even the EU, despite having a relatively high share of recycling of end-of-use products, only reaches a circularity of 13%, suggesting that more ambitious measures are required to increase circular use of resources. Recent contributions[10,11] have also highlighted the need to measure not only circularity, but also the overall size of the socio-economic metabolism. This must be carried out at sufficient detail to enable identification of opportunities for improvement.

2.2 The Concept of the Circular Economy

The rapidly increasing environmental pressures noted in Chapter 1 demand urgent action and innovative approaches. The concept of the Circular Economy, which addresses some of these challenges, has gained traction in the last few years among policy-makers, businesses, and in academia. In contrast to the linear economy, the Circular Economy proposes a system where materials are retained in circulation through the life cycle (Figure 2.1), and energy comes from sustainable sources.[12–14] The concept highlights the opportunities to create value associated with, in order of priority: re-use, repair and remanufacturing of products and components, and ultimately material recycling. It proposes a "cradle to cradle" scope for the whole value chain, including products, components and materials. By circulating materials for longer, and keeping products and materials at their highest utility and value, the system is expected to contribute to reduced environmental impacts linked to resource and energy use, while generating opportunities for business development and creating jobs. Several reports have attempted to quantify the potential global financial benefits of a Circular Economy. The Ellen MacArthur Foundation estimated an annual benefit of 1.3 trillion Euros for the EU.[13] The basis of these calculations is not always clear, however, and they have been criticised for their lack of detail, unclear assumptions,[14] and for overlooking the costs of the transition.

The idea of continued circulation of resources within the economy is by no means new. In fact, its foundations are probably ancient, but it has found recent expression in the concepts of Industrial Ecology,[15] Cradle-to-Cradle,[16] Environmental Economics,[17] the Performance Economy,[18] and related fields of knowledge. These concepts have promoted preservation of energy and circulation of resources to reduce the entropy of the socio-economic metabolism, and new ways to organise patterns of production and consumption in socio-economic systems to bring this about. The first use of the term "Circular Economy" can be traced back to Pearce and Turner.[19] They refer to a Circular Economy that internalises the value and utility provided by natural systems as suppliers of natural resources, providers of life-supporting services and sinks of emissions and waste.[20] In their paper, Pearce and Turner describe the flows between natural and man-made systems, and the need to constrain action within the laws of thermodynamics. They refer back to the seminal paper authored by Boulding in 1966, which describes earth as a spaceship "without unlimited reservoirs of anything, either for extraction or for pollution, and in which, therefore, man must find his place in a cyclical ecological system",[21] much as Ellen MacArthur has, more recently, likened the earth to a sailboat. From the 1990s, the field of Industrial Ecology has emphasised the need for a perspective that encompasses the interactions between natural and industrial systems. The "industrial metabolism" is used as an analogy to describe the flows of materials and energy within and between man-made systems, and between natural and man-made systems. Although Industrial Ecology does not refer explicitly to the Circular Economy, its core principles are essentially the same. In the words of Jelinski *et al.*[22] "Industrial ecology is a new approach to the industrial design of products and processes [which] seeks to optimize the total materials cycle from virgin material to finished material, to component, to product, to waste product, and to ultimate disposal." Following Graedel and Allenby,[15] a 3-stage system typology of industrial metabolisms is proposed, depending on the degree of reliance on primary energy and materials, and on disposal of wastes to the natural environment. Type I is a completely linear system; Type II is a system with increased circularity (quasi-closed-loop), in the form of limited use of primary materials and reduced generation of waste and pollution; Type III is an ideal system composed of completely closed loops, with reliance on external energy, but very limited dependence on external resources and sinks through regenerative practices,[23] *i.e.*, the ideal of a Circular Economy.

Another relevant contribution to the normative design of the concept of the Circular Economy comes from the "cradle-to-cradle approach",[16] which promotes product design for a circular cradle-to-cradle, rather than a linear "cradle-to-grave", life cycle. The Circular Economy "Butterfly Diagram" (Figure 2.2) illustrates separate cradle-to-cradle life cycles for renewable and finite resources[12] and highlights the differing circular economy options for each system (in the left and right sides of the "butterfly"). For finite resources, circular economy options aim for infinite cycling of materials, by extending the life of products, components and materials in use; with reuse, repair and remanufacturing, and recycling at end of use. The renewables cycle applies mainly to biological materials and, while maintaining their value for as long as possible is still desirable, their ultimate degradation is assumed, and the focus is on protection of natural capital. Intervention strategies for renewable resources are directed to preserve and recover nutrients, through loops to retain material value in cascading cycles, from higher to lower value, ultimately to regenerate productive soil and biodiverse (agri)ecosystems. While the renewables cycle includes opportunities to recover energy in organic wastes, direct incineration without cascading and/or landfilling are discouraged.

Drawing on these fields of knowledge, recent years have seen a plethora of definitions of the concept that stress different dimensions of the Circular Economy. In their review paper, Kirchherr *et al.*[24] identify at least 114 definitions of Circular Economy in the academic literature. However, recent interpretations of the idea of the Circular Economy have been developed primarily by practitioners, policy-makers, and consultants, with a pragmatic outlook and a narrative that emphasises the idea of value creation from the cycling of resources. Many of these concepts stress the transition from an economic system based on extraction and transformation, to one that builds on conservation and recovery. However, as pointed out by Kirchherr *et al.*[24] any concept that has attracted such attention and is used by different stakeholders is likely to encompass a number of different conceptions, drawing on different world views, which may blur its boundaries. Though problematic for academic rigour, emphasis on the high-level approach with some flexibility in definition of the details may be an advantage in recruiting followers of the approach.

One of the leading promoters of the idea of the Circular Economy has been the Ellen MacArthur Foundation, which defines it as: "one that is restorative and regenerative by design and aims to keep products, components, and materials at their highest utility and value at all times, distinguishing between technical and biological cycles. This new economic model seeks to ultimately

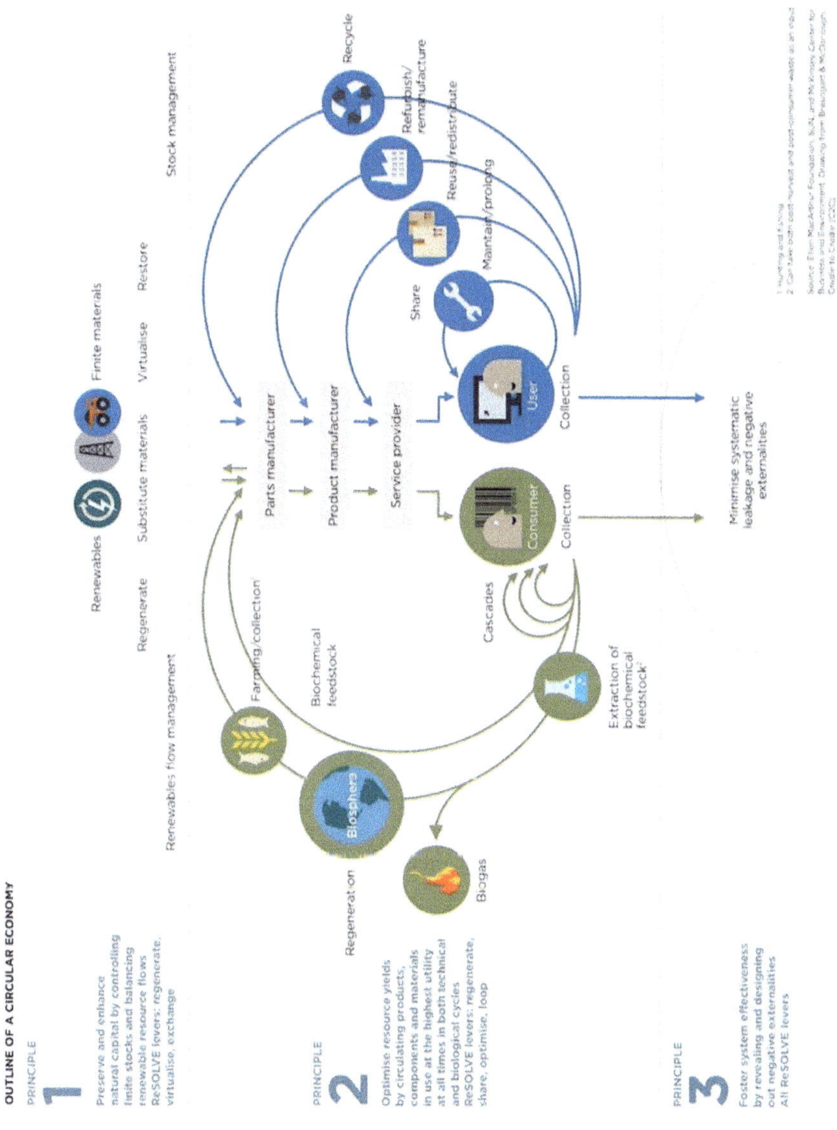

Figure 2.2 The Circular Economy 'Butterfly Diagram'.[12] Source: ref. 12, EMF. Reproduced with permission from Ellen MacArthur Foundation.

decouple global economic development from finite resource consumption."[12] This definition highlights the idea of "restoration" that responds to the need to return the Earth to a more balanced state. It implies a transition to recovery of resources and nutrients, a shift away from toxic chemical use, and a move towards renewable sources of energy. Attention is placed on the value creation potential associated with restorative activities. The definition also sees the decoupling of resource consumption from value creation as one of the main purposes of the Circular Economy.

The Circular Economy discourse has tended to emphasise scarcity as a driver for resource efficiency. An analysis of absolute and relative scarcity of different types of resources by Tucker and Ekins[25] suggests that, in the next decades, absolute scarcity concerns may only be relevant for a limited group of materials. However, regional scarcities may be a concern due to geopolitical factors; indeed, political developments are likely to continue to be influenced by stresses caused by the geographical heterogeneity of resource distribution and exploitation. In the meantime, limitations on our material consumption are already experienced globally from the output side, *i.e.*, GHG and other harmful emissions to air, land, and water; a disruptive consequence for the ability of natural systems to regenerate in line with the concept of "planetary boundaries".[26]

There have been attempts to further create linkages between the concept of the Circular Economy, and sustainability and the United Nations Sustainable Development Goals (UN SDGs), to highlight complementary elements, but also potential differences and even conflicts.[25] Some authors emphasise the need for systemic changes as part of a Circular Economy. For example, Ghisellini *et al.*[20] criticise short-sighted approaches that consider Circular Economy as an optimised way to manage waste, rather than truly capturing the potential transformative dimension, which "requires a broader and much more comprehensive look at the design of radically alternative solutions, over the entire life cycle of any process as well as the interaction between the process and the environment and the economy in which it is embedded".[20] The authors also highlight the Circular Economy for its "potential to understand and implement radically new patterns and help society reach increased sustainability and wellbeing at low or no material, energy or environmental costs".[20] To be sustainable, such systemic changes require a smooth transition from existing practice that avoids economic losses to businesses, individuals and communities, *e.g.*, through loss of existing investments ("stranded assets"), superseded employment skills, *etc.*

The Circular Economy proposes potentially disruptive restructuring of, not only the production, manufacturing, use and end-of-use stages of individual product life cycles, but also the wider socio-economic system, with an emphasis on value creation and preservation of natural systems. Its implementation thus requires a deep understanding of material flows and impacts in the whole socio-economic system, as well as sub-systems: product life cycles and value chains. Progressing to a Circular Economy requires changes in the way value is created, with an emphasis on the role of new business models capable of enabling businesses to take advantage of the economic opportunities that arise from circular and efficient use of resources. A smooth transition must be supported by appropriate policy. The following sections address some of these aspects, beginning with discussion of the Circular Economy in different sectors.

2.3 A Circular Economy in Different Industrial Sectors

2.3.1 Industrial Transitions to a Circular Economy

A key principle of the Circular Economy is the separation of cycles for finite and renewable resources, as summarised by the Circular Economy Butterfly Diagram (Figure 2.2). According to this, this book illustrates opportunities and challenges associated with the transition to the Circular Economy in different sectors, drawing on LCA case studies:

- Steel (Chapter 4) and cement (Chapters 5 and 6) manufacture are secondary industries that mainly support the construction sector. The primary raw materials in these industries are finite resources, with the potential for continuous cycling through product, component and material loops by reuse, repair, remanufacturing, and recycling, respectively.
- Plastics (Chapter 7) manufacture is also a secondary industry, with products that pervade consumer goods manufacture, including textiles (Chapter 8). Traditionally based on finite raw materials, they have a potential for continuous cycling, but also improvement of impacts through substitution by renewable bio-based alternatives.
- Agriculture, to produce crops (Chapter 11) and livestock (Chapter 12), is the primary industry that underpins the cascading cycles of secondary industries based on renewable resources (the Bioeconomy[27,28]), including paper (Chapter 10), biotic and bio-based textiles (Chapter 8), other advanced bio-based materials (Chapter 9), and biofuels (Chapter 13).
- Renewable energy (Chapter 14) is essential for operation of all production, manufacture, recovery and cascading processes, and is therefore a prerequisite for a Circular Economy, as it is for any sustainable paradigm.

Though the LCA case studies presented in this book are for primary and secondary industrial sectors, adherence to Circular Economy principles in these sectors can benefit the tertiary (service-based) and quaternary (knowledge-based) industrial sectors, through their supply chains, which are therefore also candidates for LCA.

In an attempt to streamline the Circular Economy concept, to operationalise it, and avoid falling into the trap of lack of concrete action, the Ellen MacArthur Foundation has proposed three guiding principles: "1) preserve and enhance natural capital by controlling finite stocks and balancing renewable resource flows; 2) optimise resource yields by circulating products, components and materials at their highest utility and value in both technical and biological cycles and 3) foster system effectiveness by designing out negative externalities".[13] These are shown to the left of the Butterfly Diagram in Figure 2.1, and also embedded in the ReSOLVE strategic framework for Circular Economy implementation. This framework identifies six areas of action: Regenerate, Share, Optimise, Loop, Virtualise, Exchange. There has also been recent activity in development of standards, such as BS 8001:2017 Framework for implementing the principles of the Circular Economy in organisations.

Together with the concepts of separate cycles for renewable and finite resources, these initiatives provide general guidance about how individual businesses and other organisations can go about implementing a Circular Economy. However, different industrial resources and sectors face different challenges, and require different approaches, in the transition to a Circular Economy. The following sections illustrate some of the issues in manufacturing sectors relevant to the LCA cases studies: the construction, plastics, and textiles industries.

2.3.2 The Construction Industry and the Circular Economy

The construction industry is arguably the largest global consumer of resources, including:

- 43 billion tonnes of mineral-based construction materials, which represent nearly half of all resource extraction,[29] and include more than 4 billion tonnes a year of cement,[30] as well as aggregate, asphalt, brick, plasterboard, stone, and glass
- 0.6 billion tonnes of steel,[31] among other metals such as aluminium and copper
- 0.9 billion m^3 of wood products,[32] and
- smaller, but significant, quantities of a variety of other materials, including plastics in window frames and architectural profiles, floor and wall coverings pipes, thermal and electrical insulation,[33] composite materials, advanced fabrics, adhesives, coatings, paints and varnishes.

A large proportion of these materials go into stocks, *i.e.*, homes and infrastructure, where they may remain for periods ranging from years (for internal fittings) to decades or even centuries (for well-designed structures).[34] However, there are material losses during the construction process, as well as arisings from demolition, which result in end-of-life construction and demolition waste (C&DW) equivalent in mass to nearly 40% of extraction.[29] The resulting quantity is a major fraction of total waste arisings, *e.g.*, more than double the mass of household waste in the EU.[35] Some wastes, such as asbestos, polyvinyl chloride (PVC), and adhesives, have impacts disproportionate to their relatively small mass flows. Metals are well recovered from C&DW everywhere, because of their high value, and some countries report high recycling rates for other C&DW (*e.g.*, in the EU).[35] If recovered at all, mineral-based products are generally down-cycled, with loss of embodied impacts from their production and manufacturing,[36] and this may be the most concerning effect of the linear life cycle of construction materials. For example, processing of concrete from demolition into recycled concrete aggregate results in the loss of approximately 2.6 million tonnes per year of embodied CO_2 and 2.8 million tonnes per year of embodied water in the UK alone.[37–40] While developed countries have focussed in recent decades on avoiding environmental dispersion of hazardous pollutants, and avoiding landfill, there is now a growing emphasis on moving up the waste hierarchy, to closed-loop recovery systems.[41] On the other hand, some cascading use of mineral materials could be considered desirable, to ultimately replenish soil mineral content depleted by erosion.[42]

The Circular Economy represents a potential way to mitigate the negative environmental impacts from buildings and construction, which is responsible for, *e.g.*, 36% of energy use and 39% of global CO_2 emissions in 2018, including 11% from manufacture of construction products such as cement, glass and steel.[43] Adams *et al.*[44] provide a review of the application of circular economy thinking over the building life cycle: from design; manufacture and supply of material components; construction; maintenance, use and refurbishment to end-of-life management. The Ellen MacArthur ReSOLVE framework has also been considered as an approach to conceptualising circular economy thinking in the built environment.[45]

With respect to the three guiding principles suggested by the Ellen MacArthur Foundation[13] (see Section 2.3.1), it is essential to control leakage of finite resources, *i.e.*, construction minerals and metals, from the overall system, but also to decrease demand, which is still increasing exponentially, *e.g.*, at a rate of 4% per year for construction minerals from 2002–2015.[29] A shift to cascading renewable material cycles through increased use of timber, and especially engineered timber products such as glulam and cross-laminated timber,[46] is sometimes contemplated as a solution to the emissions by the cement and steel industries. Among other practical considerations, this idea is limited by the fact that the quantities to be substituted are substantial compared with current consumption of timber in construction. There has been a net decline of 3 billion m^3 of forest stock over the past 30 years,[47] which provides an essential function as a carbon store and habitat. Management of the substitution of construction materials (and others, such as packaging; see Section 2.3.3) to encourage expansion rather than contraction of forests is therefore critical.

Design is often considered the starting point for implementation of a Circular Economy in the construction sector, as well as in others. "Design for Disassembly" (or Deconstruction; DfD) is needed to enable recovery of components, retaining their value and embodied impacts. Modular construction, in which buildings are manufactured in pieces off-site for on-site assembly, is increasingly used to improve efficiency, quality and worker safety. This approach would lend itself to incorporation of DfD. Unfortunately, the emphasis is usually on productivity, rather than enabling disassembly with material recovery, although off-site construction does reduce raw material wastage. The chief difficulty with DfD may be the longevity of many structures, which complicates recovery logistics, particularly in the context of a distant and unknown future marketplace, and disincentivises decision-makers in the present, who will not benefit from future efficiencies. Brand's concept of separating buildings into six layers with different lifetimes (from shortest to longest: stuff, space, services, skin, structure, site),[34] is helpful for enabling refurbishment (remanufacture of buildings) and recovery of materials with shorter lifespans. New business models, and accounting and procurement methods that take into account negative externalities, need to be found to encourage closed-loop building.

Within DfD, choice of materials is a critical aspect, and also the availability of systems to collect and process them, to close the loop into renewed use. The concept of Material Marketplaces, to connect producers of end-of-life materials to new users, has existed for several decades without becoming mainstream. Its potential for uptake is increasing as data management systems become more advanced, to take into account material locations in buildings (through Building Information Modelling; BIM), as well as their geographical locations (through Geographical Information Systems; GIS), and their detailed properties including value and impacts. Degradation of quality of recovered materials, over successive cycles, can be a serious issue. For example, steel and aluminium may be manufactured with a specific composition for a particular application, but without separate collection systems for different grades of metals, different grades are necessarily amalgamated into a lower grade in subsequent cycles.[48] At the same time, valuable alloy elements are lost. However, it is also true that materials are often over-specified, in quality as well as quantity, and that materials with unconventional qualities, including variability, may be acceptable in many applications. The use of blended cements and geopolymers, in which mineral wastes from other industries are used to replace the Portland cement clinker associated with 8% of global CO_2 emissions,[30] has become common;[49] indeed, it often benefits cement quality. Although the quantities of pulverised fuel ash from coal generation and blast furnace slag that are traditionally used in this practice are small relative to the overall demand for cement, further system innovations can reduce the impacts of building with cement (concrete). Commercialisation of a concrete based on biomineralisation, that sequesters CO_2, will be a disruptive innovation.[50]

Given that DfD principles were not applied in designing most of it, strategies are also required to support recovery of components and their embodied value and impacts when our existing building stock reaches end of life.[51,52] Consideration should be given to salvaging whole buildings, rather than demolishing them.

2.3.3 Plastics and the Circular Economy

Since their introduction after the Second World War, rates of plastics use have grown exponentially, reaching 350 million tonnes globally in 2017,[33] not including textiles (see Section 2.3.4). In Europe, production of plastics was around 60 million tonnes in 2018, which is around 17% of the world's plastic production. The main application for plastics in the EU is packaging (around 40%); in contrast to construction materials, their fate at end-of-use is thus substantially in the control of individual consumers. Other sectors, such as construction (around 20%), electronics, and automobiles, also represent important and growing areas of plastic consumption.[33] Plastics are cheap and versatile materials that have contributed to the rapid expansion of some sectors, and to significant price reductions of consumer commodities. The applications of plastics in sectors such as packaging have had numerous societal benefits, such as the maintenance of food quality and safety. However, is estimated that global CO_2 emissions from plastics production and the incineration of plastic waste are about 400 million tonnes a year,[7] and the rapid increase in plastics use has not been accompanied by due consideration of their end-of-life management. A large share of plastic waste corresponds to plastic packaging with a short use life, designed to protect goods in transit. While many regions have systems to collect polyethylene terephthalate (PET; *e.g.*, drinking water bottles) and high-density polyethylene (HDPE) from households, there are many plastics that are not recycled, despite the logos they bear to indicate recyclability. Some plastics, such as composite films designed to keep food fresh, or components of consumer goods, are not mechanically recyclable. They could be recovered through thermochemical processing into feedstocks, but this is not commercially practiced at scale. Evidence of the leakage of plastic waste into the natural environments and its impact on human health and ecosystems has raised concerns around sustainable use of plastics now, and in the future. Lebreton *et al.*[53] estimated uncontrolled plastic leakage to the environment in Europe at around 3.3%, with 60–90 million tonnes of plastic waste being mismanaged globally in 2015. In a business-as-usual scenario, this will grow to 155–260 million tonnes by 2060. Current traditional approaches to waste management, even in developed countries, have proven inadequate to address the problem of plastic waste, and current trends indicate that more than 1.3 billion tonnes of plastic waste will be dumped on land and in waterbodies by 2040.[34]

In Europe alone, around 25.8 million tonnes of plastic waste are generated each year, while recycling rates across European countries vary significantly, ranging from 20% to over 50%.[55,56] Recycling rates of plastics are not straightforward to calculate or compare, as plastic content by type of product differs and there is a lack of harmonised methodologies to monitor both plastic waste and its recovery. Lau *et al.*[54] report that, globally, 14% of plastic was recycled, 44% disposed of by landfilling or incineration, and 49% mismanaged (*e.g.*, by open burning or controlled dumping), in 2016. It should be noted that even modern landfill disposal (the fate of 31% of end-of-use plastics in Europe) constitutes pollution of land, even incineration with energy for recovery is problematic for PVC. Design, including DfD and material choice, can play a role in managing the technical difficulties associated with recovery of plastics from complex consumer goods, and help limit the range of plastic polymers and composites on the market. However, increasing recycling rates faces several other barriers, including low profitability due to weak demand for recyclate plastic, and high uncertainty linked to the volatile prices of primary plastics. The loss in material quality at the end of their first life has been estimated as representing an economic loss of between EUR 70 and 105 billion annually.[55] To tackle these issues, the European Commission Action Plan for a Circular Economy of 2015 highlighted plastics as a key priority area for action. The European Strategy for Plastics in a Circular Economy launched in 2018 lays down the foundations for better design, production, use and recycling of plastics.[55] It sets the target that all plastic packaging should be recyclable by 2030, and that more than half of the plastic waste generated in Europe is recycled. This was accompanied by a proposal of a new directive to target 10 single-use plastic products that account for around 70% of marine litter. It is recognised that investments to increase recycling capacity and development of a secondary market for recycled plastics is required. The European strategy also mentions the need for interventions to increase the share of more sustainable plastics, from renewable feedstocks, and/or which are biodegradable.

In the discussion of circularity of plastics, comprehensive life cycle thinking is essential to ensure that proposed solutions have an overall net positive effect and do not create a shift of burdens. Policies such as the introduction of economic charges for the use of plastic bags have proven very effective in promoting changes in consumer behaviour, shifting from single-use plastic bags to reusable plastic or textile bags. However, a recent LCA study has shown that an organic cotton bag needs to be used 150 times to break even with the impact of a single-use low density polyethylene bag with respect to climate change impacts, and 20 000 times to break even with respect to the full range of investigated impact indicators.[57] Development of biodegradable plastics has led to problems in collection and recovery systems intended for conventional undegradable plastics, or also for kitchen scraps and garden waste, since they do not decompose as readily as food waste and plant matter. While they may be suitable for cascading to land at end-of-life, it is to also important to consider the impacts of biobased plastic production, or other substitutes such as paper for plastic packaging, on the use of land, water and other resources.

2.3.4 Textiles and the Circular Economy

The textile industry used nearly 90 million tonnes of fibre in 2015, and clothing represented more than 60% of this, consuming 33 million tonnes of synthetic (plastic) fibres, nearly 14 million tonnes of cotton, and nearly 6 million tonnes of other fibres.[58] The phenomenon of "fast fashion", with rapid roll-out of new collections at low prices per garment[59] has resulted in a steady rise of the economic importance of this industry. Unfortunately, the benefits for employment and consumer choice brought by this industry also resulted in clothing waste generation of over 92 million tonnes in 2015,[60] the majority of which ended up in landfill.[58,61] The clothing industry also emits 1.2 billion tonnes of CO_2e and consumes 93 billion m^3 of water annually.[58] LCA studies have also highlighted important impacts in terms of resource depletion, water pollution and toxicity associated with the use of pesticides, fertilisers, and other chemicals, such as dyes, during the manufacturing processes.[62,63] Thirty-five per cent of microplastics released to natural waters come from synthetic textiles,[64] which has negative environmental implications for marine ecosystems.[58,65] In this sector, the mixing of finite and renewable resources represents a particular challenge, when synthetic and natural fibres are mixed as composite products, so that separation of the cradle-to-cradle cycles becomes more complex.

The fashion industry currently represents a wasteful linear system, with a high reliance on non-renewable resources and significant emissions of pollution. It is promoted by popular culture and relies on rapid loss of value of its products. Circular fashion proposes a new regenerative system with reduced reliance on non-renewable resources, extended clothing life through reuse and remanufacturing, and recovery of fibres to be used to create new clothes. The Ellen MacArthur Foundation[58] proposes four complementary pathways to disrupt current linear systems in fashion: (1) phasing out the use of chemicals of concern and control of micro-fibre release; (2) increasing utilisation rate of clothes by making more durable apparel, promoting secondary markets and exploring remanufacturing potential; (3) moving to high-quality recycling through the development of technologies which make economically feasible the separation and recovery of fibres and/or recovery of nutrients from natural fibres; and (4) moving towards the use of renewable resources, reducing water consumption and promoting regenerative practices for natural fibres. Advancing in this direction requires coordinated action across global supply chains, changes at the policy level, and changes in consumer behaviour. Companies, such as Patagonia, have explored models of product-life extension and remanufacturing, by implementing a "take back" system, that allows repurposing or reselling of some of their items. Others, such as H&M, have started to introduce recycled fibres in their collections (*e.g.*, lyocell) or creatively explore the use of waste materials to substitute for virgin materials. The British brand Elvis and Kresse uses reclaimed materials such as decommissioned hoses from the fire service or leather off-cuts to create luxury designed items that are meant to last, and illustrates opportunities for small entrepreneurs. Companies such as MUD Jeans have pioneered leasing of products, as an alternative to purchasing, that supports profitability, while providing a mechanism for end-of-use product recovery. The pervasive introduction of new technologies in fashion, such as block chain and other traceability systems to follow the fate of individual pieces of clothing, may also help in the identification of opportunities to increase efficiency, reduce wastage, and promote ways to recover materials and value through resale, remanufacturing, and/or fibre recovery.

2.4 Supporting the Transition to a Circular Economy: Assessment of Material Flows and Impacts

To make progress towards circularity, tools that provide an understanding of the system, and its environmental (and sometimes social and economic) impacts are needed. In addition to LCA (Chapter 3), the analytical tool kit for the Circular Economy usually includes Material Flow Analysis (MFA), and may include environmentally extended Input-Output modelling, which, like LCA, helps to identify opportunities for application of Circular Economy principles, and analyse the implications of their adoption.

MFA is widely used in the field of industrial ecology to describe the industrial metabolism, in terms of its flows and stock, for a defined time and space.[66] MFA quantifies the material flows of "process chains comprising extraction or harvest, chemical transformation, manufacturing, consumption, recycling and disposal of materials",[67] *i.e.*, the resource life cycle. The unit of measurement is physical units (usually tonnes), to express the inflows (inputs) and outflows (outputs) of a system or process system. In MFA waste, effluents and emissions are considered part of the output side, *i.e.*, products or unwanted residuals of industrial/societal processes. The analysis is based on the principle of conservation of matter (mass and energy) derived from the first law of thermodynamics, by which matter (mass or energy) cannot be created or destroyed, but can be converted and transformed from one form to another. Generally, depending on the focus and boundary definition, there are two main types of MFAs. Substance Flow Analysis (SFA) follows a single element or compound (*e.g.* carbon, zinc, sugar) throughout the system. System-wide MFA accounts for all materials entering and leaving the system, as well as, in some cases, "unused extraction" or displaced resources that do not enter the system as priced resources and goods. There is a continuum of methods and approaches that can be applied at multiple scales, including global, national, industrial park, region, city or business and product/process level.

In economy-wide MFA, Direct Material Input (DMI) refers to the total direct inputs of materials in an economy, including extraction and imports. Consumption-based inputs include Domestic Material Consumption (DMC), which refers to the total amount of materials directly used by an economy, or Raw Material Consumption (RMC), which is preferred, as it also captures upstream implications associated with imported goods.[68,69] Domestic Processed Output (DPO) is the most widely used output, which accounts for outflows of processed materials returned to natural systems in the form of residuals.[69]

MFA provides information about the scale of the industrial metabolism and efficiency of use of resources, and helps to identify the main areas of inefficient resource use, leakage to natural systems and dispersion, all of which represent losses to the system. A number of nations and regions have adopted MFA- and SFA-based indicators to understand opportunities to increase circularity of the system (*e.g.*, Republic of Korea, Europe, or China). Sectoral applications for construction and other resource intensive sectors are also common (*e.g.*, for steel in construction,[70] SFA for specific critical metals,[71] or substances of concern.[72]

A limitation of MFA is the aggregated character of its indicators in a mass flow, which does not capture economic and environmental impacts, or other details along the supply chain. MFA can form the basis of LCA (Chapter 3) of a process, but this technique does not lend itself to assessment of economy-wide impacts. Complex trade patterns and globalised supply chains mean that the environmental impacts caused by consumption in one country or region extend to several countries and regions in the world, in multifaceted ways which are not well captured through territorial monitoring of resource extraction and pollution.[73] Departing from the traditional concept of Input-Output Leontief matrices, which describes the flow of final and intermediate goods and services among different sectors of activity and consumers in monetary terms, Environmental Extended Input Output analysis (EEIO) includes environmental information for each sector of activity, such as resource depletion, water and land use, emissions and other burdens associated with the specific activities. Generally based on Multi-Regional Input-Output models (MRIO), EEIO provides further detail for quantifying the material footprints and associated economic transactions for all products and sectors in an economy, depicting the full complexity of supply chains[74] and can be integrated in more complex Computable General Equilibrium (CGE) models. This has the advantage of adding granularity to the material footprints associated with final products, as well as geographic detail of the origin of raw materials and intermediate products.[74] These studies, though complex, provide a comprehensive evaluation of resources embedded in products, and thus a more accurate picture of the consumption-based material footprint of different nations. Applications of EEIO modelling to assess progress to circularity and impact of policies have been undertaken in recent years. Giljum *et al.*[74] used EEIO modelling to analyse changes in the material footprint of the EU, to conclude a shift in the origin of raw materials in the composition of the final demand, with a rapidly decreasing share of materials extracted in the EU in favour of materials extracted in other regions, primarily China, and an increase in import dependency for metal ores. The study also points to the increasing growth of the material footprint of services and manufacturing products from biomass. Similarly, Koskela *et al.*[75] use EEIO modelling to identify material intensity of different sectors in Finland. The analysis points to the export sector and domestic construction sector as industrial activities with highest material footprints. Tisserant *et al.*[76] employ EEIO modelling to describe the pattern of waste generation globally and identify opportunity areas with potential to close material cycles.

These tools have been developed under different theoretical and epistemological frameworks, which is reflected in the "*scope, focus and orientation*" of the different approaches.[77] They, and other tools such as cost–benefit analysis and dynamic systems modelling can be combined into complex models for better understanding of the system, and the economic, social, and environmental implications of the transition towards the circular economy. There are a number of aspects that need to be considered:

- The system boundary needs to be clearly applied consistently across products/business models/systems under evaluation, to define the processes and flows that will be included in, or omitted from, the analysis. Although the definition of boundaries is traditionally rooted in engineering approaches, broader understanding of the institutional system that shapes those interactions needs to be considered in the definition of boundaries.[77]
- Analysis may compare different scenarios, *e.g.*, status quo or Business-as-Usual (BAU), with improved circular economy alternatives.
- Gathering data on inputs, outputs and processes is a critical and, in many cases, a limiting aspect of the analysis, as data can be fragmented, uncertain or access-restricted.
- Finally, acknowledging and reflecting on methodological limitations derived from the selection of tools and data sources allows for better interpretation of the findings.

2.5 New Business Models for Value Creation in the Circular Economy

Rooted in the pragmatic perspective promoted by practitioners of the concept, new Circular Economy business models are generally put forward as the main vehicles through which to unleash circular economy opportunities. The concept of business models refers to the core elements that define how companies and organisations generate value. Osterwalder and Pigneur[78] provide a broad definition: "A business model describes the rationale of how an organization creates, delivers, and captures value" and propose a 9-dimensional framework. The interlinked dimensions cover the following areas: key activities; value proposition; customer relationships; customers; resources; costs; channels; and revenue. The main idea behind the concept is the notion that circular economy thinking can change the way value is created in organisations, through mobilising different dimensions of the business system. Core disruption operates through changes in the revenue and resource dimensions, with cascading implications for the other dimensions, moving from a resource-intensive model to one that seeks to retain value by preserving functionality and quality of resources.

Four main principles of value creation have been identified in the Circular Economy literature.[12] The first principle is referred to as "the power of the inner circle" and highlights the potential business opportunities linked to cycling products, components and materials for as long as possible, where tighter loops are preferred to outer loops. This principle is also known as the

"3 Rs" approach, where there is a hierarchy of options, including reuse as the preferred option, remanufacturing next, and, finally, recycling. Business models thus mapped create value by reducing costs associated with materials, labour and energy while reducing impacts associated with extraction, transformation, manufacturing, and disposal of components and products.

The second principle, "the power of cycling for longer" is mostly related to extending the life of products and components through multiple lives or multiple users. This favours models that create value by preventing obsolescence and increasing product durability. Products design for durability and upgradability can prevent lost of value and create value through cycling resources in different markets (primary *versus* secondary) and for multiple users. Performance models, where the consumer pays for the performance rather than the ownership of the product (*e.g.*, MUD Jeans, see Section 2.3.4), are an example of operationalisation of this principle. This principle is also closely related to the previous one, as after the first user, reuse and remanufacturing can be additional options to extend the life of the product or its components.

The third principle of value generation refers to the "power of cascaded use". This principle builds on notions from industrial ecology and industrial symbiosis, where a waste from one activity or sector becomes a resource for another activity or sector. Whereas in renewables cycles, resources are cascaded through multiple uses to regenerate natural capital, subsequent use of finite resources between industries with different needs and specifications is also referred to as cascading. BS 8001:2017 defines cascading use as "repeated use of a resource usually at decreasing quantity/quality at each subsequent stage or cycle".[79] Food processing residues can be used by the pet and animal feed sector, or garden waste can be treated in anaerobic digestion (AD) plants for recovery of nutrients and production of biogas. Unwanted clothing can be cascaded into the second-hand markets and then into furniture or insulation applications. The cascading is possible because applications in industrial sectors are subject to different material specifications and qualities.

The fourth of the principles of value generation in the circular economy is "the power of pure circles". The intention is to avoid contamination, and remove chemicals of concern (which can accumulate in closed loops) from the system, thus reducing risks and/or costs associated with multiple reuse and closed-loop recycling systems. Contaminants may reduce the quality of materials in loops (*e.g.*, aluminium, see Section 2.3.1) or cause harm to consumers, *e.g.*, mineral oil that can accumulate in paper cycles rendering cardboard unsuitable for food-grade packaging. EU regulations on chemicals, such as REACH[80] and RoHS[81] have advanced our understanding of chemicals of concern. Eradication of contaminants can be difficult and costly, and can compromise opportunities for material recovery.

A myriad of examples of case studies of successful implementation of new business models has grown exponentially in recent years.[12,82] Other authors, such as Urbinati *et al.*[83] have highlighted that there remains a lack of understanding of how organisations can adapt their internal and external processes to follow new circular business models. They propose a taxonomy that considers: (1) how the value proposition to the customer is articulated to incorporate circular economy principles; and (2) how the value chain of the organisation is modified as a result of adapting circular economy principles.

BS 8001:2017[79] proposes a flexible 8-stage process for the practical implementation of Circular Economy principles in organisations. The first stage is the "framing", which refers to understanding of the relevance of the circular economy for their business, mapping of stakeholders and leadership. The second stage is "scoping", which implies better understanding of the current system, identifies linkages between the company current vision and the Circular Economy, and defines a Circular Economy vision and high level strategy. Stage 3 is about "idea generation" and exploring different options to increase circularity. Stage 4 "feasibility" involves undertaking feasibility studies, reviewing different alternatives and identifying preferred options. In stage 5, a "business case" of selected options is developed to understand opportunities and risks. Stage 6 is "piloting and prototyping" and thus points towards small-scale experimentation to test operationalization of the proposed Circular Economy business model. Stage 7 is "delivery and implementation" and proposes the roll-out and upscale of proven approaches to circular economy in the organisation. Finally, stage 8 "monitor, review and report" recommends a tracking and monitoring system to understand and evaluate implications of the circular economy interventions.

Recent years have seen several taxonomies of new business models that try to organise circular economy business practice.[12,82] BS 8001:2017[79] proposes a categorisation system that includes six different types of new business models: (1) on-demand; (2) dematerialisation; (3) product-life extension-reuse; (4) recovery of secondary materials/by-products; (5) product as a service (product-service system); and (6) sharing economy and collaborative consumption. These types of business models can be combined in different hybrid forms in their implementation by businesses, and not all may be suitable for different types for activity.

Interestingly, while the literature on successful business model adoption is prolific, there is still little understanding of unsuccessful stories. Understanding the barriers and hurdles of businesses, from niche to mainstream practices, is an underexplored research area. The following section, looking at the framework conditions defined by policy and regulatory frameworks may provide some insights into some of the dimensions that may require profound changes to enhance viability of Circular Economy business models.

2.6 Policy Support for the Circular Economy

The emergence and development of new business models to transition towards the circular economy requires a supportive policy framework. China and the EU have led the process of policy-making in this area. In Europe, much of the coordinated action around the circular economy has been operationalised by the EU Action Plan for the Circular Economy,[84] which builds on the pillars of the Europe 2020 strategy and the flagship initiative on resource efficiency and associated roadmap. The package has a strong focus on eco-innovation and puts the emphasis on Circular Economy solutions, covering all phases of product, component, process and material from design to disposal and recovery/recycling. Although most of the proposals from the package are non-binding, the package also includes a legislative proposal for the update of the Waste Directives, which includes more stringent targets and better harmonisation across countries. The EU Circular Economy Package also includes 54 specific actions across different areas, including circular economy requirements in the eco-design Directive, guidance on Best Available Technique Reference documents (BREFs) across sectors; guidance on substitution of hazardous substances; revised waste legislative framework; promotion of collaboration among stakeholders; promotion of secondary markets; and a plastic strategy, including a legislative proposal on single use plastic among others measures. The 54 actionable items have been implemented, which does not necessarily mean that the EU is circular, but rather that some of the policy and legislative pillars to promote the transitions towards the circular economy, including a monitoring framework, have been introduced. It is also important to understand the synergies and potential trade-offs, between policy efforts towards the Circular Economy and other well-established areas of policy action at the EU level.[85] An example of this is the case of Environmentally Harmful Subsidies or environmental implications of some of the transport and infrastructure policies. While most of the proposed changes and policy measures can indeed contribute positively to further uptake of the Circular Economy concept, further policy action and incentivisation will be required, such as the much-discussed green tax reform.[85]

China has also been extremely active in introducing the Circular economy at the core of its policy making. The 11th Five-Year Plan (2006–2010) included a full chapter on Circular Economy, with emphasis on the opportunities for more efficient use of resources and reduction of pollution, as a cornerstone of the policy of the "ecological civilisation". In fact, the Plan proposed ambitious mechanisms to promote Circular Economy, including the Circular Economy Law, which includes a range of regulatory measures to ensure efficient use of energy, resources and minimisation of waste and emissions.[69] The 12th Five-Year Plan (2011–2015) saw the Circular Economy acquiring the status of a national development strategy.[81] Emission reduction targets for certain pollutants have been extended in the 13th Five-Year Plan and have been accompanied with targets in waste reutilisation and industrial symbiosis, *e.g.*, 72% of industrial solid waste needs to be recovered and resource productivity should increase by 15%.[86] The implementation of the Circular Economy in China is structured following the spatial division of competencies at the micro, meso and macro levels.[69] At the micro level, the emphasis is put on resource efficiency and cleaner production, and better accountability of environmental performance, while the meso level focuses on improvements of overall resource efficiency, waste reduction and waste reutilisation in industrial parks.[69] Finally, the national level focuses on the overall coordination of Circular Economy policies, the establishment of a monitoring framework and coordination of large-scale demonstrations projects. An example of this is the "10–100–1000 strategy", which includes 10 major programmes for recycling industrial waste and transforming industrial areas; 100 demonstration cities where Circular Economy principles are embedded at different levels from infrastructure, to mobility and waste; and 1000 demonstration projects targeting individual companies and industrial estates.[86]

2.7 Drivers and Challenges for the Transition to a Circular Economy

The linear system has been successful in supporting rapid industrial growth, but this has been achieved at the expense of shrinking the natural systems that act as the providers of resources and life-supporting systems.[14] A different system is thus needed to accommodate the needs of an increasing population and to achieve the UN SDGs.[87] An intriguing question is why the Circular Economy, based on well-established fields such as environmental economics or Industrial Ecology, has captured the interest of business and policy stakeholders in the last decade. Contextual factors for the rapid uptake of the concept include: the increased volatility of commodity markets; risks associated to the supply of critical materials; and population dynamics and rapid development of emergent economies. The combination of these megatrends has resulted in uncertainty regarding security of supply and price of primary raw materials, that combined with policy efforts to reduce GHG emissions and increase resource efficiency have contributed to create a conducive framework for the transition to more circular systems. At the same time, information technologies, "big data", and "block-chain" have transformed information capture and management and provided opportunities to increase traceability and recovery of products, components and materials (*e.g.*, BIM, see Section 2.3.2).

Transition processes are complex however, and not exempt of difficulties. Socio-technical systems consist of network of actors, institutions and socio-technical norms, as well as material artefacts and knowledge that in interaction define the behaviour of the system and their outcomes.[88] Systemic changes such as those proposed by the circular economy concept require changes at all system levels, and by all actors and institutions. This entails overcoming path-dependencies, system inertia and socio-technological lock-ins. The concept of "the web of constraints" captures well the set of dynamic and interrelated barriers that constrain action towards the circular economy, and that include technical and technological barriers to behavioural, institutional and cultural aspects, trapped in linear loops.[89] Altering the dynamics of the system will generate costs, including learning costs, and readjustment of influence between actors in the system and, therefore, change may face resistance.

Finally, and connected to the above, it is important to work towards a better understanding of circular economy actions at different levels (micro, meso and macro), to fully monitor the progress towards circularity of the system as a whole. More work is required to fully understand the resource (and energy) implications of circular economy actions at all levels. This includes better understanding and mapping of resource flows, with an emphasis on improved accountability of embodied carbon and other impacts through the use of LCA and enhanced monitoring of global resource use and distribution patterns. Increasing resource circulation may not ensure development within "planetary boundaries"[10,26] unless the scale of the industrial metabolism is also monitored, including acknowledging aspects such as the "CE rebound effect"[90] and additions to the socio-economic stock both in developing and emerging economies.

References

1. UNECE, https://www.unece.org/fileadmin/DAM/energy/se/pdfs/UNFC/Anthropogenic_Resources/UNFC_Antropogenic_Resource_Specifications.pdf, (accessed September 2020).
2. F. Krausmann, C. Lauk, W. Haas and D. Wiedenhofer, From resource extraction to outflows of wastes and emissions: The socioeconomic metabolism of the global economy, 1900–2015, *Glob. Environ. Change*, 2018, **52**, 131.
3. International Resource Panel, https://www.resourcepanel.org/reports/global-resources-outlook, (accessed September 2020).
4. D. A. Vallero and G. Blight in *Waste*, ed. T. M. Letcher and D. A. Vallero, Academic Press, 2nd edn, 2019, p. 129.
5. S. Bringezu and R. Bleischwitz, *Sustainable Resource Management*, Routledge, London, 2017.
6. IPCC, https://www.ipcc.ch/report/ar5/wg1/, (accessed September 2020).
7. Ellen MacArthur Foundation, https://www.ellenmacarthurfoundation.org/publications/the-new-plastics-economy-rethinking-the-future-of-plastics-catalysing-action, (accessed September 2020).
8. J. M. Allwood, T. G. Gutowski, A. C. Serrenho, A. C. H. Skelton and E. Worrell, *Philos. Trans. R Soc. A*, 2017, **375**, DOI: 10.1098/rsta.2016.0361.
9. Material Economics, https://materialeconomics.com/publications/the-circular-economy, (accessed September 2020).
10. W. Haas, F. Krausmann, D. Wiedenhofer and M. Heinz, *J. Ind. Ecol.*, 2015, **19**(5), DOI: 10.1111/jiec.12244.
11. A. Mayer, W. Haas, D. Wiedenhofer, F. Krausmann, P. Nuss and G. A. Blengini, *J. Ind. Ecol.*, 2019, **23**(1), 62.
12. Ellen MacArthur Foundation, https://www.ellenmacarthurfoundation.org/circular-economy/concept/infographic, (accessed September 2020).
13. Ellen MacArthur Foundation, https://www.ellenmacarthurfoundation.org/assets/downloads/publications/EllenMacArthurFoundation_Growth-Within_July15.pdf, (accessed September 2020).

14. J. Korhonen, A. Honkasalo and J. Seppälä, *Ecol. Econ.*, 2018, **143**, 37.
15. T. E. Graedel and B. Allenby, *Industrial Ecology*, Prentice Hall, Inc, 1995.
16. W. McDonough and M. Braungart, *Cradle to Cradle: Remarking the Way We Make Things*, North Point Press, New York, 2002.
17. H. E. Daly, *Steady-State Economics*, Island Press, Washington, DC, 1991.
18. W. Stahel, *The Performance Economy*, Palgrave Macmillan, Basingstoke, 2010.
19. D. W. Pearce and K. R. Turner, *Economics of Natural Resources and the Environment*, Harvester Wheatsheaf, Hemel Hempstead, Herts., UK, 1990.
20. P. Ghisellini, C. Cialani and S. Ulgiati, *J. Clean. Prod.*, 2016, **114**, 11.
21. K. E. Boulding, in *Environmental Quality in a Growing Economy*, ed. H. Jarrett, John Hopkins University Press, Baltimore, 1996, p. 3.
22. L. W. Jelinski, T. E. Graedel, R. A. Laudise, D. W. McCall and C. K. Patel, *Proc. Natl. Acad. Sci.*, 1992, **89**(3), 793.
23. R. Lifset and T. E. Graedel in *A Handbook of Industrial Ecology*, ed. U. A. Robert and W. A. Leslie, Edward Elgar Publishing, Cheltenham, 2015.
24. J. Kirchherr, D. Reike and M. Hekkert, *Conserv. Recycl.*, 2017, **127**, 221.
25. A. Tukker and P. Ekins, *Ecol. Econ.*, 2019, **155**, 36.
26. W. Steffen, K. Richardson, J. Rockstrom, S. E. Cornell, I. Fetzer and E. M. Bennett, *et al.*, *Science*, 2015, **347**(6223), 1259855.
27. Publications Office of the European Union. https://op.europa.eu/en/publication-detail/-/publication/1f0d8515-8dc0-4435-ba53-9570e47dbd51, (accessed September 2020).
28. P. Stegmann, M. Londo and M. Junginger, *Resour., Conserv. Recycl.: X*, 2020, **6**, 100029.
29. F. Krausmann, C. Lauk, W. Haas and D. Wiedenhofer, *Glob. Environ. Change*, 2018, **52**, 131.
30. R. M. Andrew, *Earth Syst. Sci. Data*, 2018, **10**(1), 195.
31. J. M. Cullen, J. M. Allwood and M. D. Bambach, *Environ. Sci. Technol.*, 2012, **46**(24), 13048.
32. Forestry Research, https://www.forestresearch.gov.uk/tools-and-resources/statistics/forestry-statistics/forestry-statistics-2019/international-forestry/, (accessed September 2020).
33. Plastics Europe, https://www.plasticseurope.org/application/files/9715/7129/9584/FINAL_web_version_Plastics_the_facts2019_14102019.pdf, (accessed September 2020).
34. S. Brand, *How Buildings Learn: What Happens After They're Built*, Penguin, London, 1994.
35. European Environment Agency, https://www.eea.europa.eu/themes/waste/waste-management/construction-and-demolition-waste-challenges/download.pdf.static, (accessed September 2020).
36. The Green Construction Board, https://www.constructionleadershipcouncil.co.uk/wp-content/uploads/2016/05/ZAW-Report-Final-Draft-25-February-2020.pdf, (accessed September 2020).
37. MPA, www.concretecentre.com/Publications-Software/Publications/Whole-life-Carbon-and-Buildings.aspx, (accessed September 2020).

38. WRAP, www.wrap.org.uk/sites/files/wrap/CRWP-Demolition-Report-2009.pdf, (accessed September 2020).
39. MPA, mineralproducts.org/documents/Contribution_of_Recycled_and_Secondary_Materials_to_Total_Aggs_Supply_in_GB.pdf, (accessed September 2020).
40. MPA, https://www.concretecentre.com/TCC/media/TCCMediaLibrary/Products/MB_8thPerformanceReport_SCF_2014data.pdf, (accessed September 2020).
41. S. Van Ewijk and J. A. Stegemann, *J. Clean. Prod.*, 2016, **132**, 122.
42. D. L. Evans, J. N. Quinton, A. M. Tye, Á. Rodés, J. A. C. Davies and S. M. Mudd, *et al.*, *SOIL*, 2019, **5**(2), 253.
43. UN environmental programme, https://www.unenvironment.org/resources/publication/2019-global-status-report-buildings-and-construction-sector, (accessed September 2020).
44. K. T. Adams, M. Osmani, T. Thorpe and J. Thornback, *Proc. Inst. Civil Eng. Waste Resour. Manag.*, 2017, **170**(1), 15.
45. Ellen MacArthur Foundation, https://www.ellenmacarthurfoundation.org/assets/downloads/Built-Env-Co.Project.pdf, (accessed September 2020).
46. C. Rose, D. Bergsagel, T. Dufresne, E. Unubreme, T. Lyu and P. Duffour, *et al.*, *Sustainability*, 2018, **10**(11), 4118.
47. FAO, http://www.fao.org/documents/card/en/c/ca8753en, (accessed September 2020).
48. J. M. Cullen and J. M. Allwood, *Environ. Sci. Technol.*, 2013, **47**(7), 3057.
49. C. Shi, B. Qu and J. L. Provis, *Cem. Concr. Res.*, 2019, **122**, 227.
50. W. De Muynck, N. de Belie and W. Verstraete, *Ecol. Eng.*, 2010, **36**(2), 118.
51. C. M. Rose and J. A. Stegemann, *Sustainability*, 2018, **10**(1), 229.
52. C. M. Rose and J. A. Stegemann, *Sustainability*, 2019, **172**(3), 129.
53. L. Lebreton and A. Andrady, *Palgrave Commun.*, 2019, **5**(1), 6.
54. W. W. Y. Lau, Y. Shiran, R. M. Bailey, E. Cook, M. R. Stuchtey and J. Koskella, *et al.*, *Science*, 2020, **369**, 1455.
55. EC, https://ec.europa.eu/environment/circular-economy/pdf/plastics-strategy-brochure.pdf, (accessed September 2020).
56. Plastics Europe. https://www.plasticseurope.org/application/files/5715/1717/4180/Plastics_the_facts_2017_FINAL_for_website_one_page.pdf, (accessed September 2020).
57. Ministry of Environment and Food of Denmark, https://bioplasticsnews.com/wp-content/uploads/2019/10/Life-Cycle-Assessment-of-grocery-carrier-bags.pdf, (accessed September 2020).
58. Ellen MacArthur Foundation, https://www.ellenmacarthurfoundation.org/publications/a-new-textiles-economy-redesigning-fashions-future, (accessed September 2020).
59. M. Koszewska, *Autex Res. J.*, 2018, **18**(4), 337.
60. Global Fashion Agenda, https://globalfashionagenda.com/wp-content/uploads/2017/05/Pulse-of-the-Fashion-Industry_2017.pdf, (accessed September 2020).

61. WRAP. http://www.wrap.org.uk/sites/files/wrap/UK_Textiles_EPR.pdf, (accessed September 2020).
62. G. Baydar, N. Ciliz and A. Mammadov, *Resour., Conserv. Recycl.*, 2015, **104**, 213.
63. European Parliament, https://www.europarl.europa.eu/RegData/etudes/BRIE/2019/633143/EPRS_BRI(2019)633143_EN.pdf, (accessed September 2020).
64. Institution of Mechanical Engineers, https://www.imeche.org/docs/default-source/1-oscar/reports-policy-statements-and-documents/imeche-engineering-out-fashion-waste4daedc8d54216d0c8310ff0100d05193.pdf?sfvrsn=2, (accessed September 2020).
65. B. Henry, K. Laitala and I. G. Klepp, *Sci. Total Environ.*, 2019, **652**, 483.
66. P. H. Brunner and H. Rechberger, *Practical Handbook of Material Flow Analysis*, CRC press, Boca Raton, 2014.
67. S. Bringezu and Y. Moriguchi in *A Handbook of Industrial Ecology*, ed. R. Ayres and L. Ayres, Edward Elgar Publishing, Cheltenham, 2015.
68. Eurostat, https://ec.europa.eu/eurostat/documents/3859598/9117556/KS-GQ-18-006-EN-N.pdf/b621b8ce-2792-47ff-9d10-067d2b8aac4b, (accessed September 2020).
69. Y. Geng, J. Fu, J. Sarkis and B. Xue, *J. Clean. Prod.*, 2012, **23**(1), 216.
70. J. Davis, R. Geyer, J. Ley, J. He, R. Clift and A. Kwan, *et al.*, *Resour., Conserv. Recycl.*, 2007, **51**(1), 118.
71. K. Nansai, K. Nakajima, S. Kagawa, Y. Kondo, S. Suh and Y. Shigetomi, *et al.*, *Environ. Sci. Technol.*, 2014, **48**(3), 1391.
72. L. Li and F. Wania, *Environ. Int.*, 2016, **94**, 674.
73. K. Stadler, R. Wood, T. Bulavskaya, C.-J. Södersten, M. Simas and S. Schmidt, *et al.*, *J. Ind. Ecol.*, 2018, **22**(3), 502.
74. S. Giljum, H. Wieland, S. Lutter, M. Bruckner, R. Wood and A. Tukker, *et al.*, *J. Econ. Struct.*, 2016, **5**, 17.
75. S. Koskela, T. Mattila, R. Antikainen and I. Mäenpää, *Resources*, 2013, **2**(3), 151.
76. A. Tisserant, S. Pauliuk, S. Merciai, J. Schmidt, J. Fry and R. Wood, *et al.*, *J. Ind. Ecol.*, 2017, **21**(3), 628.
77. E. Iacovidou, J. Millward-Hopkins, J. Busch, P. Purnell, C. A. Velis and J. N. Hahladakis, *et al.*, *J. Clean. Prod.*, 2017, **168**, 1279.
78. A. Osterwalder and Y. Pigneur, *Business Model Generation – A Handbook for Visionaries, Game Changers and Challengers*, John Wiley and Sons, Inc., New Jersey, 2010.
79. British Standard Institution, *BS 8001:2017. Framework for Implementing the Principles of the Circular Economy in Organizations – Guide*, The British Standards Institution, London, 2017.
80. EC, https://ec.europa.eu/environment/chemicals/reach/reach_en.htm, (accessed September 2020).
81. EC, https://eur-lex.europa.eu/legal-content/EN/TXT/?qid=1511965370860&uri=CELEX:32017L2102, (accessed September 2020).

82. Accenture, https://www.accenture.com/t20150523t053139__w__/us-en/_acnmedia/accenture/conversion-assets/dotcom/documents/global/pdf/strategy_6/accenture-circular-advantage-innovative-business-models-technologies-value-growth.pdf, (accessed September 2020).
83. A. Urbinati, D. Chiaroni and V. Chiesa, *J. Clean. Prod.*, 2017, **168**, 487.
84. EC, https://eur-lex.europa.eu/legal-content/EN/TXT/?uri=CELEX%3A52015DC0614, (accessed September 2020).
85. T. Domenech and B. Bahn-Walkowiak, *Ecol. Econ.*, 2019, **155**, 7.
86. J. A. Mathews and H. Tan, *Nature*, 2016, **531**(7595), 440.
87. UN, https://www.undp.org/content/undp/en/home/sustainable-development-goals.html, (accessed September 2020).
88. J. Markard, R. Raven and B. Truffer, *Res. Policy*, 2012, **41**(6), 955.
89. M. Dijk, J. Backhaus, H. Wieser and R. Kemp, *Sustainability Sci. Pract. Policy*, 2019, **15**(1), 62.
90. T. Zink and R. Geyer, *J. Ind. Ecol.*, 2017, **21**(3), 593.

CHAPTER 3

Life Cycle Assessment as a Metric for Circular Economy

X. C. SCHMIDT RIVERA,*[a] P. BALCOMBE[b] AND M. NIERO[c]

[a] Institute of Energy Futures, College of Engineering, Design and Physical Sciences, Brunel University London, Kingston Lane, Uxbridge UB8 3PH, UK; [b] Division of Chemical Engineering and Renewable Energy, School of Engineering and Material Science, Queen Mary University of London, E1 4NS, UK; [c] Aalborg University, Department of Planning, Research group "Sustainable Design and Transition, A.C.", Meyers Vænge 15, 2450 Copenhagen, Denmark
*Email: ximena.schmidt@brunel.ac.uk

3.1 Introduction

A core component of successful implementation of a circular economy (CE) strategy is the quantification of improvements or changes to environmental impacts, resource usage, waste and/or economic costs. Life cycle methodologies enable this quantification, to assess and define the environmental performance of circular product designs, as well as large-scale industrial changes such as the implementation of different waste management options.[1] According to Iacovidou *et al.*,[2] "LCA is the best known and commonly used tool for assessing the environmental impacts of a product's life from raw material extraction to end of use, disposal and end of life management, making capable and useful comparisons between products, processes and systems." However, there are several ways to conduct a Life Cycle Assessment (LCA) and incorporate LCA into CE strategies, as well as several challenges relating to data collection, uncertainty and expertise development.

Life Cycle Assessment: A Metric for the Circular Economy
Edited by Aiduan Borrion, Mairi J. Black and Onesmus Mwabonje

Published by the Royal Society of Chemistry, www.rsc.org

This chapter outlines where LCA is incorporated into circular economy metrics and frameworks, how LCA can be used to define many of the CE metrics and indicators, followed by the taxonomy of a LCA, how to conduct one and why they are important for CE strategies. The plethora of guidelines associated with different methods or stages of LCA are described and differentiated. Key challenges and barriers associated with incorporating LCA are described, before conclusions are given.

3.2 Frameworks Implementing Circular Economy Using LCA

Several decision support frameworks for implementing circular economy (CE) strategies, both at the product and organization level, include LCA. For example, LCA plays a key role in the decision support framework developed by Niero *et al.*,[3] in which LCA is combined with the Cradle to Cradle® (C2C) certification program, in the context of packaging. The aim of the 4-step framework is to support the development of continuous loop packaging systems, where LCA is used in the third step to assess the environmental impacts of alternative scenarios of C2C certification. The alternative LCA scenarios encompass different improvement strategies, such as changes in material composition, the use of renewable energy in product manufacturing and supply chains,[4] increased recycled content, and recycling rates.[5] The application of the framework combining LCA and C2C certification allows the identification of actions to be prioritized to achieve a continuous material loop for beverage packaging, both from an environmental and an economic point of view, therefore targeting the development of circular product systems.

Another framework, named BECE (Backcasting and Eco-design for the Circular Economy), helps businesses to tackle CE holistically by embedding the concept into corporate decision making and by bringing operational and systems thinking together, thus increasing the likelihood of successful implementation. Key elements are backcasting and eco-design: backcasting is "a strategic business planning tool that aims to achieve a particular desirable future state, by exploring alternative non-predictive pathways toward it by developing different scenarios";[6] and eco-design is a tool that helps incorporate environmental considerations into product design, with the aim of minimizing life cycle environmental impacts.[6] The BECE framework consists of 10 steps, where LCA plays a key role in "product selection and evaluation" (step 5). In this context, LCA helps to address potential issues associated with burden shifting, to ensure that circularity is not achieved to the detriment of other impacts, and to identify further opportunities for improvement through eco-design.[6]

At the product level, Rönnlund *et al.*[7] developed an eco-efficiency indicator framework based on LCA, covering ten important issues of product environmental sustainability, including some key aspects for implementing CE, such as material efficiency and reutilization of secondary raw materials. This indicator framework enables an environmental sustainability benchmarking of products, and has been conceived within the metallurgical industry.

Niero and Hauschild[8] recommend using the Life Cycle Sustainability Assessment (LCSA) framework intended as a combination of LCA, Environmental Life Cycle Costing (LCC) and Social LCA (S-LCA)[9] to evaluate circular economy strategies, since it is the most comprehensive framework, and is effective at preventing burden shifting between different life cycle stages and impact categories in the value chain.

As far as the implementation of CE at the organization level is concerned, only a limited number of frameworks are available. The British Standard,[10] *BS 8001:2017 Framework for implementing the principles of the circular economy in organizations*, was developed to provide guidance to organizations in the transition toward more circular and sustainable modes of operation. It is based on an 8-stage flexible framework to assist organizations to develop a road map for continual and transformational improvement. LCA (and S-LCA) are mentioned as quantitative assessment tools as part of the feasibility assessment phase (step 4). Niero and Schmidt Rivera[11] propose to combine the BS 8001:2017 8-step framework with the LCSA framework, thus providing an operative tool to prioritize the selection of the most feasible options to implement the CE vision of an organization.

3.3 Circular Economy Metrics and Indicators Using LCA

CE strategies can be applied at the macro (*i.e.* region, nation, sector), meso (*i.e.* eco-industrial parks) and micro (*i.e.* product and organization) levels. Most attempts to define indicators for measuring CE strategy have so far addressed the macro and meso levels, and only a limited number of indicators are available at the product-level scale.[12,13] According to the European Environment Agency[14] these macro- and meso-level indicators "give insight into material flows in the economy as a whole, but are unable to capture the mechanisms behind these flows. Life-cycle approaches offer promising possibilities at the product group level but remain far from operational".[14] On the other hand, there are studies showing that LCA plays a key role in the development of meaningful indicators for the CE with regard to the product-level assessment.[12,15] A number of reviews on CE metrics have recently been performed, and are described below.

According to Bocken *et al.*[16] CE metrics are categorized into three groups: (1) narrowing resource flows (or resource efficiency), aimed at using fewer resources per product; (2) slowing resource loops, *i.e.* extending and/or intensifying the utilization period of products, thus resulting in a slowdown of the flow of resources; and (3) closing resource loops, *i.e.* aiming at a circular flow of resources, by closing the loop between post-use and production through recycling. LCA has already been used as a metric to assess the environmental sustainability of all three CE categories: *e.g.* Huysman *et al.*[17] for narrowing resource flows, Bakker *et al.*[18] for slowing resource flows and Niero and Olsen[19] for closing resource flows.

Elia *et al.*[12] analyzed the current literature on CE assessment and proposed a reference framework comprised of five metric categories for the monitoring phase of a CE strategy: reducing input and use of natural resources; reducing emissions; reducing valuable material losses; increasing share of renewable and recyclable resources; and increasing the value durability of products. Of the five CE requirements identified, LCA allows quantification of all except "increasing the value durability of products". Whilst LCA is evidently a useful tool for these CE metrics, barriers to implementation include data availability and uncertainty, time intensiveness, and ease of understanding for non-practitioners. Elia *et al.*[12] also proposed a taxonomy of index-based methodologies for measuring the adoption of CE based on two factors: (1) the index-based method typology (either single synthetic indicator or set of multiple indicators); and (2) the parameter(s) to be measured (*i.e.* material and energy flow, land use and consumption, and other life cycle-based criteria).

Saidani *et al.*[20] performed a systematic literature review considering both academic and grey literature and identified 55 sets of circularity indicators, developed by scholars, consulting companies and governmental agencies. Furthermore, inspired by existing taxonomies of eco-design tools and sustainability indicators, and in line with the CE characteristics, they developed a taxonomy of circularity indicators including 10 categories. These categories include: the levels of CE implementation (*e.g.* micro, meso, macro); the CE loops (maintain, reuse, remanufacture, recycle); the performance (intrinsic, impacts); the perspective of circularity (actual, potential) they are taking into account; or their degree of transversality (generic, sector-specific). Although most of the identified circularity indicators address material and/or resource circularity, some of the studies combine them with LCA results. The combination of circularity indicators with LCA results has revealed potential trade-offs *e.g.* between the goals of resources circularity and reducing environmental burden.[15,21,22]

After a critical appraisal of the BS 8001:2017, Pauliuk[23] proposes a CE indicator dashboard for the quantitative assessment of CE for product systems and organizations. This list is based on different methods, *i.e.* material flow analysis (MFA), material flow cost accounting (MFCA) and LCA, and thus includes different categories of indicators, which measure both physical circularity, monetary value, and potential environmental impacts. As far as LCA is concerned, it plays a central role in the category "Climate, energy & other", which includes the reduction of GHGs, energy demand and water, land, and material use as goals. The Global Warming Potential[24] (GWP), Cumulative Energy Demand[25] (CED) or Cumulative Exergy Demand[26] (CexD) and water, land, material footprints, or a combination thereof are suggested as indicators. Table 3.1 summarizes the categories, indicators and the life cycle thinking tool recommended.

Table 3.1 CE Indicators aided by life cycle thinking methodologies – Life Cycle Assessment (LCA), Life Cycle Costing (LCC) and Social Life Cycle Assessment (S-LCA) (Adapted from ref. 23, https://doi.org/10.1016/j.resconrec.2017.10.019, under the terms of the CC By 4.0 license, https://creativecommons.org/licenses/by/4.0/).

Group	Category	CE Indicator	Life cycle thinking methodologies[a]
Circular Economy (BS 8001:2017)	Regenerate	Footprint of regenerative flows	LCA
Life cycle	Maintain financial value	Ratio between recirculated and total economic product value	LCC
Resource Efficiency	Natural resource conservation	Amount and type of primary resource that is replaced by the CE activity/strategy	LCA
		Resource footprint or mineral depletion indicator	LCA
Climate, Energy and others	Reduce greenhouse gases	Quantification of greenhouse gas emissions and improvement opportunities offered by CE activity/strategy	LCA
	Reduce energy demand	Cumulative energy demand (CED), cumulative exergy demand	LCA
	Reduce water, land and material use	Water, land, material footprints	LCA
	Reduce exposure to critical materials	Quantification of vulnerability to supply restriction and supply risk and the environmental impacts associated to it.	LCA
	Address social indicators	Social life cycle indicators	S-LCA
	Reduce cross-impacts	Assessment of CE strategies that provide the highest environmental impact improvement in the relevant categories	LCA
		Ratio of economic costs of reducing environmental damage over product price	LCA, LCC

[a]Other tools such as material flow analysis (MFA) and material flow cost accounting (MFCA) might be used encompass the life cycle thinking tools.

LCA can also be used as an input to conduct cost-environmental impact multi-objective optimization, for example, to answer the question: "which CE strategies have the highest impact reduction potential in the relevant categories?" The Eco-cost Value Ratio (EVR) by Scheepens *et al.*[27] can be used to measure the costs of reducing environmental damage over product price. To address social indicators, social life cycle indicators (*i.e.* employment, work safety, transparency, supplier relations, *etc.*) by Benoit *et al.*[28] are suggested. Furthermore, LCA may be combined with MFA to address the following goals:

- Supply chain footprint of regenerative flows;
- Natural resource conservation, *i.e.* by answering to the question: "What and how much primary resource does the circular economy activity replace?";[29]
- Resource footprint, or mineral depletion indicator.

Parchomenko *et al.*[30] used the method of Multiple Correspondence Analysis (MCA) to assess 63 CE metrics and 24 features relevant to CE, such as recycling efficiency, longevity and stock availability. They identified three main clusters of metrics addressing: (1) resource-efficiency; (2) materials stocks and flows; and (3) product-centric perspective.

Based on a material flow perspective, Helander *et al.*[31] (2019) showed that most available indicators do not capture environmental pressures related to the CE activities they address. They found that many indicators focus on a single CE activity or process, which does not necessarily contribute to increased environmental sustainability overall, and thus they suggested that a material- and resource-based footprint approach, accounting for major environmental inputs and outputs, is necessary—while not sufficient—to assess the environmental sustainability of CE activities. Footprints can indeed be used at different scales, depending on the CE activity they aim to assess, *e.g.* they can be used by companies, cities, and national and supranational bodies for monitoring and evaluation to support resource governance and waste management.

LCA-based metrics included in the resource-efficiency cluster are frequently combined with other metrics, such as the assessment of Industrial Symbioses (IS-LCA), food waste management systems (FW-LCI), the evaluation of business models (LCA-EVR),[27] or for the identification of more sustainable supply chain partners (SSCN).[32] This reveals the flexibility of the LCA approach to be combined with other metrics, which makes it more challenging to identify clearly opposing CE elements for LCA-based metrics. Some elements, such as system stability and longevity tend to appear less likely reflected using LCA-based metrics. Moreover, the results showed poor integration of resource-efficiency and product-centric perspectives, while the product-centric and system-dynamic perspectives are least frequently assessed.[30]

3.4 Life Cycle Assessment Methodology

LCA is a standardized methodology that was first created to study the environmental impacts of systems, products and services[33] with a system approach *i.e.* the whole life cycle of the system from the extraction of raw materials, manufacturing and processing, distribution and transport, as well as end-use and waste management. Hence, this tool can be used to identify potential improvements across each life cycle stage and to support

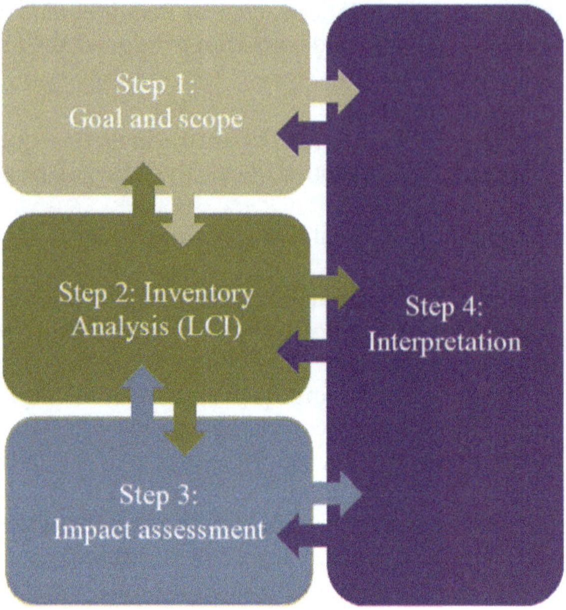

Figure 3.1 Life Cycle Assessment Framework.[34,35]
Reproduced from ref. 34 and 35 with permission from BSI. Permission to reproduce extracts from British Standards is granted by BSI Standards Limited (BSI). No other use of this material is permitted. British Standards can be obtained in PDF or hard copy formats from the BSI online shop: www.bsigroup.com/Shop.

decision-makings processes considering environmental aspects affecting ecosystems such as air, soil and water, and resources and human health. Additionally, LCA is used to support marketing strategies in the case of developing product declarations and other environmental certifications.[34,35] The great value associated with LCA methodology lies within its qualities as a systematic, transparent, comprehensive, and holistic approach.

The LCA methodology is part of the ISO management systems standards, specifically ISO 14040/44.[29,30] The method consists of four steps: the definition of goal and scope; the life cycle inventory (LCI) analysis; the impact assessment; and the interpretation of the results. Figure 3.1 summarizes the 4-step framework, while each step is described below.

Step 1: Definition of Goal and Scope

Given the value of LCA as a comprehensive and transparent tool for environmental performance assessment, the clear definition of the goal and scope of the study is vital. Defining the goal must illustrate the purpose of the study, who will receive the benefit of it and the decision-context of the study (*i.e.* support decision of government policy). The scope relates to the goal of the study, study boundary, functional unit, data requirements and assumptions, impact assessment method, allocation methods, limitations of the study and critical review.[34,35] Critical review provides general quality assurance, support transparency, and depending on the intention of the study, it could be a compulsory step for certification.[36] The goal and scope of the study may change during the development of the study, due to data limitations or in light of preliminary findings or systematic iterations, which is one of the key characteristics of LCA.

The functional unit (FU) defines the unit of analysis and must be quantifiable (measured) and clear: it is the reference flow that the LCA will be performed on. This also helps with comparisons with other studies, as well as validation. The FU usually refers to units of mass, volume, area and energy, but could relate to a specific condition and purpose.

Once the purpose and audience of the study is defined and, in particular, the decision-context is clarified, the LCI method or modelling framework must be identified. There are two LCI methods, namely attributional (A-LCI) and consequential (C-LCI). Attributional is the traditional style of assessment and as defined by Curran *et al.*:[17] "A-LCA considers the flows in the environment within a chosen temporal window", *i.e.* it considers the potential environmental impact associated with the specific inputs and outputs flows used in the systems and its life cycle in a specific time frame. Curran *et al.*[37] state that the C-LCI "considers how the flows may change in response to decisions", and hence how the potential environmental impacts directly and indirectly associated to the inputs and outputs flows of a system and its life cycle would be affected by changes or stimulus in the system or its life cycle. A-LCI is typically used for understanding the impacts associated with a product and its life cycle while C-LCI is usually use to assess the implications of broader changes in system pathways, such as with changes in policy. Hence, C-LCI is usually used for decision- and policy-making purposes.[38]

The system boundary also depends on the goal and scope of the study and must be clearly defined. The inclusion and exclusion of stages and systems should be clarified, and inputs/outputs included for each process involved. Additionally, cut-off criteria, *e.g.* mass, energy and environmental significance, must be defined, as well as data quality and allocation issues (see step 2 for details). As defined by ILCD[36]: "In principle all quantitatively relevant activities that can be attributed to a system (or a result of the consequences, in the case of consequential modelling) should be included in the system boundaries unless they are quantitatively irrelevant, applying the cut-off criteria. The need for inclusion and the possibility of exclusion of activities can only be decided for the given case in view of the required completeness and precision of the results."

Assumptions relating to data manipulation and aggregation must be described and documented in a consistent manner across the study, to ensure transparency, accuracy, completeness and precision, as well as to allow study repetition by other practitioners.[34–36]

Finally, the selection of the impact assessment method and indicators should be selected accordingly with the goal of the study. The most popular impact assessment methods are ReCiPe,[39] CML 2002,[40] ILCD,[41] Eco-indicator 99[42] and IMPACT 2002+.[43] Some impact assessment methodologies are developed for specific categories of indicators. For example, USEtox,[44,45] which focuses on toxicity-related aspects. A comprehensive analysis and comparison of most of the current impact assessment methods is given by EC-JRC[46] for information, and further assessment of the effects of using different assessment methods is given in Renou *et al.*[47]

Step 2: Life Cycle Inventory Analysis

The definition of the goal and scope provides a structure to develop the Life Cycle Inventory (LCI). LCI first involves the collection, calculation and validation of the data; at this point, the redefinition of the system boundary could be assessed if difficulties are encountered while building the LCI. Decisions should be made after performing sensitivity analysis and considering issues such as the cut-off criteria defined in the scope.

The quality of data input governs the representativeness of the LCA results and may involve an array of different sources. Up-to-date primary data sources (*e.g.* from recent industrial or governmental sources) should be prioritized, particularly for direct emissions sources or those that are expected to dominate results. However, this data may not be available, in which case historical or secondary source data is used instead. Secondary data include: well-known databases (*i.e.* Ecoinvent,[48] GaBi Database,[49] Open LCA Database,[50] Life Cycle Data Network);[51] government reports; sector or industry best available technology (BAT) reports; and peer-reviewed academic, government and industry publications. The data collection process may be iterative: the potential importance of certain data sources may not be known until output results are analysed. At this point, an assessment of data quality and the importance of the result will drive the collection of more up-to-date/better-quality information.

In terms of representativeness and quality of data, the geographical and time context are key aspects and must be considered when comparing and validating results. In the case of primary data, confidentiality may restrict use and publication. If so, both confidentiality and transparency must be addressed when building the inventory and reporting, which may be *via*:

- The inclusion of a separate complementary report, not publicly available, containing the confidential information for reviewers (who sign confidentiality agreements) to assess the analysis and ensure quality, robustness and reproducibility of the study; or
- The aggregation and synthesis of confidential data to reduce confidential characteristics whilst enabling reproducibility.

The final step of the LCI is the incorporation of the LCI modelling framework and allocation or system expansion/substitution approaches previously defined in step 1. Allocation refers to: "partitioning the input or output flows of a process or a product system, between the product system under study and one or more other product systems".[34,35] Allocation is a vital consideration for multi-output processes, where the embodied emissions must be fairly accounted for. A clear differentiation between products and waste is also an important consideration. The ISO 14044[35] standard describes three steps to deal with allocation:

1. If possible, allocation should be avoided; some suggestions to achieve this are:
   a. Divide the unit process and allocate it in sub-processes;
   b. Expand the product system to include the functions related to the co-product.
2. If allocation cannot be avoided, the inputs and outputs of the system should be distributed within its products or functions, considering a faithful representation of the physical relationships between them.
3. If physical relationships cannot be defined or do not fully represent the system, the inputs and outputs of it should be assigned within its products or functions considering an accurate representation of other relationships. For instance, a well-considered characteristic is the economic values of co-products;[30,47] recently, new considerations combining characteristics such as biophysical characteristics have been considered.[52]

Step 3: Life Cycle Impact Assessment (LCIA)

The life cycle impact assessment (LCIA) starts with the definition of the impact assessment methods and indicators categories from the goal and scope. LCIA is formed by mandatory and optional elements. The mandatory elements comprise the selection of the indicators, impact assessment method and the characterization models. The selection criteria of these elements should be justified and documented, aligning with the goal and scope of the study. The optional elements include normalization, grouping, weighting and data quality analysis,[37] where the assumptions and information required to perform these steps must be documented and clearly explained.

There are two impact assessment-modelling approaches: midpoints and endpoints. The midpoint approach refers to "a point in the cause-effect chain (environmental mechanism) of a particular impact category, prior to the endpoint, at which characterization factors can be calculated to reflect the relative importance of an emission or extraction in a Life Cycle Inventory (LCI)".[53] For example, a mid-point climate impact is the global warming potential, defined in terms of radioactive forcing and atmospheric half-lives of emissions, rather than an eventual damage caused by the emission. The endpoint approach considers the potential damage to ecosystems, human health and resources that an emission would have; hence, the endpoint converts (and aggregates) midpoint categories into damage-related impact categories.[54]

The endpoint approach results in higher uncertainties due to the broader damage assessment and additional required assumptions and data,[55] but offers a potentially more tangible and policy-focused metric. The endpoint approach requires the aggregation of impacts into single units corresponding to ecosystems, human health and resources, making it much simpler to understand the trade-off between impacts, compared with midpoints. The midpoint approach has lower uncertainty, but is not always easy to understand for non-practitioners, and is difficult to communicate effectively.

Step 4: Life Cycle Interpretation

The interpretation of results should stem from the goal and scope of the study. The interpretation first recognizes and discusses the relevant issues found in the results. Key aspects to include in the interpretation of results involve an assessment of the key contributing life cycle stages, and of the consistency and comprehensiveness of the study. Sensitivities relating to underlying assumptions, methodological aspects relating to assessment methods and indicators, as well as system boundaries are typically considered in parallel to the results. The development of conclusions and recommendations, and the discussion of the limitations of the study is the final aspect of interpretation.[34,35]

3.5 Life Cycle Assessment Standards and Guidelines

The ISO Standard 14040/44 framework has served as a basis for the development of environmental assessment guidelines, focusing on the analysis of products and organizations, countries and sectors. Due to the worldwide concerns about climate change, most of the guidelines focus on the assessment of greenhouse gas emissions (GHG), also known as carbon footprinting. However, growing interest in resource management due to increased pressures, *e.g.* on water use, has also led to the development of guidelines for water footprinting. Figure 3.2 summarizes some of the most common and recent guidelines that show their connection with the LCA framework. The following section briefly describes them.

Carbon footprint refers to the assessment of the GHG emissions associated to a product, system or organization, accounting for its whole life cycle. Most guidelines focus on products, but frameworks at organization levels have been also developed.[56] Although the ISO 14040/44 is the main basis for almost all the carbon footprint guidelines, the ISO 14060 series[57] are the main international standard series that provide a common, comprehensive, transparent and consistent framework to help quantify, monitor, report and verify GHG accounting. Additionally, they aim to help develop GHG management strategies, plans and mitigation actions. These guidelines also enable performance and progress assessments related to GHG mitigation actions. Depending on the scope, the guidelines are divided on product level (ISO 14067) and organizational level (ISO 14069). Furthermore, country-specific and sector-specific

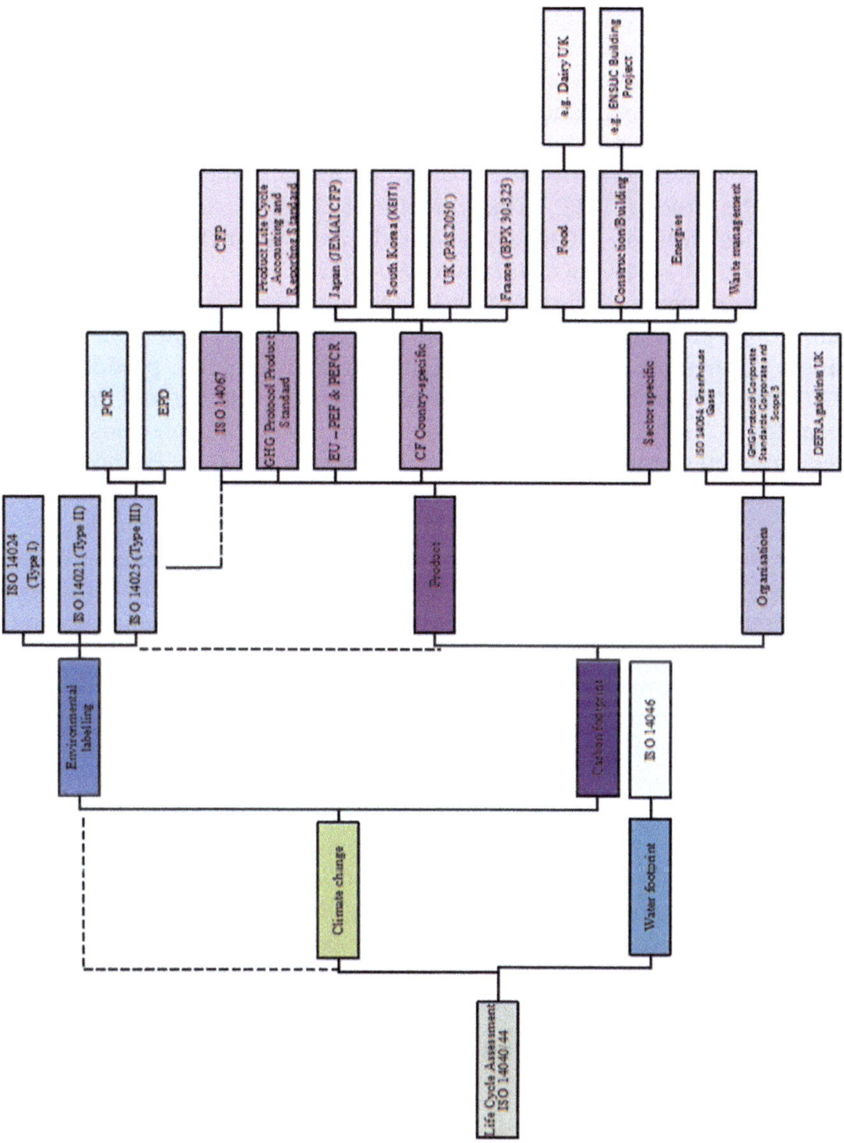

Figure 3.2 Environmental assessment guidelines based on LCA ISO Standard.

guidelines have been also developed. The following section describes some of the most widely used guidelines.

3.5.1 Carbon Footprint of Products

In the last decade, several guidelines, frameworks and standards have been developed to help accounting and reporting GHG emissions, particularly at the product level. Numerous studies have assessed the commonalities and differences across this plethora of guidelines.[58–60] At product level, the ISO 14067 2013/2018 is the main current guideline that defines the requirements, principles and framework to carry out and communicate the carbon footprint of a product (CFP). This guideline follows the ISO 14040/44 and also complies with the 14020 series related to environmental product declaration (*e.g.* ISO 14020, ISO 14024 and ISO 14025).

Prior to the release of the ISO 14067, other guidelines were developed by organizations such as the World Resources Institute (WRI) and the World Business Council on Sustainable Development (WBCSD), who produced the *GHG Protocol Product Standard: Product Life Cycle Accounting and Reporting Standard*[61] in 2011. This protocol complements two other previously published guidelines produced by WRI/WBCSD, namely, the *GHG Protocol Scope 3* and the *GHG Protocol Corporate Standard*, which together deliver a comprehensive and systematic framework to assess and manage value chain GHG for businesses and organizations from all economic sectors. These guidelines provide, not only a carbon footprint framework, but also examples and case studies, guidance for developing the GHG inventories and recommendations to set emission targets.

The European Union also released guidelines to contribute to the environmental assessment of products through the *Product Environmental Footprint (PEF),* which "is a multi-criteria measure of the environmental performance of a good or service throughout its life cycle".[62] The Product Environmental Footprint Category Rules (PEFCR) in the context of the PEF pilot initiative aims to ensure that LCA is carried out, focusing on significant contributors (foreground and background processes) in life cycle stages, and relevant impact categories. This set of guidelines provides a detailed framework that helps to harmonize the process of developing a product carbon footprint, with the further aim of developing a product declaration.

3.5.2 Environmental Labeling

The ISO 14020 series provides guidelines for environmental labeling and communication. Each of these guidelines provides specific frameworks to help businesses and organizations to communicate the environmental characteristics of their products and services. For instance, ISO 14024 (Type I label) provides guidelines for developing the voluntary declaration of environmental labeling, guiding the selection of product category, functions and environmental criteria to demonstrate compliance and get certification.

The ISO 14021 (Type II label) specifies how to communicate, in writing or verbally, the environmental characteristics of products; the guidelines are not restricted to any specific environmental characteristics. Finally, the ISO 14025 (Type III label), which follows the LCA framework (ISO 14040/44), delivers a detailed methodology to assess and report the environmental impacts of products from similar groups of characteristics. This standard introduces the concepts of Product Category Rules (PCR) and Environmental Product Declaration (EPD). PCR is a set of guidelines and requirements for each specific product group, which aid the development of EPD in a transparent manner and enable future comparison within products.[63]

EPD is a voluntary environmental impact declaration that allows the documentation of the environmental impacts of a product. These declarations are independently assessed and they aim to aid the transparent and comparable communication of the life cycle environmental impacts of products.[64]

3.5.3 Country Specific Guidance

Most of the country-specific carbon footprinting guidelines were either developed before the ISO 14060 series development, or they have adapted the ISO standard to national requirements. Most of the guidelines also comply with the labeling standard 14020 series. For instance, in the UK the "PAS 2050: Specification for the assessment of the life cycle greenhouse gas emissions of goods and services"[65] was first created in 2008 to help industries and other organizations to assess the GHG emissions of their products and services, offering a systematic, operational and comprehensive framework to implement the IPCC 2007 guidance. Although PAS 2050 was created in the UK, it has been recognized and used worldwide.[65,66] In 2012, Japan also released its own carbon footprint guidelines called the JEMAI CFP Program,[67] which is also based on ISO 14040/44. This initiative—which started in 2008 and is led by the Japanese Ministry of Economy, Trade and Industry—aims to help organizations to apply the LCA standard, focusing on communication and labeling aspects. The JEMAI CFP Program has a complete Japanese database to assess the carbon footprint of products, which complement their own product certification framework, called Ecoleaf,[68] which also complies with the ISO 14025. Similar efforts were seen in South Korea, where the Korean Environmental Industry and Technology Institute (KEITI) presented their own carbon-labeling scheme in 2009. The interest in carbon footprinting and labeling led to the creation of the Asia Carbon Footprint Network (ACFN), which aims to develop a common framework for carbon footprinting and labeling across Asia. Furthermore, in 2015, The Chinese Manufacturers' Association of Hong Kong (CMA) developed a carbon footprint and labeling scheme aiming at businesses in Hong Kong.[66] In Europe, France has also developed its own carbon footprint and labeling guidelines, having product-specific communication guidelines as an extra feature.[69]

3.5.4 Sector-specific Guidance

As with country specific guidelines, sector- or subject-specific frameworks were born from the current ISO 14040/44 standard and product carbon footprint frameworks. For instance, with help from Carbon Trust, in 2010 the UK Dairy Industry developed its own "Guidelines for the Carbon Footprinting of Dairy Products in the UK".[70] The dairy sector has adapted the PAS2050 framework to the unique issues associated with the dairy industry. For the construction sector, the European Commission Intelligent Energy for Europe Programme and nine European organizations developed the framework—ENSLIC Building Project[71]—to promote the "use of LCA in the European building and property industry". The energy sector is one the most widely studied, however a standardised framework has not been agreed yet. There are several guidelines and policy mechanisms that have been developed, overviews of the research carried out in the field,[72,73] and methodologies applied to specific energy generation such as photovoltaic.[74] The EU Renewable Energy Directive (RED 2008/28/EC, and now RED II 2018/2001/EU)[75] stipulates national renewable energy targets for each member state combined to reach 20% by 2020 and 32% by 2030. Each member state estimates their contribution to renewable energy and to GHG emissions reduction as per the directive *via* emission factors. The use of liquid biofuels accounts for estimates of the impact of indirect land use change (ILUC) to encourage the use of more benign biofuel production.

The UK Renewables Obligation[76] is a mechanism to increase large-scale renewable electricity producers in the UK. An obligation is placed on electricity producers to obtain a specified quantity of Renewables Obligation Certificates (ROCs),[76] which are given to renewable electricity producers, per unit of electricity produced. Each type of renewable electricity is given a certain quantity of certificates per MWh of electricity produced, based on their life cycle emissions and broader sustainability characteristics.

For transport, the UK's renewable fuel transport obligation (RFTO)[77] mandates that a proportion of transport fuel must be derived from renewable sources, and stipulates a methodology for the reporting of each fuel's "renewability" factor. These factors are derived within the guidance for different biogenic and non-bio renewable fuels, to account for differences in their life cycle GHG impact.

The application of the LCA standards in the field of waste management has been also widely studied,[78] although the lack of consensus is still present, with several guidelines and methodologies.[79–82]

3.5.5 Organization Level

Similar to product carbon footprint guidelines, the assessment of GHG emissions at an organizational level has also seen the development of several guidelines and protocols.[60] For instance, the Organizational Environmental Footprint (OEF), developed by the European Commission, is defined as "a multi-criteria measure of the environmental performance of a product-providing organization from a life cycle perspective".[83]

The ISO/TS 14072, published in 2014 by the International Organization for Standardization, ISO/TS 14072 "Environmental management—Life cycle assessment—Requirements and guidelines for organizational life cycle assessment" is defined as a "compilation and evaluation of the inputs, outputs, and potential environmental impacts of the activities associated with the organization as a whole or portion thereof adopting a life cycle perspective". It is a global initiative, which can be utilized by any organization with interest in applying LCA, and extends the application of ISO 14040-44 to all the activities of an organization.

In 2015, the United Nations Environment Programme (UNEP) and the Society of Environmental Toxicology and Chemistry (SETAC) partnership Life Cycle Initiative (UNEP/SETAC Life Cycle Initiative) published a "Guidance on Organizational Life Cycle Assessment (O-LCA)", hereafter UNEP Guidance.[84] This Guidance attempts to align with ISO/TS 14072, and therefore adopts the same definition of O-LCA as ISO/TS 14072, but utilizes the term "O-LCA". It is also a global initiative, and is intended to be used by organizations of all sizes. It complements the ISO/TS 14072 standard, but aims to be a more detailed accompanying document, and thus includes different case studies that involve the use of environmental multi-impact assessments with life cycle approaches.

The ISO 14064 standard also provides guidelines to help assess GHG emissions, providing a tool to help quantify, monitor, report and validate emissions for businesses, governments, regions and other organizations. This framework guides issues related to the quantification of GHG emission and mitigation opportunities, to help with the development, implementation and monitoring of action and activities related to GHG management. Hence, this standard also provides guidelines related to inventory development and quality management, auditing and verification.

3.5.6 Water Footprint

Water footprint refers to the total water use and changes of its quality across the life cycle of a system, as well as the potential impacts associated with this resource. Water footprinting follows the LCA standard (ISO 14044) and regulated by the ISO Standard 14046[85] "Environmental management—Water footprint—Principles, requirements and guidelines (ISO 14046:2014)". Although water footprinting can be developed as a stand-alone assessment (related to water only), it is a complement to an LCA study as it assesses in detail, the impacts associated with water use. Additionally, water footprint incorporates important aspects that LCAs often lack, *i.e.* regional or local characteristic and temporal dimensions.

3.5.7 Environmental Life Cycle Costing

Thus far, there is no official standard guideline for the economic LCA of products. However, the environmental life cycle costing (LCC) methodology, developed by SETAC[86] has been commonly used. To align with the LCA methodology, a code of practice has been developed that builds on the ISO 14040/44. Hence, LCC requires the development of the four-step framework and uses a life cycle thinking approach, considering all the stages of system (*e.g.* production of raw materials, manufacturing, *etc.*).

LCC accounts for all costs associated with the life cycle of a product,[86] considering all life cycle stages,[87] from the production of materials or components, processing or manufacturing, the use and maintenance, and end-of-life management. Additionally, research and development can be also considered.[86]

Hence, LCC accounts for all costs, which relates all the mass and energy flow from LCA with their respective monetary flows, associated to the whole the life cycle of a product;[88,89] for example, LCC of a ready-made meal[90] is described as:

$$LCC = C_{RM} + C_M + C_D + C_T + C_U + C_{E\text{-}o\text{-}L}$$

where:

LCC	Total life cycle cost of the ready-made food product
C_{RM}	Cost of raw materials (cultivation and processing of ingredients)
C_M	Cost of manufacturing (ready-made food product)
C_D	Cost of distribution (retailer and wholesalers)
C_T	Cost of transport
C_U	Cost of use phase
$C_{E\text{-}o\text{-}L}$	Cost of disposal (unless recycling when the cost is subtracted)

Other economic indicators have been also used, depending on the system under study. For instance, value added, which represents the profit margin (sales minus the costs of raw materials and services, has been commonly used in the assessment of food goods,[90,91] housing and construction.[92]

LCA and LCC can be used in combination, to identify the contribution from the main actors of the value chain, *i.e.* consumers, producers and waste management operators in the case of a comparative analysis of aluminum cans.[8] When the economic aspect was considered in the comparative LCA-LCC analysis of aluminum cans production, use, collection and recycling in two different cities (Bologna and Copenhagen), a trade-off emerged, since the best option from an environmental point of view (closed loop recycling) is also leading to higher costs. Another example of combined application of LCA and LCC is in the case of aluminum cans.[93]

3.5.8 Social Life Cycle Assessment

The incorporation of the social aspects of sustainability under a life cycle perspective has been a long-term challenge. Some early efforts to develop social LCA (S-LCA) started with the development of social indicators. Some examples are the ones developed by Azapagic and Perda[94] and the IChemE

Sustainability Metrics focusing on industries. For example, social aspects include those related to workplace (*e.g.* Promotion rate, Lost time accident frequency), society (*e.g.* Indirect community benefit per unit value added) and others (*e.g.* Discrimination). Metrics were also developed by the AIChE *via* the Sustainability Index,[95] although not under a life cycle perspective.

Currently, the methodology most widely used is the latest guideline released by UNEP/SETAC,[96,97] which develops a *S*-LCA that follows the LCA methodology approach. This methodology divides the stakeholders in five groups, namely: workers, consumers, local community, society and value chain actors. Each group or subsector has a set of aspects that are critical to consider, which are accounted for by quantitative, semi-quantitative or qualitative indicators. For instance, for the subcategory workers, one of the aspects is "hours of work" and one of the indicators is "Number of hours effectively worked by employees (at each level of employment)".

3.5.9 Integration of Environmental, Economic and Social: Sustainability Assessments

Holistic Life Cycle Sustainability Assessments (LCSA) integrate the environmental, economic and social dimensions of sustainability. The framework developed by UNEP/SETAC[97] combines the life cycle assessment (LCA) methodology for the environmental assessment, the economic assessment through the life cycle costing (LCC) assessment methodology and the social life cycle assessment (S-LCA) for the respective social considerations. Within the guidelines, UNEP/SETAC[98] stated that "although normalization, aggregation and weighting are optional steps according to ISO 14040, any aggregation and weighting of results of the three techniques used are not recommended because of the early stage of LCSA research and implementation and because the individual aims of each of the techniques applied are not directly comparable to the other".

However, the integration of multiple metrics and perspectives presents an important challenge: how to consider the trade-offs across different impacts or indicators and the three dimensions of sustainability. This is exacerbated by the difficulties in integrating a large number of indicators for supporting decision-making processes. In the case of LCA, the framework does consider weighting and normalization as a mechanism to integrate environmental indicators, however, this necessitates a highly subjective decision to quantify the relative importance of the different impacts (*e.g.* weighing global warming potential against depletion of resources).[99]

Although there are several methodologies and approaches to integrate indicators and sustainability dimensions, there is no consensus. For instance, to address the weighting issue in the three dimensions of sustainability, Hofstetter *et al.*[100] developed the "Life Cycle Sustainability Triangle", which is based on the graphic representation of chemical mixtures that allow weighting three parameters. There is also the "Life Cycle Sustainability Dashboard"

(LCSD) developed by Traverso *et al.*[101] and adapted from the "Dashboard of Sustainability" methodology.[102,103] This methodology and its correspondent software allow the integration of indicators and groups them into categories defined by the user. The indicators are then ranked and the performance is evaluated. LCSD has been used by UNEP/SETAC, owing to its transparency and reader-friendly characteristics,[104] but this "does not mean that other emerging methods are neither appropriate nor transparent nor user-friendly".

From an engineering perspective, multi objective optimization (MO) has been mentioned as a tool that has been mainly used to integrate environmental and economic indicators.[105] However, multi-criteria decision analysis (MCDA) is the method most broadly used, as it allows the integration of indicators and sustainability dimensions, including stakeholders' perspectives, by providing a single score index.[105,106]

3.6 Benefits and Challenges of Using LCA in CE

Inevitably, there are various challenges and barriers to incorporating LCA into CE strategies. Indeed, in their analysis of corporate sustainability reports in the Fast Moving Consumer Goods (FMCG), Stewart and Niero[107] outlined a lack of references to sustainability performance indicators or assessment methodologies with regard to CE activities. Regarding packaging, food and beverage, household goods and textiles, and personal care and household products, only a limited number of companies present a dedicated set of key performance indicators for their CE approach. Such indicators are based on metrics addressing either material efficiency, LCA results or use of the C2C certification program.[107] One of the outcomes of the 63rd discussion forum on life cycle assessment, "How can LCA support the circular economy?", held in Zurich in 2016,[1] was that "LCA and other assessment tools should be used to evaluate options for CE solutions to ensure a positive balance of efforts and benefits in both new product designs and increased recycling."

But there are several drawbacks and challenges associated with the implementation of LCA. Some such drawbacks are listed in Table 3.2, alongside benefits.[108–110] Particular challenges include the representativeness of the study, given the difficulty in obtaining relevant data of sufficient resolution and scope. Thus, many simplifications may be required to complete the study, including 'steady-state' assumptions (*e.g.* for attributional LCA) and the application of linear relationships for indicators.

How to handle the trade-off issue is key for the success of CE. Trade-offs exist first in terms of material circularity and environmental impact, as highlighted by Geissdoerfer *et al.*[111] in their analysis of the linkage between sustainability and the CE concept. Furthermore trade-offs exist with regard to different sustainability aspects (environmental, economic and social), as highlighted by Figge *et al.*[112] and Niero and Kalbar,[15] who encourage further research on the combination between circularity measures and life cycle sustainability.

Table 3.2 The benefits and challenges associated with implementing LCA in CE.[108–110]

Benefits	Challenges
– Allows the understanding of the environmental impacts of a product or service including its whole life cycle. – Provides a quantitative assessment that supports decision-making processes. – A scientific-based method that uses a holistic approach (life cycle). – Allows a consistent assessment of the environmental impacts generated in each stage, process and activity included in the life cycle of the system studied. – Provides opportunities for indirect supply chain management. – Avoids "problem shifting", which means that the possible solutions include the understanding of the effects across the supply chain; hence, it will help to avoid improving impacts in one stage by causing unexpected problems in another stage. – Allows the comparison among products, processes and designs. – Supports decision-making for businesses (*e.g.* environmental performance of products/services) and governments (*e.g.* waste management facilities, regulations and policies).	– Usually performed in a steady-state basis, so difficult to understand the impacts of changes in the future (*e.g.* new technologies, *etc.*) – Does not provide the understanding of local impacts. – Due to the complexity of representing the whole life cycle of the system, several simplifications and assumptions are made, which will affect the results of the assessment. However, the ISO standard provides a framework to help with the transparency of these assumptions and to avoid arbitrary decisions. – LCA is mainly based on physical characteristics and less so in economic ones, hence it is not easy to include effects such as market changes, new technology development, among others. – LCA assesses the potential impacts of a system based on a specific functional unit, which could be defined arbitrary, and with lack of time and space definition. – LCA predominantly uses a linear approach. – Data availability is a critical issue, in terms of quality and representativeness.

In order to advance the assessment of CE strategies at the product level, Niero and Kalbar[15] propose to couple material circularity and LCA indicators by means of the Multi Criteria Decision Analysis (MCDA) and recommend the use of Technique for Order of Preference by Similarity to Ideal Solution (TOPSIS) (or similar methods) to make sense of the complementary CE indicators, namely material circularity and LCA indicators.

3.7 Summary

In summary, LCA is a vital tool for the quantification and assessment of effectiveness and impacts associated with CE strategies. Incorporating LCA allows a comprehensive and transparent assessment of the product, service or organization, and can help to identify any potential unintended consequences associated with a change in process. The implementation of LCA can appear complicated for non-practitioners, but there exist clear guidelines on how to conduct LCAs, as described in this chapter. There are several frameworks for the incorporation of LCA into CE strategies, as described here and LCA may play a key role in the development of meaningful indicators for CE with regard to the product level assessment. Furthermore, LCA allows the environmental assessment of CE strategies in relation to design and product life cycle, which provide clear and quantifiable evidence to support decision-making. The environmental indicators can be also complemented with social and economic ones, as the LCA framework also allows the development of economic and social assessment under a life cycle perspective, which complement the development of indicators associated to these aspects.

Finally, the holistic and system approach characteristics embedded in the LCA, LCC and *S*-LCA framework enable understanding of the impacts and effects of each CE strategy across the whole life cycle, which helps with avoiding shifting impacts across stages and impacts categories. This chapter includes an assessment of the challenges associated with implementing LCA in CE, which include the difficulty in managing numerous indicators in parallel, and the lack of available data. However, the benefits associated with transparency, consistency, comparability and systems approach highlight its complementarity with a CE strategy.

References

1. M. Haupt and M. Zschokke, *Int. J. Life Cycle Assess.*, 2017, **22**, 832–837.
2. E. Iacovidou, C. A. Velis, P. Purnell, O. Zwirner, A. Brown, J. Hahladakis, J. Millward-Hopkins and P. T. Williams, *J. Cleaner Prod.*, 2017, **166**, 910–938.
3. M. Niero, M. Z. Hauschild, S. B. Hoffmeyer and S. I. Olsen, *J. Ind. Ecol.*, 2017, **21**(3), 742–753.
4. M. Niero, S. I. Olsen and A. Laurent, *J. Ind. Ecol.*, 2018, **22**, 760–772.
5. M. Niero, A. J. Negrelli, S. H. Boas, S. I. Olsen and M. Birkved, *J. Cleaner Prod.*, 2016, DOI: 10.1016/j.jclepro.2016.02.122.
6. J. M. F. Mendoza, M. Sharmina, A. Gallego-Schmid, G. Heyes and A. Azapagic, *J. Ind. Ecol.*, 2017, **21**, 526–544.

7. I. Rönnlund, M. Reuter, S. Horn, J. Aho, M. Aho, M. Päällysaho, L. Ylimäki and T. Pursula, *Int. J. Life Cycle Assess.*, 2016, **21**, 1719–1748.
8. M. Niero and M. Z. Hauschild, in *Procedia CIRP*, vol. 61, 2017.
9. W. Kloepffer, *Int. J. Life Cycle Assess.*, 2008, **13**, 89–95.
10. British Standard, *BS 8001:2017 Framework for implementing the principles of the circular economy in organizations*, 2017.
11. M. Niero and X. C. Schmidt-Rivera, in *Procedia CIRP*, vol. 69, 2018.
12. V. Elia, M. G. Gnoni and F. Tornese, *J. Cleaner Prod.*, 2017, **142**, 2741–2751.
13. M. Linder, S. Sarasini and P. van Loon, *J. Ind. Ecol.*, 2017, **21**, 545–558.
14. European Environment Agency, EEA Report No 6/2017, 2017, pp. 1–52.
15. M. Niero and P. Kalbar, *Resour., Conserv. Recycl.*, 2019, **140**, 305–312.
16. N. M. P. Bocken, I. de Pauw, C. Bakker and B. van der Grinten, *J. Ind. Prod. Eng.*, 2016, **33**, 308–320.
17. S. Huysman, S. Sala, L. Mancini, F. Ardente, R. A. F. Alvarenga, S. De Meester, F. Mathieux and J. Dewulf, *Resour., Conserv. Recycl.*, 2015, **95**, 68–76.
18. C. Bakker, F. Wang, J. Huisman and M. den Hollander, *J. Cleaner Prod.*, 2014, **69**, 10–16.
19. M. Niero and S. I. Olsen, *Resour., Conserv. Recycl.*, 2016, **114**, 18–31.
20. M. Saidani, B. Yannou, Y. Leroy, F. Cluzel and A. Kendall, *J. Cleaner Prod.*, 2018, **207**, 542–559.
21. G. Lonca, R. Mugg, T. Hugue, S. Bernard and M. Margni, in *Designing Sustainable Technologies, Products and Policies*, ed. E. Benetto, K. Gericke and M. Guiton, 2018, pp. 33–42.
22. S. Walker, N. Coleman, P. Hodgson, N. Collins and L. Brimacombe, *Sustainability*, 2018, **10**, 1–14.
23. S. Pauliuk, *Resour., Conserv. Recycl.*, 2018, **129**, 81–92.
24. IPCC, Climate Change 2013: The Physical Science Basis, Working Group I Contribution to the Fifth Assessment Report of the Intergovernmental Panel on Climate Change, Cambridge University Press, Cambridge, United Kingdom and New York, NY, USA, 2013, p. 1535.
25. M. A. J. Huijbregts, L. J. A. Rombouts, S. Hellweg, R. Frischknecht, A. J. Hendriks, D. van de Meent, A. M. J. Ragas, L. Reijnders and J. Struijs, *Environ. Sci. Technol.*, 2006, **40**, 641–648.
26. M. E. Bösch, S. Hellweg, M. A. J. Huijbregts and R. Frischknecht, *Int. J. Life Cycle Assess.*, 2007, **12**, 181–190.
27. A. E. Scheepens, J. G. Vogtländer and J. C. Brezet, *J. Cleaner Prod.*, 2016, **114**, 257–268.
28. C. Benoît, G. A. Norris, S. Valdivia, A. Ciroth, A. Moberg, U. Bos, S. Prakash, C. Ugaya and T. Beck, *Int. J. Life Cycle Assess.*, 2010, **15**, 156–163.
29. R. Geyer, B. Kuczenski, T. Zink and A. Henderson, *J. Ind. Ecol.*, 2015, **20**, 1010–1017.

30. A. Parchomenko, D. Nelen, J. Gillabel and H. Rechberger, *J. Cleaner Prod.*, 2018, **210**, 200–216.
31. H. Helander, A. Petit-Boix, S. Leipold and S. Bringezu, *J. Ind. Ecol.*, 2019, **23**, 1278–1291.
32. H. Winkler, *CIRP J. Manuf. Sci. Technol.*, 2011, **4**, 243–246.
33. H. Baumann and A.-M. Tillman, *The hitch hiker's guide to LCA: an orientation in life cycle assessment methodology and application*, 2014.
34. ISO, *ISO 14040-environmental management – life cycle assessment - principles and framework*, Geneva, 2006.
35. ISO, *ISO 14044-environmental management – life cycle assessment - requirements and guidelines*, Geneva, 2006.
36. ILCD, *General Guide For Life Cycle Assessment – Detailed Guidance*, 2010, https://publications.jrc.ec.europa.eu/repository/bitstream/JRC48157/ilcd_handbook-general_guide_for_lca-detailed_guidance_12march2010_isbn_fin.pdf.
37. M. A. Curran, M. Mann and G. Norris, *J. Cleaner Prod.*, 2005, **13**, 853–862.
38. T. Ekvall, A. Azapagic, G. Finnveden, T. Rydberg, B. P. Weidema and A. Zamagni, *Int. J. Life Cycle Assess.*, 2016, **21**, 293–296.
39. M. Goedkoop, R. Heijungs, M. Huijbregts, A. De Schryver, J. Struijs and R. Van Zelm, ReCiPE 2008: A life cycle impact assessment method which comprises harmonised category indicators at the midpoint and the endpoint level, 2008.
40. J. B. Guinée, *Handbook on Life Cycle Assessment: Operational Guide to the ISO Standards*, Kluwer Academic Publishers, Dordrecht, Boston, 2002.
41. European Commission, *Characterisation factors of the ILCD Recommended Life Cycle Impact Assessment methods. Database and Supporting Information*, Joint Research Centre, Institute for Environment and Sustainability, Luxembourg, 1st edn, 2012.
42. M. Goedkoop and R. Spriensma, Eco-indicator 99, a damage-oriented method for lifecycle impact assessment, methodology report (update April 2000), PRe Consultants BV, Accessed from: www.pre-sustainability.com/download/EI99_annexe_v3.pdf, 2000.
43. O. Jolliet, M. Margni, R. Charles, S. Humbert, J. Payet, G. Rebitzer and R. Rosenbaum, *Int. J. Life Cycle Assess.*, 2003, **8**, 324.
44. R. K. Rosenbaum, T. M. Bachmann, L. S. Gold, M. A. J. Huijbregts, O. Jolliet, R. Juraske, A. Koehler, H. F. Larsen, M. MacLeod, M. Margni, T. E. McKone, J. Payet, M. Schuhmacher, D. van de Meent and M. Z. Hauschild, *Int. J. Life Cycle Assess.*, 2008, **13**, 532.
45. M. Huijbregts, M. Hauschild, O. Jolliet, M. Margni, T. McKone, R. K. Rosenbaum and D. Van de Meent, USEtox User manual, Accessed from: www.usetox.org/sites/default/files/support-tutorials/user_manual_usetox.pdf, 2010.
46. European Commission, *ILCD Handbook: Analysis of existing Environmental Impact Assessment methodologies for use in Life Cycle Assessment Joint Research Centre*, Institute for Environment and Sustainability, 2010.

47. S. Renou, J. S. Thomas, E. Aoustin and M. N. Pons, *J. Cleaner Prod.*, 2008, **16**, 1098–1105.
48. Ecoinvent, https://www.ecoinvent.org/, 2020.
49. GaBi Software, GaBi LCI Database, http://www.gabi-software.com/international/databases/gabi-databases/, 2020.
50. Open LCA, https://nexus.openlca.org/databases, 2020.
51. LCDN, https://eplca.jrc.ec.europa.eu/LCDN/, 2014.
52. F. Ardente and M. Cellura, *J. Ind. Ecol.*, 2012, **16**, 387–398.
53. S. G. Mackenzie, I. Leinonen and I. Kyriazakis, *Int. J. Life Cycle Assess.*, 2017, **22**, 128–137.
54. J. C. Bare, P. Hofstetter, D. W. Pennington and H. A. U. de Haes, *Int. J. Life Cycle Assess.*, 2000, **5**, 319.
55. I. Alyaseri and J. Zhou, *Heliyon*, 2017, **3**, e00268.
56. M. Z. Hauschild, M. Goedkoop, J. Guinée, R. Heijungs, M. Huijbregts, O. Jolliet, M. Margni, A. De Schryver, S. Humbert, A. Laurent, S. Sala and R. Pant, *Int. J. Life Cycle Assess.*, 2013, **18**, 683–697.
57. European Commission, *Analysis of Existing Environmental Footprint Methodologies for Products and Organizations: Recommendations, Rationale, and Alignment*, Ispra, Italy, Accessed from: http://ec.europa.eu/environment/eussd/pdf/Deliverable.pdf, 2011.
58. ISO, 14067:2018(en), Greenhouse gases — Carbon footprint of products — Requirements and guidelines for quantification, Accessed from: www.iso.org/obp/ui/#iso:std:iso:14067:ed-1:v1:en, 2018.
59. S. Manfredi, K. Allacker, N. Pelletier, E. Schau, K. Wolf, R. Pant and D. Pennington, *Comparing the European Commission product environmental footprint method with other environmental accounting methods*, 2015.
60. T. Gao, Q. Liu and J. Wang, A comparative study of carbon footprint and assessment standards, *Int. J. Low-Carbon Technol.*, 2014, **9**, 237–243.
61. European Commission, *ILCD Handbook. Recommendations for Life Cycle Impact assessment in the European context – based on existing environmental impact assessment models and factors*, Joint Research Commission, Accessed from: http://lct.jrc.ec.europa.eu, 2011.
62. P. Bhatia, C. Cummis, A. Brown, L. Draucker, D. Rich and H. Lahd, *Product Life Cycle Accounting and Reporting Standard*, World Resources Institute, Accessed from: www.ghgprotocol.org/sites/default/files/ghgp/standards/Product-Life-Cycle-Accounting-Reporting-Standard_041613.pdf.
63. European Union, 2013/179/EU: Commission Recommendation of 9 April 2013 on the use of common methods to measure and communicate the life cycle environmental performance of products and organisations, 2013.
64. EPD, The International EPD® System, Accessed from: https://www.environdec.com, 2020.
65. British Standard, PAS 2050:2011 Specification for the assessment of the life cycle greenhouse gas emissions of goods and services,

Accessed from: http://shop.bsigroup.com/upload/Shop/Download/PAS/PAS2050.pdf, 2011.
66. Pre-Sustainability, *Life Cycle-Based Sustainability —Standards & Guidelines*, Accessed from: www.pre-sustainability.com/download/Life-Cycle-Based-Sustainability-Standards-Guidelines.pdf, 2012.
67. JEMAI CFP Program, *Carbon Footprint of Products (CFP)*, Accessed from: www.cfp-japan.jp/english/overview/index.html, 2012.
68. JEMAI, *Ecoleaf environmental label*, Accessed from: www.ecoleaf-jemai.jp/eng/, 2002.
69. C. Cros, E. Fourdrin and O. Réthoré, *Int. J. Life Cycle Assess.*, 2010, **15**, 537–539.
70. U. K. Dairy, *Guidelines for the Carbon Footprinting of Dairy Products in the UK*, Dairy UK, Dairy Co and Carbon Trust, Accessed from: https://dairy.ahdb.org.uk/resources-library/research-development/environment/carbon-footprinting-dairy-products-in-the-uk/#.XDvSUFz7Q2w, 2011.
71. T. Malmqvist, M. Glaumann, S. Scarpellini, I. Zabalza, A. Aranda, E. Llera and S. Díaz, *Energy*, 2011, **36**, 1900–1907.
72. A. Laurent, N. Espinosa and M. Z. Hauschild, in *Life Cycle Assessment: Theory and Practice*, ed. M. Z. Hauschild, R. K. Rosenbaum and S. I. Olsen, Springer International Publishing, Cham, 2018, pp. 633–668.
73. R. Turconi, A. Boldrin and T. Astrup, *Renewable Sustainable Energy Rev.*, 2013, **28**, 555–565.
74. IEA, *Methodology Guidelines on Life Cycle Assessment of Photovoltaic Electricity*, International Energy Agency, Paris, France, 2011.
75. EU, DIRECTIVE (EU) 2018/2001 of the European Parliament and of the Council, https://eur-lex.europa.eu/legal-content/EN/TXT/PDF/?uri=CELEX:32018L2001&from=EN.
76. DECC, UK Department of Energy and Climate Change, 2010 to 2015 government policy: low carbon technologies, https://www.gov.uk/government/publications/2010-to-2015-government-policy-low-carbon-technologies/2010-to-2015-government-policy-low-carbon-technologies#appendix-5-the-renewables-obligation-ro.
77. UK Government, *Renewable Transport Fuel Obligation (RTFO) guidance*, 2019, www.gov.uk/government/publications/renewable-transport-fuel-obligation-rtfo-guidance-2019.
78. A. Laurent, J. Clavreul, A. Bernstad, I. Bakas, M. Niero, E. Gentil, T. H. Christensen and M. Z. Hauschild, *Waste Manage.*, 2014, **34**, 589–606.
79. N. B. Chang and A. Pires, *Sustainable Solid Waste Management*, 2015.
80. European Commission, Environmental Assessment of Municipal Waste Management Scenarios: Part II Detailed Life Cycle Assessments, Accessed from: https://eplca.jrc.ec.europa.eu/uploads/waste-part-II-Detailed-LCAs.pdf, 2017.
81. European Commission, *Joint Research Council*, Accessed from: http://eplca.jrc.ec.europa.eu/uploads/waste-part-II-Detailed-LCAs.pdf, 2007.

82. G. Finnveden, M. Nilsson, J. Johansson, Å. Persson, Å. Moberg and T. Carlsson, *Environ. Impact Assess. Rev.*, 2003, **23**, 91–123.
83. European Commission, Organisation Environmental Footprint (OEF) Guide, Joint Research Centre, Institute for Environment and Sustainability, Ispra, Italy, 2012.
84. UNEP/SETAC, *Guidance on Organizational Life Cycle Assessment*, United Nations Environment Programme, Accessed from: https://www.lifecycleinitiative.org/wp-content/uploads/2015/04/o-lca_24.4.15-web.pdf, 2015.
85. BS EN ISO, 14046:2016 *Environmental management. Water footprint. Principles, requirements and guidelines*, 2016.
86. T. E. Swarr, D. Hunkeler, W. Klöpffer, H.-L. Pesonen, A. Ciroth, A. C. Brent and R. Pagan, *Int. J. Life Cycle Assess.*, 2011, **16**, 389–391.
87. G. Rebitzer and D. Hunkeler, *Int. J. Life Cycle Assess.*, 2003, **8**, 253–256.
88. W. Kloepffer, *Int. J. Life Cycle Assess.*, 2008, **13**, 89.
89. J. M. F. Mendoza, A. Gallego-Schmid, X. C. Schmidt Rivera, J. Rieradevall and A. Azapagic, Sustainability assessment of home-made solar cookers for use in developed countries, *Sci. Total Environ.*, 2018, **648**, 184–196.
90. X. C. Schmidt, Rivera and A. Azapagic, Life cycle costs and environmental impacts of production and consumption of ready and home-made meals, *J. Cleaner Prod.*, 2015, **112**(1), 214–228.
91. D. Amienyo and A. Azapagic, *Int. J. Life Cycle Assess.*, 2016, **21**, 492–509.
92. R. M. Cuéllar-Franca and A. Azapagic, *Int. J. Life Cycle Assess.*, 2014, **19**, 174–193.
93. R. Princigallo, D. Visini, A. Bonoli, S. I. Olsen and M. Niero Comparative environmental and economic assessment of production, use and recycling of aluminium cans: Bologna vs Copenhagen, in *Proceedings of the Italian LCA Network conference 2016 Life Cycle Thinking: sustainability and circular economy. Ravenna – 23–24 giugno 2016*, ed. A. Dominici Loprieno, S. Scalbi and S. Righi, ENEA, Roma; 2016, pp. 352–359.
94. A. Azapagic and S. Perdan, *Process Saf. Environ. Prot.*, 2000, **78**, 243–261.
95. AIChE, Sustainability Index, Accessed from: https://www.aiche.org/ifs/resources/sustainability-index, 2008.
96. UNEP/SETAC, *Guidelines for Social Life Cycle Assessment of Products*, United Nations Environment Programme (UNEP) and Society of Environmental Toxicology and Chemistry (SETAC), Accessed from: www.lifecycleinitiative.org/wp-content/uploads/2012/12/2009%20-%20Guidelines%20for%20sLCA%20-%20EN.pdf, 2009.
97. UNEP/SETAC, *The Methodological Sheets for Subcategories in Social Life Cycle Assessment (S-LCA)*, United Nations Environment Programme (UNEP) and Society of Environmental Toxicology and Chemistry (SETAC), Accessed from: www.lifecycleinitiative.org/wp-content/uploads/2013/11/S-LCA_methodological_sheets_11.11.13.pdf, 2013.

98. UNEP/SETAC, *Towards a life cycle sustainability assessment. Making informed choices on products UNEP/SETAC Life Cycle Initiative*, Accessed from: www.lifecycleinitiative.org/wp-content/uploads/2012/12/2011%20-%20Towards%20LCSA.pdf, 2011.
99. M. Finkbeiner, E. M. Schau, A. Lehmann and M. Traverso, *Sustainability*, 2010, **2**, 3309–3322.
100. P. Hofstetter, A. Braunschweig, T. Mettier, R. Müller-Wenk and O. Tietje, *J. Ind. Ecol.*, 1999, **3**, 97–115.
101. M. Traverso, M. Finkbeiner, A. Jørgensen and L. Schneider, *J. Ind. Ecol.*, 2012, **16**, 680–688.
102. P. Hardi and P. Semple, *The Dashboard of Sustainability—From a Metaphor to an Operational Set of Indices*, 2000.
103. J. Jesinghaus, *On the Art of Aggregating Apples & Oranges*. European Commission JRC/ISIS, I-21020 Ispra (VA), 2000.
104. S. Valdivia, C. M. L. Ugaya, J. Hildenbrand, M. Traverso, B. Mazijn and G. Sonnemann, *Int. J. Life Cycle Assess.*, 2013, **18**, 1673–1685.
105. A. Azapagic and S. Perdan, *AIChE J.*, 2014, **60**, 3998–4007.
106. G. M. Zanghelini, E. Cherubini and S. R. Soares, *J. Cleaner Prod.*, 2018, **172**, 609–622.
107. R. Stewart and M. Niero, *Bus. Strategy Environ.*, 2018, **27**, 1005–1022.
108. J. B. Guinée, M. Gorrée, R. Heijungs, G. Huppes, R. Kleijn, L. Oers, A. Wegener Sleeswijk, S. Suh, H. A. Udo de Haes, H. de Bruijn, R. van Duin and M. Huijbregts, *Life cycle assessment: An operational guide to ISO standards*, http://www.bookmetrix.com/detail/book/0bc29205-4177-48af-aaea-9348ffb3bcab#citations, 2001.
109. UNEP, *Why take a life cycle approach?* United Nations Publications Retrieved from: https://www.lifecycleinitiative.org/wp-content/uploads/2012/12/2004%20-%20Why%20take%20LCA%20-%20EN.pdf, 2004.
110. UNEP/SETAC, *Greening the Economy Through Life Cycle Thinking Ten Years of the UNEP/SETAC Life Cycle Initiative*, UNEP/SETAC, Retrieved from: https://www.lifecycleinitiative.org/wp-content/uploads/2013/03/2012_LCI_10_years_28.3.13.pdf, 2012.
111. M. Geissdoerfer, P. Savaget, N. M. P. Bocken and E. J. Hultink, *J. Cleaner Prod.*, 2017, **143**, 757–768.
112. F. Figge, A. S. Thorpe, P. Givry, L. Canning and E. Franklin-Johnson, *Ecol. Econ.*, 2018, **150**, 297–306.

CHAPTER 4

Case Study – Construction Material (Steel)

J. P. BIRAT

IF Steelman, Gâte-Chaux, Semécourt 57280, France
Email: jean-pierre.birat@ifsteelman.eu

4.1 Introduction

Construction is an important economic sector, which comprises buildings and civil engineering structures. It is active in cities, towns and villages, and in the networks of information, energy and transport that link them together and innervate them. In other words, construction is in charge of erecting the backbone of the anthroposphere.[1]

Quantitatively, construction accounts for 6% to 14% of Gross Domestic Product (GDP), depending on the geographical reference system, in either developed or emerging countries, or the world. In Europe, it accounts for 9% of GDP, 18 million jobs and a turnover of $1.6 trillion (2016). Weighted against physical parameters related to materials, energy or emissions, it accounts for 40% of energy consumption, 45% of Greenhouse Gas (GHG) emissions (Europe), 50% of steel consumption (world) and 100% of cement consumption (Europe). Finally, people spend 90% of their time inside buildings and most of the remaining 10% of their time using built infrastructure.[2,3]

Civil Engineering, or the Public Works sector, as it is also known in some countries, comprises a broad array of categories which engage materials, mostly concrete and steel, for long periods of time:[1] *e.g.* electrical grid equipment, permanent and other structures (bridges, dams, *etc.*), railroad networks, roads and highways, pipeline and other underground structures.

Life Cycle Assessment: A Metric for the Circular Economy
Edited by Aiduan Borrion, Mairi J. Black and Onesmus Mwabonje

Published by the Royal Society of Chemistry, www.rsc.org

Figure 4.1 Applications of steel in the world by economic sectors. Reproduced from ref. 4 with permission from WorldSteel.org.

Recycling at end-of-life, especially of steel or iron, is active in the sector and essential for "closing the loop" (see Figure 4.1).[4]

It is legitimate to apply Life Cycle Assessment (LCA) methodology to the construction sector, in as far as LCA has the ambition to give a full picture of the global economy. This, however, requires some accommodations with respect to a method that was historically developed to describe parts of industrial artifacts (the *functional unit*) that are usually mass-produced in large numbers and therefore can meaningfully be described by average figures.[5]

The particular features of the construction sector are described in Section 4.2, to explain what challenges they raise in terms of carrying out a LCA. Section 4.3 explains the physics of construction relevant to LCA, therefore, mostly energy issues plus a few more relevant items. LCA in construction is described from a particular case study centered on a steel building in Section 4.4. Some particular approaches specific to the construction sector and relevant for LCA are described in Sections 4.5 (Environmental Product Declaration or EPD) and 4.6 (Green Building Rating Systems or GBRSs). Perspectives, related to the major transitions that the construction sector is likely to live through in connection with the energy and ecological transitions that are currently underway, are discussed in Section 4.7. This will lead

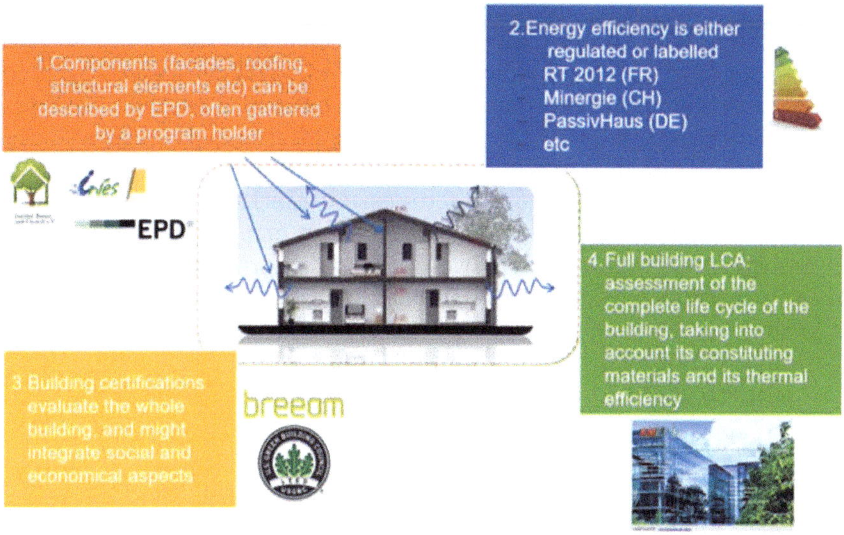

Figure 4.2 Various levels of assessment of a building including LCA. Image reproduced with permission from ArcelorMittal.

to the concept of the social value of construction artifacts (Section 4.8) and to the conclusion of this chapter (Section 4.9).

At a practical level, this means that the assessment of a construction artifact can be carried out from a series of complementary standpoints. They will be reviewed in the next sections of this chapter and summarized in Figure 4.2. This pictorial presentation is borrowed from a study by ArcelorMittal, part of which was included and published in the *Large Valorization on Sustainability of Steel Structures* or LVS[3] project,[6] which will be used as the core of Section 4.5.

4.2 The Main Features of Construction in the Light of LCA Methodology

Construction exhibits the following features.

- Construction units cover a wide range of sizes, from individual homes to buildings, including high-rise buildings; from modest pedestrian bridges to airports, sea harbors, dams, wind turbine farms, high-speed train tracks or long motorways. In comparison, consumer goods, commonly studied in LCAs, are small but more numerous. Doing an LCA – a costly endeavor that ought to have meaning beyond a particular study – is therefore justified by the size of a building and the fact that it is seldom unique.[†] In the case of individual homes, for example, many similar ones may be built.

[†]The Eiffel tower is unique and was 130 years old in 2019. Doing an LCA for the Eiffel Tower would be possible, although rather meaningless! As would discussing its end-of-life or recycling.

- Construction units are made from an assembly of different structural materials, and almost always include concrete and steel, but also glass, copper (for wiring and pipes), aluminum (window frames), mineral materials (for insulation), wood and traditional materials (*e.g.* adobe brick) in some parts of the world. In construction, structural materials cooperate rather than compete with each other.
- Construction exhibits long lifetimes (50 years or more), often with intermediary refurbishment (buildings) or maintenance (civil engineering infrastructure). Long lives raise special problems with regards to LCA methodology, because ordinary LCA is a static methodology, where only one aspect of time is taken on board, the time of the life-cycle: outputs are thus spread out over long periods of time and adding them together, as LCA does, is not "physical".
- Today, because of these long lifetimes, the use-phase (or life-in-use) of a building ought to be more important, in terms of environmental footprints, than the construction phase or the end-of-life phase (the use-phase amounts to roughly 80–90% of the total life cycle impacts, *cf.* Section 4.5). Building elements and building materials are often recycled or reused, and this is important because of the sheer mass of materials involved: thus, buildings are an integral part of the Circular Economy, even if the loop is closed over long periods of time. Moreover, steel, which is recycled to a high degree, averages out the low footprints related to the production from scrap in Electric Arc Furnaces (EAF), with the higher footprints of the initial production stage in an Integrated Steelmill from iron ore and coal/coke, because steel recycling rate is high (80%) and because steel is indefinitely recyclable.[7]
- The major part of the environmental footprints of construction is therefore related to the energy used (heating and cooling of buildings, energy used by electrical devices) and emissions during the life-in-use, mainly GHG emissions. There are important health issues related to indoor air pollution, but also to the contamination of buildings with bacteria [sick-building syndrome (SBS) and legionellosis], which is related to many causes, but is often due to the improper management of moisture. However, LCA is not the preferred or the best-suited tool to deal with these matters.

4.3 The Physics of Construction Relevant to LCA Issues

This section aims to give some general and simple elements related to the physics of buildings, in order to explain what an LCA delivers. The physics of buildings deals with both structural and non-structural performance of a building, the latter being what matters for LCA.

The function of a building is to create a boundary – the envelope – between a living space and the outside, in order to ensure the safety and the comfort of the people inside.

Comfort is usually understood in terms of temperature, moisture and availability of fresh air inside the building. It is therefore related to the *hygrothermal performance* of the building, thus to the heat transfer between the building envelope and its environment, under the condition that air and moisture are allowed to circulate in addition to energy/heat.

The complexity of analyzing the heat transfer in buildings is due to the complexity of the buildings themselves, and of the boundary conditions due to daily and annual cycles, which lead to unsteady-state conditions. Only measurements or sophisticated computer modeling can give access to the corresponding information.

Let us simply mention a few important issues:

1. Heat transfer through the envelope is mainly related to thermal conduction through the walls, floor, roof and windows under the gradient of temperature between inside and outside. The basic physics is related to two coefficients that describe the thermal behavior of the envelope: the diffusivity $\left(\frac{\lambda}{\rho c}\right)$ and the effusivity $\left(\sqrt{\lambda \rho c}\right)$ – where λ is the thermal conductivity, ρ is the volumetric mass density and c is the heat capacity – which are characteristics of steady-state and transient heat transfer, respectively. Because steady-state is a first and rough approximation of heat transfer through an envelope, building experts use the *U-value* (actually, a heat transfer coefficient), which is directly related to the ratio $U = \frac{\lambda}{L}$, in W m^{-2} K, where λ is the conductivity and L the thickness of the envelope. Engineering solutions consist of implementing *thermal insulation* as a first approximation to reduce conduction heat transfer through the envelope wall.
2. Heat is not simply transferred through the envelope, it is also accumulated in it: one then talks about the *thermal mass* of the building. This represents, for example, the heat stored in the walls or the floors. Thermal mass regulates the dynamics of temperature evolution and is an important part of the feeling of comfort of the people inside the building. Thermal mass is directly related to the heat capacity c of the material. Using phase-change materials in the envelope can increase its thermal mass.
3. *Moisture* is also an important parameter, because the humidity in the air depends on temperature, in this case inside and outside temperatures. Thus, cooling will precipitate liquid water, while heating will evaporate some of it. The amount of latent heat involved in these transformations is of the same order of magnitude as the heat flux by conduction through the walls.
4. *The circulation of air* is also essential to help humidity circulate and to ensure that air quality inside the building is safe (accumulation of CO, radon, volatile organic emissions, *etc.*) This also leads to energy exchanges that are of the same order of magnitude as conduction and moisture management.

Ultimately, heat transfer converts into annual *energy consumption* and heating/cooling costs. It is the time-integrated consequence of the interplay of the phenomena outlined earlier. This integrated value is what is captured in LCA, with the particular filters that LCA uses to take energy on board of its analysis (not the detailed mechanisms of the heat balance of the building). Emissions are mostly related to this energy consumption, either GHG emissions or others related to upstream emissions due to the fuel used.

The physics of buildings today is embedded in simulation tools, which are readily used by the major players of the construction sector, foremost of which are architecture and engineering design firms, but also material providers to the sector. An example of such a model, called AMECO3, is mentioned in Section 4.4.

A further level of sophistication is represented by the *Building Information System* (BIS), which is a kind of database related to a building, which in effect can create a virtual building when connected to CAD tools:[8] this makes archiving easier than with old-fashioned drawings, facilitates maintenance and transformation-refurbishment of buildings and makes it easier to carry out an LCA, by delivering a *bill of materials* as well as other parameters, automatically. It is related to *Industrie 4.0*[‡] or to *Industry of the Future.*[§] [9]

Health is a different matter, although, as mentioned in Section 4.2, it is related to physics, *i.e.* to the circulation of air and of moisture. LCA, however, does not take this very much into account in its present impersonation, which is one reason why "beyond-LCA" approaches were developed.

The physics of infrastructure obey completely different rationales from that of buildings. Structural performance and durability are essential, although the lifetimes of bridges or dams may be higher than those of buildings, while those of roads are shorter (*e.g.* 30 years). Heat transfer is mostly a non-issue in these cases, but the use-phase can be related, for example, to traffic (roads, bridges) and, therefore, some of the traffic-related environmental burden is passed on to the infrastructure, in full or with an allocation rule; the core of the matter is that the road surface quality influences fuel consumption. These topics will not be reviewed here, as they are very specialized, and the relevant LCA methodology is still under construction.[10–12]

There is a rich literature of textbooks, encyclopedias and treatises related to the construction sector.[13,14]

‡*Industrie 4.0* refers to the *4th industrial revolution* as applied to industry, *i.e.* the ongoing change of technological episteme that takes on board the present level of automation in which data are exchanged automatically by *cyber-physical systems* (CPS), that include Internet of Things, (IoT) cloud computing, artificial intelligence, digital twins, *etc.*

§*Industry of the Future* was the name used to designate the new industrial paradigm, in the US and France. In Germany, where the expression *Industrie 4.0* was coined, the latter expression is preferred.

4.4 LCA of a Building: A Case Study

An example of a LCA devoted to a building is presented in this section. It is an office building designed by ArcelorMittal and included in the LVS^3 project.[6]

The functional unit of this LCA study is a whole building. Two versions were designed and modeled in terms of energy needs: one is a building fully made of steel-reinforced concrete, while the other is purely steel, except for the foundations, the concrete floor and the composite floor made of concrete on a steel decking (this version is called *eco-optimized composite structure* in the LVS^3 report). Both are shown in Figures 4.3 and 4.4, and in

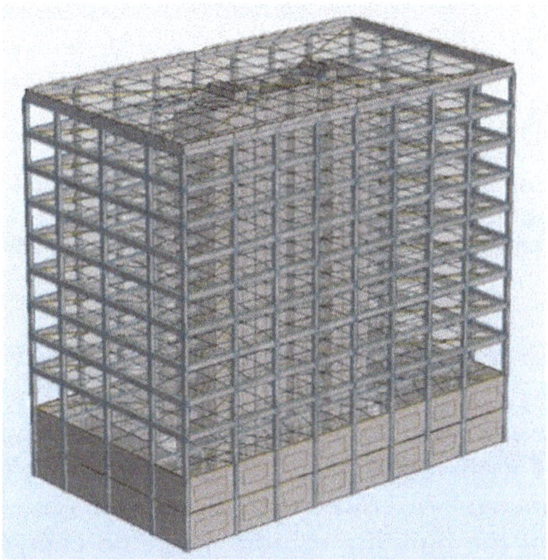

Figure 4.3 The "mostly concrete building" used as one of the versions of the functional unit in Section 4.4.
Image reproduced with permission from ArcelorMittal.

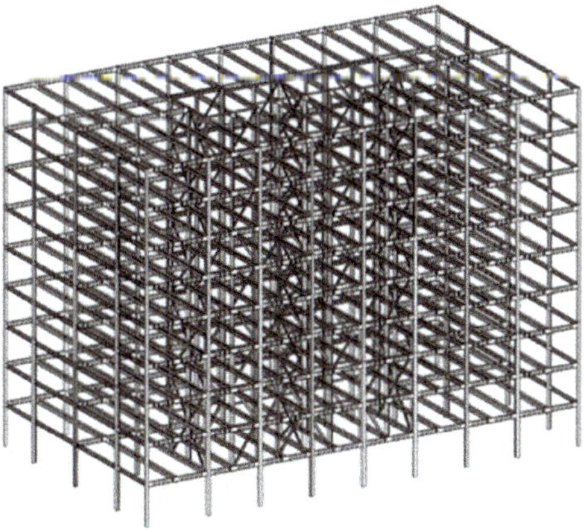

Figure 4.4 The "mostly steel building" used as one of the versions of the functional unit in Section 4.4.
Image reproduced with permission from ArcelorMittal.

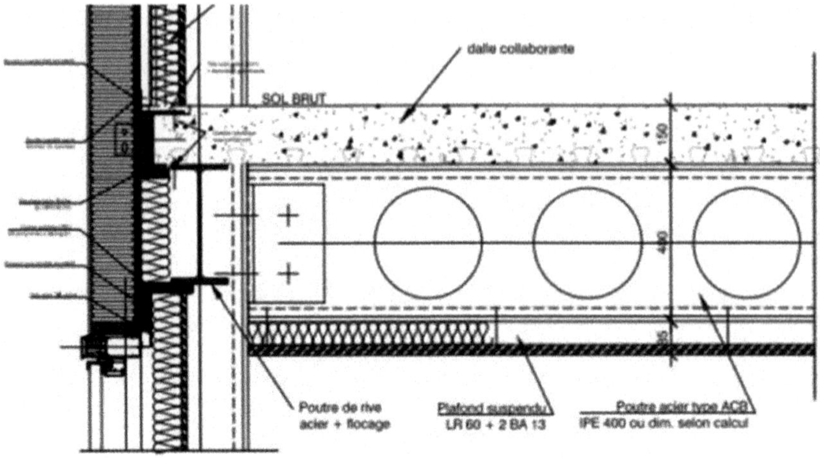

Figure 4.5 Details of design of the wall and floor connection in the "mostly steel building" (eco-optimized composite structure).
Image reproduced with permission from ArcelorMittal.

the design features of wall and floor connection in the steel-building version in Figure 4.5.

The point of the study was to use LCA to compare the two solutions, which differ in terms of their major structural material and, therefore, they are called "mostly concrete" or "mostly steel" solutions in what follows.

Both versions of the building exhibit the same energy demand, as they follow the same environmental regulations, so that both use-phases are identical: in each version, the façade is made of light steel panels, with 50 mm of extruded polystyrene (XPS), the roof is insulated with 18 cm of expanded polystyrene (EPS) and the windows are double-paned with solar protection where this is made necessary by the building orientation. The winter temperature (heating) is set at 20 °C, while the summer temperature (cooling) is set at 26 °C. Hot water is delivered by an electrical boiler (DHW). Mechanical ventilation renews internal air and ensures heating and cooling. The building is located in Paris, France. Its lifespan is assumed to be 50 years.

The design generates a *bill of materials*, which is essential to implement the LCA. The energy balance is calculated over a period of one year by a *heat transfer model* included in the AMECO3[¶] software.

The software calculates LCA outputs, *i.e.* the following environmental impacts: Global Warming Potential (GWP in CO_{2eq}), Ozone Depletion Potential (ODP in t_{CFCeq}), Acidification Potential (AP in t_{SO2eq}), Eutrophication

[¶]Two methods of analyzing heat transfer were used and compared: a fully dynamic calculation and one quasi-static calculation based on monthly averages. The average gap between them was 7.2%, which strengthens the argument in favor of the simpler, quasi-static model.

Potential (EP in t_{PO4eq}), Photochemical Ozone Creation (POC in $t_{Etheneeq}$) and Abiotic Depletion Potential, both for elements (ADP-e in t_{Sbeq}) and fossil fuels (ADP-ff in GJ NCV). Depletion of resources and generation of waste is also part of the calculations. Most materials are recycled at end-of-life, depending on their nature, while some building elements are reused and some energy is recovered: this brings in a credit in the life-cycle impacts.

The model calculates the impacts according to 4 modules: module A deals with the production of materials and the construction process; module B deals with the use-phase; module C deals with the end-of-life and module D calculates various credits (or burdens) related to reuse, recycling or the valorization of waste.

A detailed analysis of the life cycle of materials is given in ref. 5. It shows that a material has a life-cycle inventory, which is included as background information in the LCA of the building. For example, while GWP is due mostly to the production of steel, because most steel production is energy-intensive and generates GHG emissions, either in the integrated route, where primary raw materials (iron ore) are used, or in the Electric Arc Furnace route, where secondary raw materials (steel scrap) are used. The other indicators are mostly influenced in the upstream part of materials production, *i.e.* mining of iron ore, where applicable, and transportation of raw materials.

The GWP results for the two building versions is given in Figure 4.6. It shows very clearly that the use-phase is always the most important phase (module B) – it actually amounts to 90% of the total life-cycle GWP. Then comes the production and construction phase (module A), which represents somewhat less than 10% (module C) and the end-of-life, about 1%. The credits for recovering energy or materials at end-of-life are still less (module D).

The GWP is mostly determined by the use-phase and, therefore, improvements in terms of GHG emissions should be focused on that phase. That does not mean that materials have no significant connection with the overall GWP, as will be seen later, but it is of a second order of magnitude compared to the use-phase. The end-of-life is even of less importance; which, however, does not mean either that reuse and recycling should be ignored.

A finer analysis shows that air conditioning is the major contributor to the GWP, followed by the production of hot water. Heating is the 3rd largest contributor, and is slightly more important than the production and construction phase.

Comparing Figures 4.7 and 4.8 shows that materials have a fairly large influence on module A's GWP. The difference in GHG emissions between the steel and concrete buildings is due to the fact that the less specific footprint of concrete is overcome by its larger mass. High-strength steel, per unit of function in the building, performs better than concrete.

Very often, the main focus of the LCA analysis is Global Warming Potential, but this is ignoring LCA's power to convey a broader message in terms of more environmental indicators.

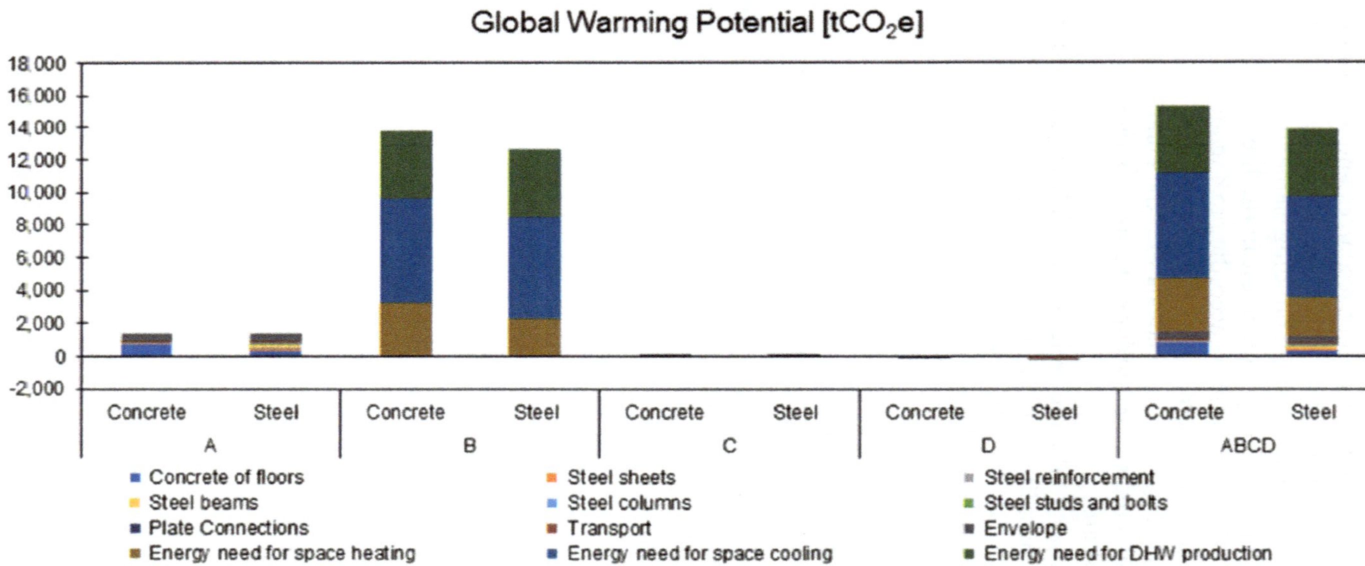

Figure 4.6 Life-cycle impacts on Global Warming Potential (GWP) for the steel and concrete solutions. Image reproduced with permission from ArcelorMittal.

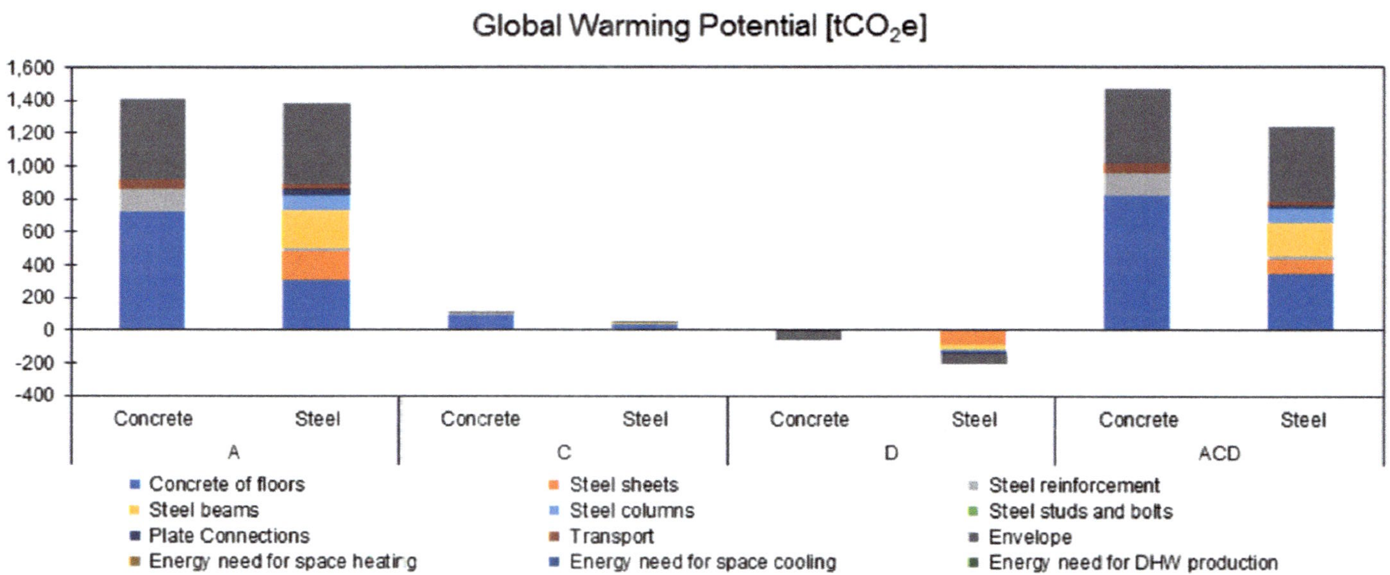

Figure 4.7 Life-cycle impacts on Global Warming Potential (GWP) for the steel and concrete solutions – focus on materials (*i.e.* module B is excluded).
Image reproduced with permission from ArcelorMittal.

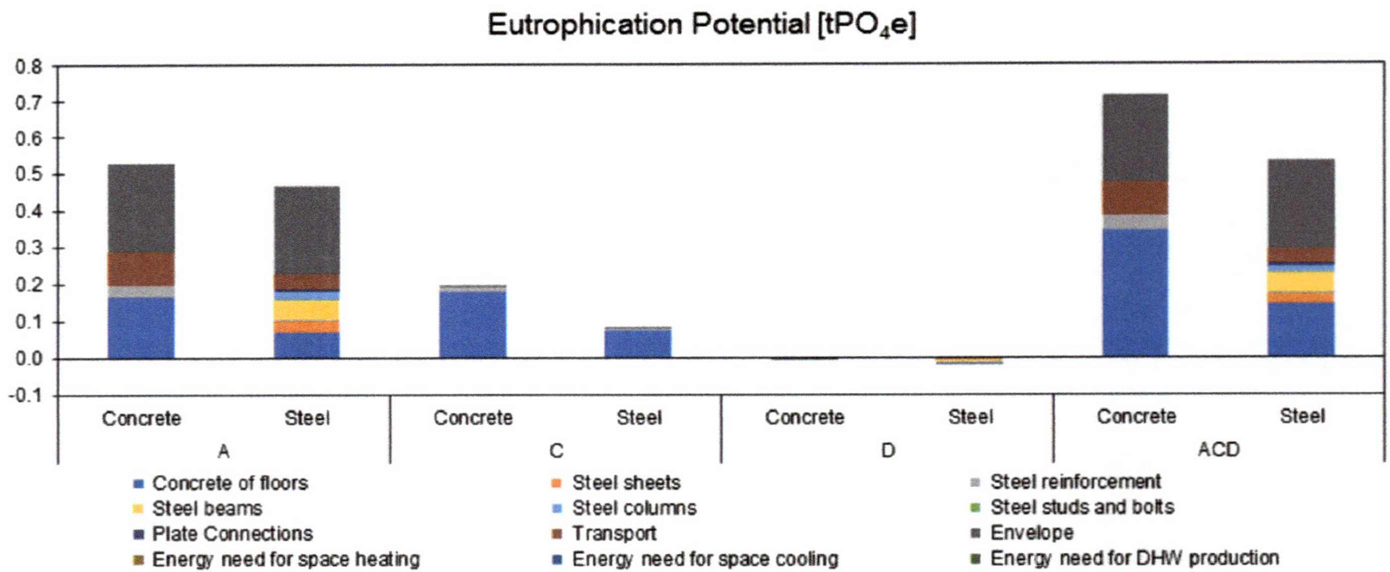

Figure 4.8 Life-cycle impacts on Eutrophication Potential (EP) for the steel and concrete solutions – focus on materials (*i.e.* module B is excluded).
Image reproduced with permission from ArcelorMittal.

Figure 4.8 shows the eutrophication indicator (EP) in a presentation similar to that of the previous figure. Eutrophication is caused by the release of plant nutrients, which foster algae growth in fresh or sea water, and the indicator focuses on P (phosphates) and N (nitrides). They originate from agriculture (fertilizers, discharge from animal farming and aquaculture), industry (wastewater discharge) or cities (municipal water discharge) but also the combustion of fossil fuel (ash). In the present LCA, this is due to the transport of materials to the construction site and then of waste from the demolition site at end-of-life. The mass of concrete is larger than that of steel and this explains the difference between both building versions. There are also similar effects prior to the construction step (transport within the production chain of materials), which are included in the background data and thus are not immediately "visible", as such. The steel building slightly outperforms the concrete building for this indicator, as it did for the GWP.

The next figures show other indicators calculated in a standard LCA. The results for steel and concrete buildings are shown side by side in the figures, for easy comparison. All life-cycle phases are reported (modules A, B, C and D), as well as the total of the 4 phases. ODP is shown in Figure 4.9, AP in Figure 4.10, POCP in Figures 4.11 and 4.12, ADP-e in Figure 4.13, and ADP-ff in Figure 4.14.

For all the new indicators, the use-phase (module B) far exceeds the other phases, except for Ozone Depletion. In phase A, best and worse results toggle back and forth between steel and concrete solutions, depending on the indicator. In phase B, steel systematically outperforms concrete. The overall, LCA results aggregated over the four phases show a slight advantage for the steel solution, the difference remaining small.

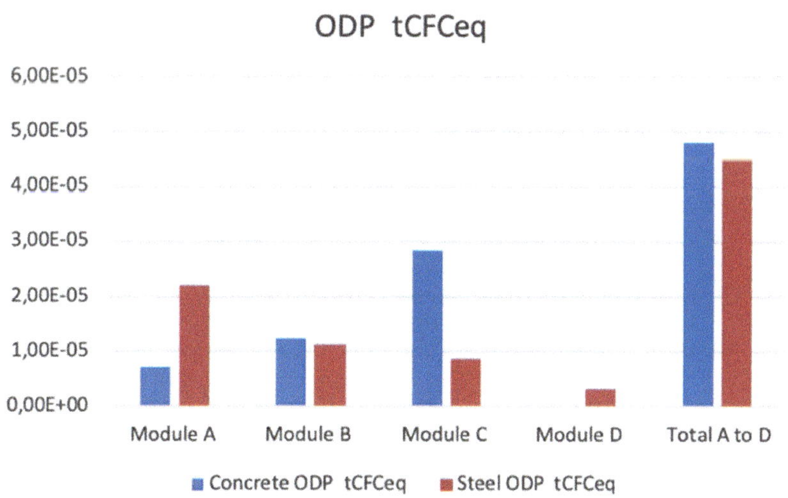

Figure 4.9 Life-cycle impacts on Ozone Depletion Potential (ODP) for both steel and concrete solutions.
Image reproduced with permission from ArcelorMittal.

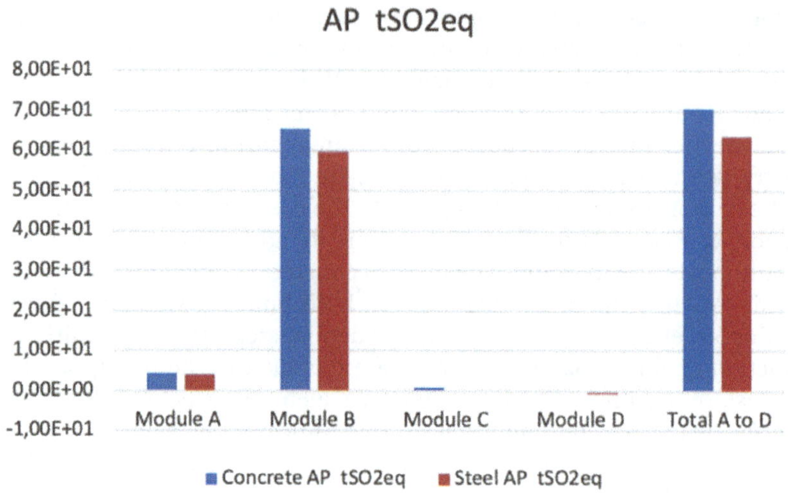

Figure 4.10 Life-cycle impacts on Acidification Potential (AP) for both steel and concrete solutions.
Image reproduced with permission from ArcelorMittal.

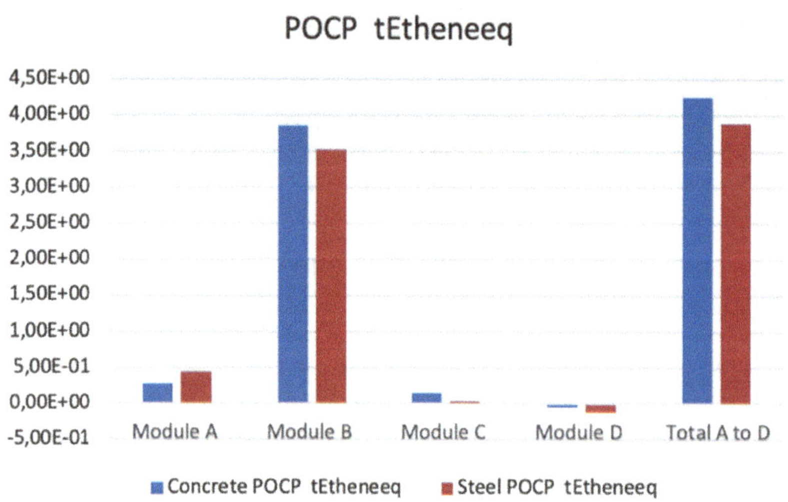

Figure 4.11 Life-cycle impacts on Photochemical Ozone Creation Potential (POCP) for both steel and concrete solutions.
Image reproduced with permission from ArcelorMittal.

However, this is a provisory conclusion as, in the future, the importance of the use-phase should decrease and that of the production phase-A increase: this should rekindle the competition between materials (*cf.* Section 4.5).

The analysis can be elaborated further by moving *beyond LCA*. The following criteria were proposed in the Steligence project[15] (the comparison is between a steel and a concrete scenario, somewhat similar to what was done

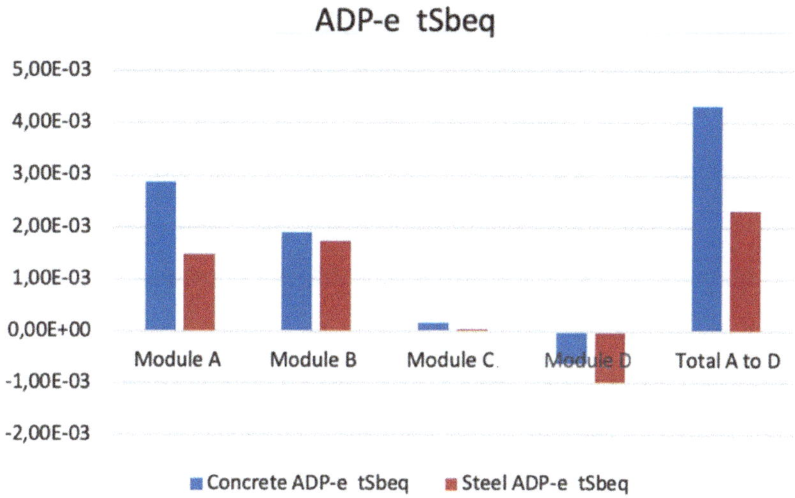

Figure 4.12 Life-cycle impacts on Abiotic Depletion Potential for elements (DAP-e) for both steel and concrete solutions.
Image reproduced with permission from ArcelorMittal.

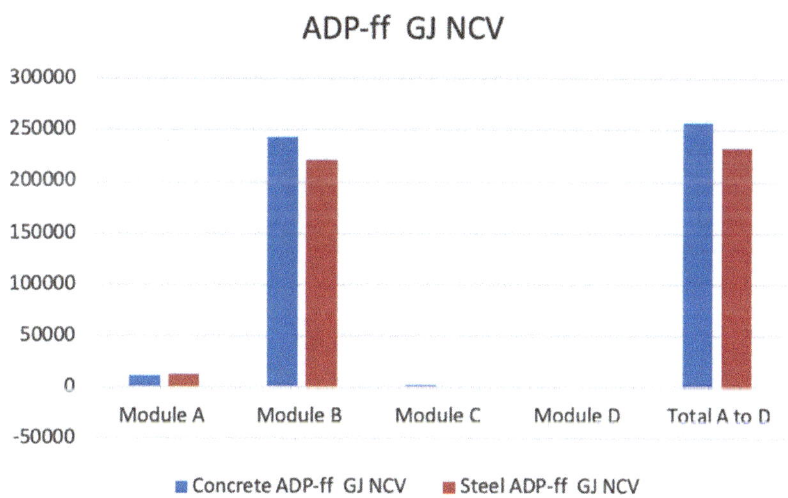

Figure 4.13 Life-cycle impacts on Abiotic Depletion Potential for fossil fuels (ADP-ff) for both steel and concrete solutions.
Image reproduced with permission from ArcelorMittal.

in the LVS[3] study) and the results of the comparison are given in brackets (steel *vs.* concrete buildings):

- *Mass* of the building (50% reduction)
- *Compactness* of the building (30% reduction in height)
- *Truck traffic around the construction site* (30% reduction)

Figure 4.14 An environmental product declaration document (HISTAR steel sections of ArcelorMittal).
Image reproduced with permission from ArcelorMittal.

- *Speed of construction* (24% reduction in costs, due, for example, to the design without intermediary columns, made possible by the use of long-span light beams)
- *Environmental ratings*, see Section 4.6 (3 extra points in BREEAM, 3 extra points and a Platinum rating *vs.* a Gold rating in LEED)
- *Costs* (reduction of 7% in construction cost, 20% in cost of ownership)
- *Comfort* is a more complex parameter: it is indeed related to the thermal mass of the building and, as such, depends heavily on design; in a multistory office building, an important part of the thermal mass is located in the floors and some of the solutions involving composite steel and concrete floors can outperform a thicker, more massive concrete slab.

There are limits, however, to what the quantitative analyses available today can bring, even with indicators going beyond LCA.[16] Indeed, *comfort* would probably need to be analyzed further, as well as *noise,* which is also an element in the feeling of comfort. Air quality, in terms of *indoor pollution* and *bacterial contamination*, is also an important matter, which still mostly eludes these methods.

4.5 Environmental Product Declarations (EPDs)

Environmental Product Declarations (EPD, FDES in French, for *fiche de déclaration environnementale et sanitaire*) are a kind of standardized and simplified LCA describing a product, either a final product such as a window, or an intermediary product such as a material. In this particular case, it is similar to the life cycle inventory (LCI) of steel or concrete, for example. EPDs were particularly developed in the case of construction to provide design offices with environmental information relative to the very many materials that are used in a building: the EPD proposes a format defined by standards.[17] At the international level, EPDs are based on standards ISO 14025 and EN 15804, in addition to the ISO 14040/44 LCA standards.

An example of the EPD of a steel grade of ArcelorMittal, with the commercial name "HISTAR" (the material from which the steel building presented in Section 4.4 is made), is shown in Figure 4.14 and Figure 4.15. It looks formally like an LCA without a use-phase (module B); the functional

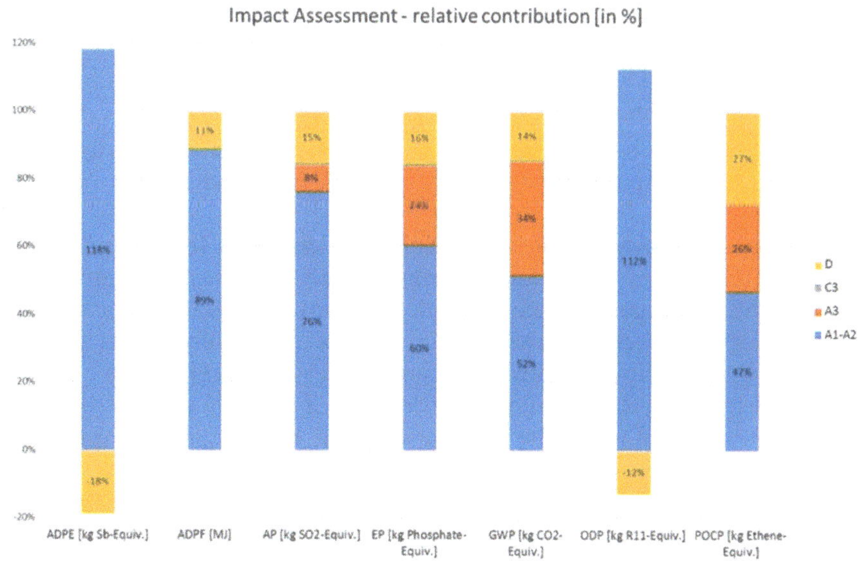

Figure 4.15 Indicators from an Environmental Product Declaration document (HISTAR steel sections of ArcelorMittal).
Image reproduced with permission from ArcelorMittal.

unit would be, for example, 1 meter of a heavy section; indeed, the EPD data were obtained by carrying out an LCA. The document also shows a long list of regulations and standards with which the material and the EPD itself comply.

Historically, when EPDs were introduced, attempts were made to find a way to include the use-phase in the EPD. The rationale was that materials are not neutral as far as the use-phase is concerned, but that some of them can bring more energy savings and/or reduced emissions than others: this would mean allocating the burdens of Module B to each of the constituting materials, with an allocation rule to be imagined. Because the use-phase is case specific and because the building designers meet the minimum standard requirements for energy consumption anyway, no matter what materials solutions are chosen, such an extension of an EPD was never fleshed out.

EPDs provide the environmental information that is needed at the time of final design and erection, to be integrated in the BIM (Building Information Modeling) file and thus used to carry out the LCA of the building. LCA software, relying on LCA databases, can achieve the same result, but the EPD empowers the material producer to communicate personally, on the environmental issues related to their own products. EPDs are centrally published by IBU (*Institut Bauen und Umwelt,* Germany).[18]

4.6 Green Building Rating System

As was pointed out earlier, a LCA does not provide all the information one would wish to have with regards to a building: there is a need for non-LCA or beyond-LCA information.

The construction sector has developed a series of methods to deal with this: they are called *Green Building Rating Systems (GBRS)* or *Environmental Certification of Buildings (ECB)* and several of them are in common use.[19,20] They also purport to present LC information in a "user-friendly" way, where LCA is complex and technical. They are thus the exact contrary of EPDs.

ECBs usually provide an *ecolabel.* They participate in the contemporary trend for businesses to voluntarily comply with a set of environmentally related rules, packaged by an independent organization, rather than simply meeting regulatory requirements in a kind of anonymity. ECBs provide businesses with visibility, with a convenient way to communicate environmentally friendly practices, and they also make it possible for businesses to maintain a distance with "the law" by showing an autonomous approach. ECBs also have an ethical dimension.[21] Moreover, they also step in when regulations are weak, in international contexts for example, and bring in order and rationale for green practices, where there are few regulatory guidelines (*cf.* NGOs such as *the Forest Stewardship Council* or the *Carbon Trust*). There are international standards, which define ECBs and certification practices, namely ISO 140021 and ISO140024.

Because they refer to regulations, ECBs tend to be organized regionally: BREEAM‖ in the UK, HQE** in France and SEED†† in the US are some of the ECBs, with an international reach for the first two.

The HQE label is taken in what follows as an example of the ECB approach. In 2002, the label set 14 targets organized in four families to be met at the "highest" possible level compared to the best of the state-of-the-art technology:

- *Ecoconstruction:* integration of the building in its physical environment; integrated choice of construction processes and products; low-impact construction site
- *Ecomanagement:* energy, water and waste (operational and maintenance waste) management
- *Comfort:* hygrothermal, noise, visual and olfactory
- *Health:* health quality of inside spaces, of inside air and of water.

That approach was reformulated in 2015, with more contemporary words to designate the families of indicators: quality of life, economic performance, environmental friendliness and socially responsible management replaces the older categories. Mostly environmental and health matters are included in the indicators. An example of an HQE building is the ICADE Premier House, in Munich, Germany.[22]

4.7 Perspective on Construction and LCA Related to the Energy and Ecological Transitions

The agenda for fighting climate change is focused on a series of strategies, which address the various activities of the anthroposphere. They point to two major "transitions", the *energy transition* and the *ecological transition,* distinctly but complementarily. These transitions address both the overall energy system and the various anthropogenic activities responsible for GHG emissions (transport, buildings, industry, *etc.*): the levers consist in greening the energy and in redesigning anthropogenic systems into leaner ones. The best solution, however, probably cannot be purely technical. A collective effort to favor greener and leaner lifestyles will, most likely, also be necessary.

The construction sector itself should also be responsible for greening its practices, alongside other anthropogenic activities that contribute to climate change. The major step is to reduce energy consumption, which would entail a reduction in GHG and other emissions (aiming towards zero-emissions), and the restriction of energy use to renewable energy, generated on site (*e.g.* solar panels or photovoltaic cells integrated in the envelope of the building).[23]

‖Building Research Establishment Environmental Assessment Method. BRE is the parent organization.

***Haute qualité environnementale.*

††Social, Economic, and Environmental Design.

The first societal response for decarbonizing construction consists of cutting heat losses by better insulating buildings (low-consumption buildings) and, presently, existing buildings in Europe are being refurbished as part of an effort to move in this direction.

The second stage consists of proposing a paradigm shift in building design, by positing that zero-energy/zero emissions or negative energy/emissions are possible in a building, and then by exhibiting technical solutions to arrive at these targets. Prototypes and concept buildings have been designed, modeled and eventually built. Today, regulations for future buildings imagine that such zero- or negative-footprint buildings will become the norm, for example, in 2020 in the EU.[24] Several expressions are commonly used to designate such a building: *zero-energy building, ZNE (Zero Net energy building), NZEB (near-zero energy building)*. They refer to buildings which are, at the same time, consumers of energy from the electricity grid and generators of renewable energy that can be sold back to the grid: the annual exchange of energy with the grid ought to be zero to deserve the zero-energy building designations. If the on-site generation of energy exceeds its needs, then one speaks of a *positive energy* or *energy-plus-house*. One also sometimes uses the expression of *passive building* or *active building*. An NZEB will save 50–70% more energy than typical buildings today.[25]

Note that the concept of NZEB is focused on the use-phase of the building. A full life-cycle definition of the energy used in the building *is not part* of this vision. *Mutatis mutandis,* this is similar to the emission standards for automobiles, which, today, focus only on tail-pipe emissions and thus also on the use-phase of the vehicle.[26] Provision is made, however, to expand the definition of NZEB to a Life Cycle Thinking framework.[27]

The shift to NZEB will change LCAs significantly, as is schematically illustrated in Figure 4.16, which shows the decline in importance of the use-phase. The construction phase, called "materials" in the figure, will shrink accordingly. The literature, sometimes, makes the distinction between *embodied energy* and *operational energy* to distinguish the contributions of the materials and of the use-phase.

For a more precise LCA see, for example, ref. 28 where an attempt was made to check if the concept of a life-cycle-thinking, zero-energy building (LCT-NZEB) makes sense: "it was found that an overall emissions balance

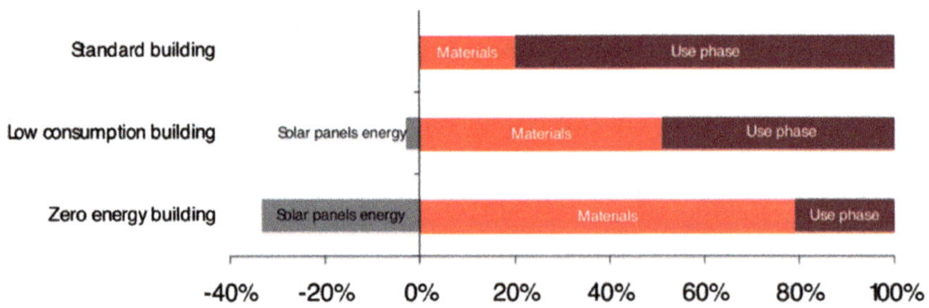

Figure 4.16 Comparison between a standard, present-day building and low-consumption and zero-energy buildings (figures are only illustrative). Image reproduced with permission from ArcelorMittal.

(the paper is focused on CO_2 emissions) including both operational and embodied energy is difficult to reach and would be unobtainable in a scenario of low carbon electricity from the grid". This conclusion is contingent to the kind of zero-energy technology studied in the article and further research might exhibit solutions that match the LCT-NZEB criterion.

4.8 Social Value of Construction

The construction sector has developed a rich set of methods to describe the sustainability of construction: LCA and LCT hold the upper hand, but they are complemented by beyond-LCA indicators and approaches, which focus on environmental and health issues left fallow in LCA. Is this enough? How, in particular, is the social dimension taken on board? This is an open question, as indeed, the social dimension is mostly left aside. These matters are currently being discussed and explored within the framework of the *Society and Materials Conferences.*[29]

The simple example of the Eiffel Tower, an emblematic building, can help raise the main questions related to the concept of social value as applied to construction. The Eiffel Tower was built in 1889, for the World Fair, which was held in Paris in 1889. It was supposed to be destroyed at the end of the event. Its initial value was covered by the price paid by the Eiffel company to erect it (8 million French francs in 1890). The Tower is still standing tall over the Paris skyline 150 years later. It has become a monument, part of the cultural heritage of France and of the world (the Seine Banks are a UNESCO world heritage site). Everything that the tower has brought to society over its 150-year life has therefore added extra value to it, which can be called its *social value.*[28] The social value of the Eiffel Tower includes its role as a major tourist attraction in the city, as the site of two famous restaurants, as a communication tower, as an image sold in millions of copies as postcards or sculptures, as a semaphore of Paris at night, as an important element in firework shows, *etc.* At a more abstract level, it is a symbol of the industrial revolution, of the strength, social importance and modernity of steel (actually puddled iron) as a material, *hic et nunc.* Its lattice tower structure shows the building materials, which, in ordinary construction, are mostly hidden, to the passer-by and the user.

Note also that much of the social value of the Eiffel Tower has not been monetized (paid): it is conceptually similar to the Eco-System Services (ESS) provided by biodiversity – which is why the extra services brought by an anthropogenic artifact are called Eco-Social-System Services (ESSS).[30]

4.9 Conclusions

Life Cycle Assessment is today commonly used in the building sector. For example, at the European level, a series of standards was developed (or is under development) to ensure consistency in the "assessment of the sustainability aspects of new and existing construction works (buildings and civil engineering works)" by the European Committee for Standardization (CEN) through Technical Committee (TC) 350, *i.e.* CEN/TC350. Within CEN/TC350,

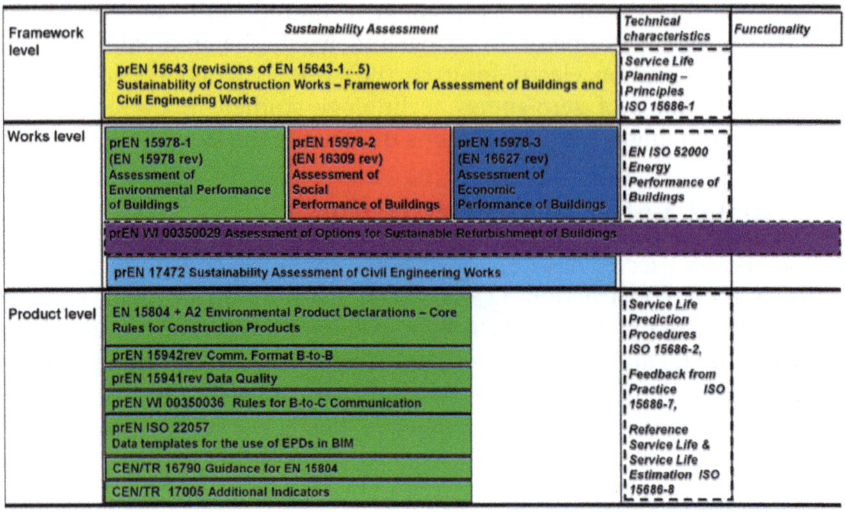

Figure 4.17 Overview of standards from CEN/TC350.
Reproduced from ref. 30 with permission from AFNOR.[32]

eight working groups deal with building products, environmental, economic and social performance of buildings, civil engineering works as well as refurbishment. An overview of CEN/TC350 activities is shown in Figure 4.17. Similar activities are ongoing both at national and global levels (within ISO), the International Organization for Standardization, ISO/TC 59 SC17.[31]

As in all the fields where LCA is used, LCA in construction needs to evolve beyond its present achievements and to improve further. The subtler features such as comfort, health issues and the social value of constructions will continue to foster research in methodology development within LCA *per se*, but also beyond LCA. But more general issues, like how energy is handled in LCA (mostly indirectly, through a series of complementary indicators), may have to be revisited.

Acknowledgements

The author would like to acknowledge the help given by ArcelorMittal Research in the supporting content of this chapter, and in improving it beyond the initial draft. He is particularly grateful to Roberto Turconi and Anne-Laure Hettinger.

References

1. J.-P. Birat, P. Russo, B. Gros, J. Dalsheimer and J. L. Phelouzat, *Waste Recycling in the construction sector*, Steel in Sustainable Construction, IISI World Conference 2002, 2002, Luxembourg, Proceedings, pp. 293–298.

2. Construction production (volume) index overview, Eurostat, 2018, https://ec.europa.eu/eurostat/statistics-explained/index.php/Construction_production_(volume)_index_overview.
3. Worldsteel in figures, worldsteel, 2018, 30 pages, https://www.worldsteel.org/en/dam/jcr:f9359dff-9546-4d6b-bed0-996201185b12/World+Steel+in+Figures+2018.pdf.
4. Steel, the permanent materials in the circular economy, worldsteel, 2018, http://circulareconomy.worldsteel.org.
5. J. P. Birat, Life Cycle Assessment (LCA) and Related Methodologies, in *Treatise on Process Metallurgy*, ed. S. Seetharaman, Elsevier, 2013, ch. XII, 25 pages.
6. J. F. Demonceau, O. Vassart, K. Jarmai, P. García, L. Cascini, M. Netušil, V. Ungreanu, M. Piasecki, M. Founti, P. Elguezabal, P. Može, I. Talvik, B. Åstedt, P.-O. Martin, J.-P. den Hollander, V. Huet, H. Gervásio; A. K. Kvedaras and R. Siebers, *Large valorisation on sustainability of steel structures (LVS3) - Final report – Study*, EU2791EN, 2014, 59 pages.
7. J. P. Birat, N. Prum, K. Yonezawa and L. Aboussouan, The value of recycling to society and its internalization into LCA methodology, *Rev. Metall.*, 2006, 50–61.
8. C. Eastman, P. Teicholz, R. Sacks and K. Liston, *BIM Handbook A Guide to Building Information Modeling for Owners, Managers, Designers, Engineers, and Contractors*, John Wiley, 2008.
9. A. J. Ross, *The Industries of the Future*, Simon Schuster, 320 pages, 2016.
10. C. Milachowski, T. Stengel and C. Gehlen, *Life Cycle Assessment for Road Construction and Use*, EUPAVE, Brussels, 18 pages, 2011.
11. A. L. Hettinger, J. P. Birat, O. Hechler and M. Braun, Sustainable Bridges – LCA for a composite and a concrete bridge, in *Economical Bridge Solutions Based on Innovative Composite Dowels and Integrated Abutments*, Springer Vieweg, 2015, pp. 45–56, 179 pages, DOI: 10.1007/978-3-658-06417-4_3.
12. CEN/TC350 WG6, AFNOR, Paris, on-going work, 2018, http://portailgroupe.afnor.fr/public_espacenormalisation/centc350/index.html.
13. A. Picon (sous la direction de), *L'Art de l'ingénieur, Constructeur, Entrepreneur, Inventeur*, Centre Pompidou – Le Moniteur, 598 pages, 1997.
14. *Sustainable Steel Buildings: A Practical Guide for Structures and Envelopes*, ed. B. Hauke, M. Kuhnhenne, M. Lawson, M. Veljkovic and R. Siebers, Wiley Blackwell, 362 pages, 2016.
15. *Steligence, The intelligence construction choice*, White Paper, An Objective Methodology for Assessing the Sustainability Impacts of Building Construction Solutions, ArcelorMittal, 2018, 38 pages, available from https://steligence.arcelormittal.com.
16. J. P. Birat and J. S. Thomas, *Beyond Life-Cycle Thinking: the SOVAMAT initiative and the SAM seminars*, ECOBALANCE, 2008, International Conference on Ecobalance, Tokyo, Japan.
17. La déclaration environnementale des produits du bâtiment, Ministère de la cohésion des territoires et des relations avec les collectivités territoriales, France, published on 8 April 2014.

18. Institut Bauen und Umwelt (IBU), https://ibu-epd.com/en/.
19. Guide to Sustainable Building Certifications, Realdania and the Dreyer Foundation, 2018, 154 pages.
20. R. J. Cole and M. Jose Valdebenito, The importation of building environmental certification systems: international usages of BREEAM and LEED, *Build. Res. Inf.*, 2013, **41**(6), 662–676.
21. J. P. Birat, The environment, from the standpoints of philosophy, social sciences and political activism, *Mater. Tech.*, 2018, DOI: 10.1051/mattech/2018067.
22. *New Office Design for the ICADE Premier House in Munich*, Landau + Kindelbacher, Archdog, the best social network for architects, http://arcdog.com/portfolio/new-office-design-for-the-icade-premier-house-munich/.
23. PHOSTER project, Photovoltaic Steel Roof, ArcelorMittal, accessed 28 December 2018, http://ds.arcelormittal.com/construction/france/news/Life-Phoster/language/EN; ARCIGS-M project, accessed 28 December 2018, https://www.arcigs-m.eu.
24. On the energy performance of buildings, European Union DIRECTIVE 2010/31/EU, 19 May 2010.
25. US Department of Energy, *What is a Zero Energy Building*, 2017, YouTube, https://www.youtube.com/watch?v=FysJKq5yCfg.
26. J. S. Thomas and J. P. Birat, Methodologies to measure the sustainability of materials – focus on recycling aspects, *Rev. Metall.*, 2013, **110**(1), 3–16.
27. A. Hermelink, S. Schimschar, T. Boermans, L. Pagliano, P. Zangheri, R. Armani, K. Voss and E. Musall, *Towards nearly zero-energy buildings, Definition of common principles under the EPBD Final report*, Ecofys, 467 pages, 2013.
28. L. Georges, M. Haase, A. Houlihan Wiberg, T. Kristjansdottir and B. Risholt, Life cycle emissions analysis of two nZEB concepts, *Build. Res. Inf.*, 2015, **43**(1), 82–93.
29. J. P. Birat, A. Declich, S. Belboom, G. Fick, J.-S. Thomas and M. Chiappini, Society and Materials, a series of regular seminars based on a dialogue between soft and hard sciences, *Metall. Res. Technol.*, 2015, **112**, 501.
30. J. P. Birat, Philosophy, social sciences and holistic views about materials, in *Environment and Society, Sustainability Materials Science*, EDP Science, 2019, ch. 17.
31. https://www.iso.org/committee/322621.html.
32. CEN/TC350 standard, AFNOR, http://portailgroupe.afnor.fr/public_espacenormalisation/centc350/index.html.

CHAPTER 5

Case Study – Construction Materials (Cement, Aggregates and Concrete Products)

MD. UZZAL HOSSAIN AND CHI-SUN POON*

The Hong Kong Polytechnic University, Department of Civil and Environmental Engineering, Hung Hom, Kowloon, Hong Kong
*Email: cecspoon@polyu.edu.hk

5.1 Introduction

Over time, sustainability has increasingly been given importance and priority in most of the industrial sectors, including the construction sector, due to its increasing materials usage, consumption of energy and water resources, waste generation, and their environmental consequences. Cement and concrete are the key construction materials, and are used globally. The consideration of environmental impacts of the production of aggregates, cement and concrete are thus of significant importance for promoting sustainability in the construction industry.[1] Cement production accounts for about 8% of total global anthropogenic CO_2 emissions.[2] The global production of cement is expected to increase by 25% by 2020 and 32% by 2030, based on 2010 production levels (the total global production is expected to reach 4.8 billion metric tonnes by 2030).[3] In addition, the annual consumption of concrete is around 25 gigatonnes (Gt) globally, amounting to over 3.5 tonnes *per capita*.[4] Although the environmental impacts of aggregate production are comparatively lower, it also consumes significant amounts of natural resources. For example,

Life Cycle Assessment: A Metric for the Circular Economy
Edited by Aiduan Borrion, Mairi J. Black and Onesmus Mwabonje

Published by the Royal Society of Chemistry, www.rsc.org

construction aggregates demand was 53.2 billion metric tons (Gt) in 2017, which is expected to increase to 66.2 Gt by 2022.[5]

When there is resource scarcity and shortage in supply, industrial symbiosis[†] can play a significant role in lowering environmental impacts and promoting green economic growth, *i.e.* the system can help link industrial development and carbon reduction.[6] The symbiosis encourages more circular and sustainable practices in the manufacturing sector,[7] by turning waste/by-products into valuable materials from one industry, to be reused as feedstock for another.[8,9] More importantly, it creates a cooperative network to exchange material resources, energy, water and/or by-products.[10] Therefore, resources productivity will be the key focus in the construction industry in the future. Circular economy (CE) is an economic system that aims to reduce resource use and minimize waste through the use and re-use of materials in closed cycles.[11] Industrial symbiosis is already recognized as a practical approach to CE, and is already embedded in the EU CE package, through the ratification of an EU law in 2018.[7] It could contribute significantly to global sustainability by increasing the circularity of materials in the manufacturing sector, reducing resources depletion and minimizing waste from the systems.[7,10] One such symbiosis network has led to the development of eco-industrial parks of different sizes throughout the world, especially in Europe, USA, South Korea, Japan and China.

There are three major trends in sustainable cement production, including: (1) the modification of plants/manufacturing processes; (2) the use of alternative fuels (mainly fuel derived from waste materials or biofuel); and (3) the use of alternative raw materials (mainly replacement of clinker materials by industrial wastes and by-products, and other waste materials). Also, two major sustainable trends can be identified for concrete production, including the use of: (1) recycled aggregates; and (2) different alternative cementitious materials to partially or fully replace Portland cement.[12,13] These solutions promote cleaner production by reducing emissions and minimizing different waste materials, based on the industrial ecology principle for saving natural resources.

Life cycle assessment (LCA) is a recognized method[14] that has been extensively used in evaluating the environmental impacts of construction materials and products. Based on its significance, numerous LCA studies have focused on evaluating the sustainability performance of cement production from different points of view, including assessing different methods for technological improvements and plant variability,[15] the use of alternative materials (*e.g.* fly ash and granulated blast furnace slag),[16–18] types of cement[19] and the use of alternative fuels.[20–23] Similarly, the environmental

[†]"An industrial symbiosis constitutes a local partnership where, partners provide, share and reuse resources to create shared value. The purpose of industrial symbiosis is to create loops of technical or biological materials while minimising the leakage and waste in the loops – demonstrating some key parts of a circular economy, at a local scale." The Ellen MacArthur Foundation (2019).

performance assessment of concrete products and concrete production has been carried out in several studies, which mainly focus on comparing conventional concrete production[24] and the use of alternative materials to substitute cement, such as fly ash,[24,25] silica fume,[26] blast furnace slags,[27] glass powder[28] *etc.*, for different applications.

This chapter aims to evaluate the environmental impacts of commonly used construction materials and products including cement, aggregates, concrete products and concrete. The chapter looks at LCA as a methodology for assessing environmental impacts, with case-specific analyses for the utilization of alternative materials and fuels in production processes within the CE principle.

5.2 Life Cycle Assessment

5.2.1 Framework

According to the International Organization for Standardization (ISO), environmental life cycle assessment deals with the environmental aspects and potential environmental impacts (*e.g.* use of resources and environmental consequences of releases) throughout a product's life cycle, from raw material acquisition through production, use, end-of-life management, recycling and final disposal.[29] It is considered to be an effective, systematic and quantitative decision support tool to evaluate the environmental burdens of a product/process, and to evaluate alternatives for environmental improvement.[29]

The principle and framework for conducting environmental LCA has been provided by the ISO in its guidelines: *ISO-14040: Environmental Management – Life Cycle Assessment – Principles and Framework* (Figure 5.1).[29]

5.2.1.1 Goal and Scope Definition

The first step of a LCA is goal and scope definition, which includes defining the reasons for conducting the study, defining and describing the product,

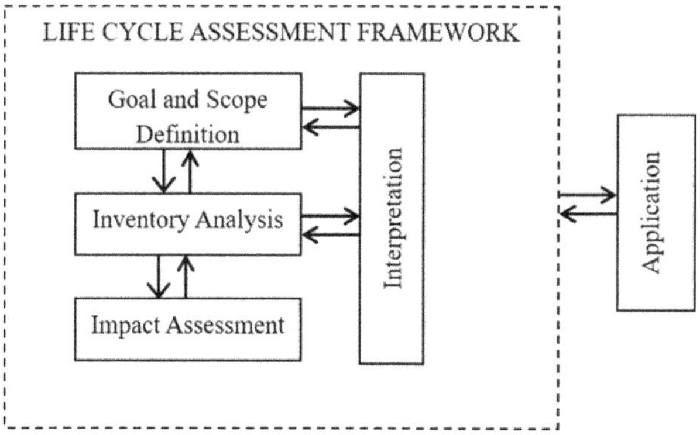

Figure 5.1 Life cycle assessment framework.[29]

process or activity, establishing the context in which the assessment is to be carried out and identifying the system boundaries, functional unit, assumptions, impact assessment methodology and environmental impact categories, *etc.* In the case of LCA of construction materials, system boundaries would be cradle-to-gate, cradle-to-site/use, *etc.*, while the functional unit may be 1 tonne or 1 m^3, depending on the type of materials. Making appropriate assumptions is often inevitable in the LCA study. The selection of system boundary, functional unit, assumptions *etc.* is provided in the case-specific examples in later sections of this chapter.

5.2.1.2 *Life Cycle Inventory Analysis*

The second step is life cycle inventory analysis (LCI), which includes data collection and calculation procedures to quantify relevant inputs and outputs of a product or process system. In LCI, data should be energy inputs, raw material inputs, ancillary inputs, other physical inputs, products, co-products and waste, releases into the air, water and soil, and other environmental aspects that should be collected within the system boundary and goal of the study. The collected data are utilized to quantify the inputs and outputs of a unit process.[29]

5.2.1.3 *Life Cycle Impact Assessment*

The life cycle impact assessment (LCIA) step aims to evaluate the magnitude and significance of potential environmental impacts of a product or a system identified in the inventory analysis. According to ISO 14044,[29] three mandatory elements are associated with the LCIA step, including:

1. Selection of impact categories, category indicators and characterization models;
2. Assignment of LCI results to the selected impact categories (classification); and
3. Calculation of category indicator results (characterization).

Optional elements of a LCIA include normalization, grouping, weighting and data quality analysis (*e.g.* gravity analysis, uncertainty and sensitivity analysis), which are often used in impact assessment.

5.2.1.4 *Life Cycle Interpretation*

This is the last step of LCA, which evaluates the results of the inventory analysis and impact assessment. It includes the identification of the significant issues based on the results of the LCI and LCIA steps of LCA, an evaluation that considers completeness, sensitivity and consistency; and conclusions, limitations and recommendations resulting from the assessment.[29]

5.2.2 Methodology

The LCIA step intends to estimate the magnitude and significance of all the environmental impacts obtained in the LCI phase. This step combines several basic elements including characterization, normalization, damage assessment and single score of the LCI results into an indicator result. In the characterization step, the LCI results are converted to LCIA results through characterization modelling, using a list of predefined factors. Using characterization factors, LCI results are converted to indicators of impact categories. In the normalization step, an indicator result is converted to a normalized value, according to a normalization factor. The factor can be defined according to a country, a region or the world.[30] In damage assessment, a number of impact category indicators are combined into an 'end-point' damage category. The last step of the impact assessment consists of converting the end-point categories into a single eco-point based on the standardization factors. The eco-point is known as 'single score', which is a dimensionless figure, measured in units of milli-points (mPt), which indicates the potential number of people affected by the environmental impacts in a period of one year. Further details of impact calculation can be found in the literature.[31]

Different environmental impact assessment methods have been developed, including IMPACT 2002+,[32] CML 2001,[33] Eco-indicator 99,[34] EDIP 2003,[35] LIME,[36] ReCiPe 2008,[37] TRACI[38] *etc.* A number of tools, especially software packages, are available to facilitate environmental impact assessment. Some of these are fully compliant with ISO standards for LCA including SimaPro, GaBi, Umberto, eBalance, EcoCalculator *etc.*[39] The use of standard LCI databases is a common practice in LCA studies.[40] Due to the high temporal variability of LCI databases, significant variations of LCA results are observed, and thus regional LCI databases are preferred in the LCA community.[41] Therefore, some databases are developed from global/regional perspectives (*e.g.* Ecoinvent; CPM LCA database), while some are based on local perspectives (*e.g.* USLCI database; CLCD). In addition to case-specific local or regional studies, Chinese Life Cycle Databases (CLCD) were mostly used in the case studies presented in this chapter. However, other global and regional databases (*e.g.* Ecoinvent, USLCI, *etc.*) were also used where local literature/databases were not available. For the case studies presented, two selected mid-point environmental impacts, including global warming potential (GWP) and non-renewable energy consumption were assessed by SimaPro 8.1.0 software using the IMPACT 2002+ method.[32]

5.3 Circular Economy: Concept and Approach

The concept of circular economy (CE) is drawing increasing attention from both researchers and practitioners, as CE aims to save the maximum value of materials through recycling, preserving resources and eliminating waste generation. CE is defined as "economy that is restorative and regenerative by design, and which aims to keep products, components and materials at their highest utility and value at all times, distinguishing between technical and biological cycles".[42] Thus, materials should enter the CE framework during their useful service lifetime.[6] However, the guidelines for CE implementation and choosing CE indicators are still in the early stages of development. Few studies have proposed guidelines for implementing CE principles. For example, whether CE indicators should be based on LCA, or material flow analysis (MFA) and material flow costing.[6] Another study implies that a "cradle-to-cradle (C2C)" approach should be considered for CE adoption instead of a "cradle-to-grave" LCA approach.[43] Dieterle *et al.*[44] also highlighted that the cradle-to-grave approach, with the credits for substituted materials, is not fully suitable for meaningful interpretation for CE setting, and thus suggested that C2C LCA approach should be better aligned with CE settings.

For the purpose of utilizing alternative materials from secondary resources, recycling rates can be good performance indicators for CE settings, as the recycling rates (for both open- and closed-loop recycling systems) can provide useful information in quantifying the circulated materials.[45] At its infancy stage of development, only a few studies can be found to interlink CE with LCA in cement and concrete production. For example, Smol *et al.*[11] reviewed the possible use of sewage sludge ash for the production of building materials like cement, brick, ceramics, aggregates, *etc.* by adopting the CE concept in the construction industry.

Thus, two approaches including the closed-loop and open-loop recycling systems can be adopted in CE, depending on the type of materials and their potential reutilizations. In CE principles, when waste materials are transformed into valuable resources by designing the products to be easily repairable, recoverable, and reused as primary materials for the same industry, it is named as a closed-loop system (where the products are redirected to the manufacturer). Another system, named open-loop recycling, is where the recovered products are used in the manufacturing of other products with industrial symbiosis (through recycling, downcycling and upcycling).[46,47] For example, Niero and Olsen[46] pointed out that the closed-loop recycling system of aluminum cans has lower climate change impacts over the other recycling scenarios. In contrast, the use of mixed waste glass as alternative cementitious materials in an open-loop circular approach can reduce the overall environmental impact in concrete production.[47] Some case-specific examples will be highlighted in this chapter for construction materials production within the CE principle, where different waste materials are adopted, whether by the open-loop or the closed-loop system.

5.4 Case Studies of Production of Construction Materials in Hong Kong

5.4.1 LCA of Cement Production

5.4.1.1 The Production Process

The production of ordinary Portland cement (OPC), a key constituent in concrete or concrete products, requires a huge amount of natural resources and energy.[1] The production process of OPC requires raw materials (primarily limestone and clay) acquisition, and then grinding and mixing of them, to produce a raw mixture that is then pyroprocessed in a rotary kiln at temperatures approaching 1480 °C to form the cement clinker. The clinker is then ground and blended with various additives and/or alternative cementitious materials and gypsum to form OPC (Figure 5.2). However, the equipment used in cement production facilities can vary depending on the age, size and design of the plants.[48]

In addition to the commercially available OPC, the cement plant in Hong Kong also produces an eco-glass (Eco-GC) cement, which is a new type of cement with the incorporation of different percentages of waste glass powder, and a Portland-fly ash cement (PFC). The constituents of the three different types of cement are given in Table 5.1. The compressive strength of OPC and Eco-GC-1 are similar (strength class: CEM I 52.5 N; conforms to BS EN 197-1:2000 specification).[49] However, the compressive strength of Eco-GC-2 is slightly lower than the OPC and PFC at 28 days, but they are comparable at 90 days. The details of the manufacturing processes of different types of cement can be found in our previous study.[49]

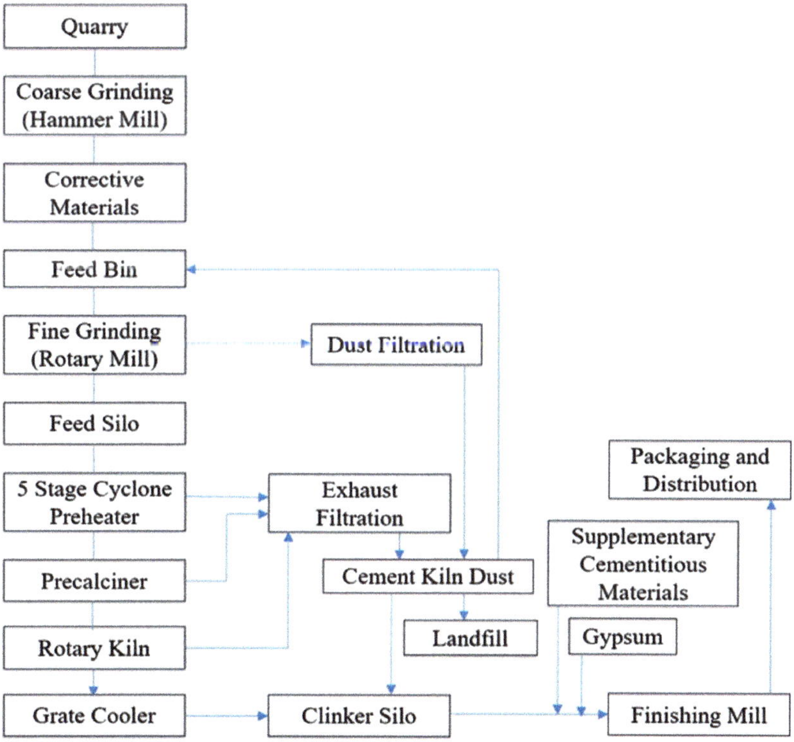

Figure 5.2 The cement production process.
Adapted from ref. 48 with permission from Elsevier, Copyright 2016.

Table 5.1 Compositions of different types of cements.[49]

Materials	Type of cement: Ordinary Portland cement (OPC)	Portland-fly ash cement (PFC)	Eco-glass cement (Eco-GC-1)	Eco-glass cement (Eco-GC-2)
Clinker (limestone 80%, sand 9%, clay 9% and iron ore 2%)	92%	75%	90%	75%
Gypsum	5%	5%	5%	5%
Pulverized fly ash	—	20%	—	—
Limestone powder	3%	—	2%	—
Waste glass powder	—	—	3%	20%
Total	**100%**	**100%**	**100%**	**100%**

For clinker production, the principal raw materials (*e.g.* limestone, clay, sandstone and iron ore) are imported from various locations (from mainland China, Japan and the Philippines) to the manufacturing plant in Hong Kong, and they are then separately ground to prepare fine powders. A synthetic gypsum is used, which is a by-product of the flue-gas desulfurization process from local coal-fired power plants.

About 20% fly ash (FA) (by wt.) is used as clinker replacement in the PFC. FA is a by-product generated from coal-fired power plants. Therefore, the transport distance and energy requirements of FA processing are modelled separately.[49] For Eco-GC production, waste glass powder ground from glass cullet obtained from locally generated waste glass bottles, are used as a partial replacement of clinker. The recycling, collection and processing of waste glass can be energy intensive. Three possible sources of waste glass collection systems are identified in Hong Kong, such as glass bottles recovery from designated recycling bins, hotels and shops; refuse transfer stations and landfill sites. The average distance of waste glass collection and transport to the cement manufacturing site was estimated. After collection, waste glass goes through several processing steps before use in the blended cement. The collected glass bottles are then crushed to glass cullet (<3.55 mm), and ground to make fine powder. Mechanical forklifts and conveyers are used for on-site handling and transportation of the cullet to the storage bunker.[49,50] The avoided impacts for landfilling the glass bottle were considered, as landfilling is the common practice for waste glass bottles disposal in Hong Kong. The transport of materials and energy requirements with their upstream data sources/databases are shown in Tables 5.2 and 5.3.[49]

5.4.1.2 *System Boundary with Adoption of Recycled Materials*

Case-specific LCA results using recycled materials are considered for cement production in Hong Kong, with a "cradle-to-gate" system boundary.

Table 5.2 Raw materials and transport data for cement manufacturing in Hong Kong.[49]

Materials	Exported from	Transport type	Distance (km)
Limestone	Guangdong, China	Inland barge	250
Limestone	South Japan	Ocean ship	3202
Sand	Guangdong, China	Inland barge	250
Clay	Guangdong, China	Inland barge	250
Iron ore	South Japan	Ocean ship	3202
Gypsum	Local (power plant)	Trucks (18 t)	7
Limestone (for limestone powder)	Philippines	Ocean ship	1720.5
Coal	Indonesia	Ocean ship	4073
Fly ash to plant	Local	Truck (30 t)	6
Waste glass bottles	Locally generated	Truck (18 t)	55
Waste glass bottles to landfill sites	Locally generated	Truck (18 t, 30 t)	41

Two strategies, such as the use of biofuel (produced from recycled wood waste), and the use of recycled materials as a substitute for clinker, are adopted in cement production in Hong Kong. In the first case, the use of recycled (waste) wood is considered as a replacement for fossil fuel (coal) in the production system. The second case considers the adoption of recycled alternatives (*e.g.* glass powder and pulverized fly ash) to clinker as an addition to produce different types of cement (considering their strength classes). This is shown in Figure 5.3.

Adoption of alternative fuel: Waste-derived fuels are good potential sources of alternative energy supply in cement plants.[20,21,48] Due to their high energy content, biofuel (wood pellets) derived from wood wastes are also a good source of energy for co-firing with fossil fuels.[51] This alternative fuel from wood waste is produced and used in the cement plant through an open-loop recycling system. Based on our previous study on LCA of energy generation from wood pellets produced from wood wastes and coal in Hong Kong,[52] several useful data have been gathered, as below.

To produce 1 kWh of heat:

- Fuel requirement: About 184 g of coal (with 90% efficiency), and about 226.42 g of wood pellets (with 80% efficiency) is required.
- Total energy requirement: About 4.53 MJ from coal, and 1.17 MJ from wood pellets.
- GHGs emission: About 0.33 kg of CO_2 eq. for coal, and 0.06 kg of CO_2 eq. for wood pellets.

With the use of wood pellets as a substitute for coal, it is possible to save about 3.36 MJ of non-renewable energy (considered as energy saving factor, ESF) and reduce 0.27 kg CO_2 eq. GHGs emissions (considered as carbon

Table 5.3 Sources of energy for materials/processes for cement manufacturing.[49]

Materials/processes	Energy requirement	Upstream data
Limestone production	Referring to the database	USLCI[a]
Sand production	Referring to the database	Ecoinvent[b]
Clay production	Referring to the database	Ecoinvent[c]
Iron ore production	Referring to the database	Ecoinvent[d]
Gypsum production	Referring to the database	Ecoinvent[e]
Limestone powder processing (crushing and grinding for making powder)	3.51 kWh t^{-1} limestone	CLP[f]; CLCD[g]
Coal production and transport to cement plant in Hong Kong	Referring to the database	Hossain *et al.*[52]
Electricity consumption (crushing, raw materials grinding and clinker production)	87 kWh t^{-1} clinker	CLP[f]; CLCD[g]
Coal combustion during calcinations in plant in Hong Kong	Referring to the database	Hossain *et al.*[52]
Cement blending	48 kWh t^{-1} cement	CLP[f]; CLCD[g]
Transport, road	Referring to the database	CLCD[h]
Transport, sea (inland barge)	Referring to the database	CLCD[i]
Transport, sea (ocean ship)	Referring to the database	USLCI[j]
Waste glass processing	25 kWh t^{-1} glass cullet	CLP[f]; CLCD[g]
Waste glass landfilling	Referring to the database	ELCD[k]

[a]USLCI: Limestone at mine. US Life Cycle Inventory Database. National Renewable Energy Laboratory, 2012. US Department of Energy 2012.
[b]Ecoinvent: Sand (RoW), gravel and quarry operation. Ecoinvent System Processes, Swiss Centre for Life Cycle Inventories (2013).
[c]Ecoinvent: Clay (RoW), clay pit operation. Ecoinvent System Processes, Swiss Centre for Life Cycle Inventories (2013).
[d]Ecoinvent: Iron ore at mine, 46% Fe (GLO). Ecoinvent System Processes, Swiss Centre for Life Cycle Inventories (2013).
[e]Ecoinvent: Gypsum, mineral (RoW), desulfurisation of hard coal flue gas. Ecoinvent System Processes, Swiss Centre for Life Cycle Inventories (2013).
[f]CLP: Sustainability Report 2014, the China Light and Power (CLP). Hong Kong (2014).
[g]CLCD: Electricity Generation, China Southern Grid. Chinese Life Cycle Database Version 0.8. Sichuan University, IKE Environmental Technology Co., Ltd (2010).
[h]CLCD: Heavy-duty diesel truck transportation (30 t, 18 t)-CN-AP.
[i]CLCD: Bulk Cargo Shipping (2500 t)-CN-AP.
[j]USLCI: Transport, ocean freighter, residual fuel oil powered. US Life Cycle Inventory Database. National Renewable Energy Laboratory, 2012. US Department of Energy.
[k]ELCD: Landfill of glass/inert waste EU-27. ELCD (European reference Life Cycle Database), Joint Research Centre (JRC) of the European Commission, ELCDDATA 13292600296 (2013).

reduction factor, CRF) for 1 kWh of heat generation. Two equations were used to quantify the impacts of using wood pellets to replace coal, in terms of energy consumption and GHGs emissions reduction, as follows:[49]

$$\text{Energy saving (\%)} = \frac{\dfrac{\text{ESF} \times \text{\% of replacement (coal energy)}}{3.6}}{\text{Energy consumption per tonne of clinker production}} \times 100 \tag{5.1}$$

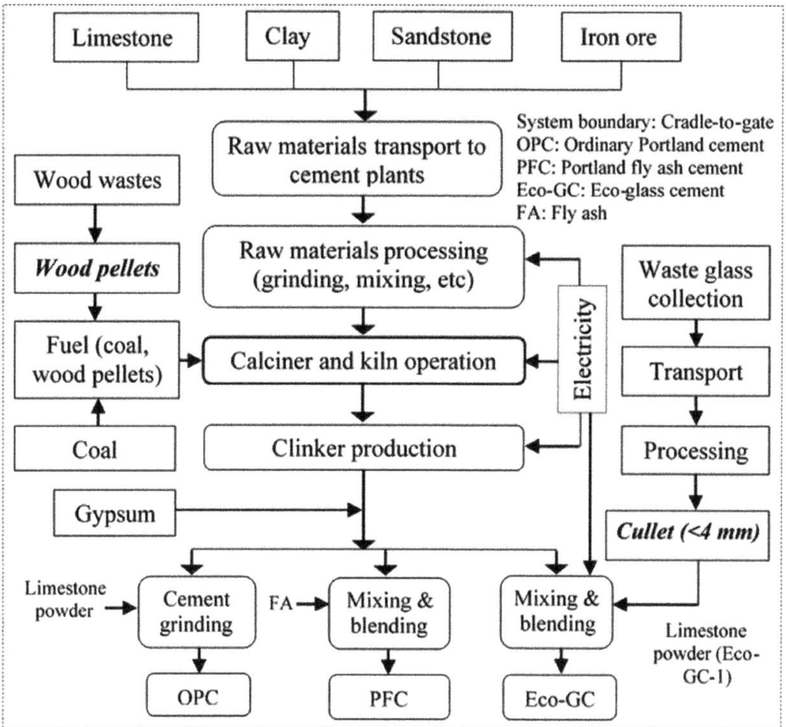

Figure 5.3 System boundary for cement production in case study.[49]

$$\text{GHGs reduction}\ (\%) = \frac{\dfrac{\text{CRF} \times \%\ \text{of replacement (coal energy)}}{3.6}}{\text{GHGs emission per tonne of clinker production}} \times 100 \tag{5.2}$$

Adoption of waste glass powder: Waste glass powder is produced from crushing and grinding waste glass beverage bottles and used in cement production through an open-loop recycling system. The collection, transport and processing of waste glass bottles to produce ground powder is described above and mentioned in Tables 5.2 and 5.3. In LCA modelling, the collection and processing-related energy (associated emissions) are considered as induced impacts, and energy associated with the collection and landfilling of the waste glass bottles are considered as avoided impact.[44]

5.4.1.3 *Environmental Impacts of Cement Production with Recycled Materials*

The use of two alternative, recycled materials, one used as fuel and the other used as clinker materials were used in these examples. The summary of the selected impact categories for 1 tonne of clinker production is shown in Table 5.4. Clinker production is an important step and intermediate material in cement manufacturing. The results show that more than 1 tCO_2 eq. GHG is emitted and more than 5.5 GJ of non-renewable energy is needed for the production of 1 tonne of clinker. It can be seen that about 54% CO_2 is emitted for the calcination of limestone, whereas 32% is emitted for burning coal, 6% for clinker materials production and transport, and 5% for electricity consumption. For non-renewable energy consumption, about 70% was associated with coal, whereas 18% was associated with electricity consumption, and 12% for the clinker materials.

The impacts of other additional materials, such as gypsum, fly ash, limestone powder and glass cullet that are used for cement production, are shown in Table 5.5. The results showed that the avoided impacts due to avoiding landfill disposal of waste glass are quite significant compared with producing glass powder. Compared with those additional materials, the final processing (*e.g.* cement blending) contributes higher impacts.

Based on the compositions of different types of cement manufactured (Table 5.1), the selected impacts categories with their strength classes are shown in Figure 5.4. The results show that about 980 kg CO_2 eq. GHG emissions with 5.7 GJ of non-renewable energy consumption are associated with OPC production. The impacts are considerably higher for OPC production in Hong Kong, as Hong Kong entirely depends on importing the raw materials, and the use of coal as fuel for clinker production. For equal compressive strength at 28 days, the results reveal that Eco-GC-1 is associated with a 2% reduction of the total GHGs emission with a 2% saving of the total energy consumption compared with the OPC.

Table 5.4 Selected impact categories for 1 tonne clinker production.

Materials/process	Processes	GHGs emission (kg CO_2 eq.)	Energy consumption (MJ)
Limestone	Quarrying, processing and transport to plant	54.11	624.79
Sandstone	Quarrying, handling and transport to plant	1.24	14.86
Clay	Quarrying, handling and transport to plant	1.12	13.60
Iron ore	Quarrying, processing and transport to plant	1.97	18.04
Coal	Production, processing and transport to plant	29.38	3868.40
	Coal combustion for clinker production	331	
Calcination	CO_2 emission during calcinations (decomposition of limestone)	551	
Electricity consumption	Processing (crushing, mixing, *etc.*)	54.81	1028.34
Total		**1024.63**	**5568.03**

Table 5.5 Selected impacts of additional materials.

Materials/ Processes	Processes	GHGs emission (kg CO_2 eq.)	Energy consumption (MJ)
Gypsum (1 kg)	Gypsum production and transport from desulfurization of hard coal flue gas	0.12	0.50
Limestone powder (1 kg)	Quarrying, crushing, grinding and transport to plant	0.04	0.56
Glass powder from waste glass bottles (1 kg)	Waste glass collection, transport and processing (*e.g.* crushing, grinding and sieving) for making powder	0.05	0.76
	Avoided impacts due to collection, transport and landfill	− 0.03	− 0.45
Fly ash (1 kg)	Induced impacts of FA due to collection, transport to plant and processing	0.006	0.11
Final processing (1 t cement)	Electricity consumption for processing (*e.g.* cement mixing, blending, *etc.*)	30.24	567.36

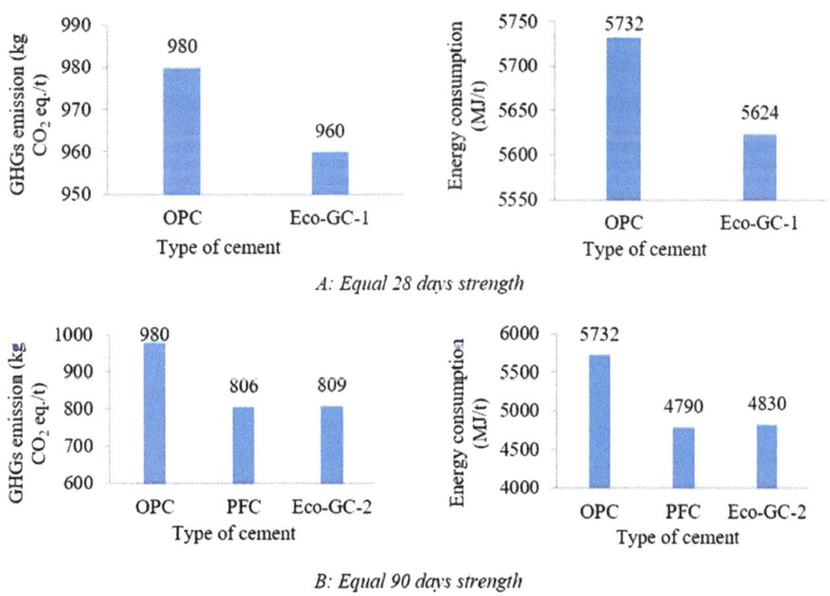

Figure 5.4 Selected impacts of different types of cement manufacturing.[49]

When 90 days equal strength is considered, considerably higher environmental impact savings are observed for alternative cement produced with by-product (FA) and waste materials (glass powder). Compared to OPC, the production of PFC and Eco-GC-2 were associated with a 17% and 16%

lower energy consumption, with a saving of 18% and 16% GHGs emission, respectively. Although the impact categories of PFC and Eco-GC-2 production are similar, a considerable amount of waste glass can be utilized for producing Eco-GC-2, which can save landfill space; this is very important in land-scarce regions like Hong Kong.

The current conventional cement production in Hong Kong has a high environmental impact. In addition, Hong Kong is facing a serious scarcity of land for waste disposal. This case study showed that waste glass has good potential to be recycled and reused as clinker replacement for blended cement production, and wood pellet produced from wood waste also has great potential to be used as a partial replacement for coal in the cement industry. The recovery and utilization of waste materials in cement production not only reduces the environmental impacts from the industry, it also can be an important outlet for waste utilization to alleviate the severe waste management problem in Hong Kong.

5.4.2 LCA of Aggregate Production

5.4.2.1 Aggregate Production

In Hong Kong, the demand for natural aggregates (NA) is mostly fulfilled by import from mainland China. For NA production, raw material extraction by mining (for crushed stone), and excavation and collection (for river sand) is the first step, and includes blasting, crushing, dredging (for river sand), transit and offloading. After screening, different types of crushers are used, depending on the aggregate size requirement. In addition, crushing/grinding machines with screeners are used to produce fine aggregates from crushed stones. Transport of aggregates includes on-site transport from the extraction site to the processing area, then to the vessel loading points using trucks, and off-site transport to Hong Kong by inland-coastal barges, and then to sites in different areas of Hong Kong using trucks.

In addition, huge amounts of construction and demolition (C&D) waste are generated in Hong Kong, and most of them are disposed of in public landfills. However, C&D waste is a good potential source of recycled aggregates production for further reuse in construction through a closed-loop system. Some information regarding LCA of aggregates production from C&D waste is shown in Table 5.6 and Figure 5.5, as an example of the system boundary of recycled aggregate (RA) production (LCI data and systems boundary are based on Hossain *et al.*[50]). The sorted, good-quality C&D waste is transported to the RA production plant and the average transport distances of C&D waste to the recycling and public fill sites are estimated. The collected C&D waste is sorted and then fed by mechanical shovels into hoppers for crushing. Different types of crushers and sieves are used, depending on the requirements of aggregate size. The produced RA is stored, and then transported to use sites.

Similar to C&D waste, a considerable amount of waste glass bottles are generated in Hong Kong, and only a small proportion are recycled. However,

Table 5.6 Materials and transport data of aggregate production.[50]

Materials	Locations	Transport type	Distance	Upstream database
Natural aggregates (RS, River sand)	Extraction sites to Dongguan, Guangdong Province, China	Trucks (30 t)	20 km	CLCD[a]
	Dongguan to Hong Kong Port	Inland barge	128 km	CLCD[b]
	Hong Kong Port to utilization sites	Trucks (30 t)	30 km	CLCD[a]
Natural aggregates (CS, Crushed stone)	Extraction sites to Dongguan, Guangdong Province, China	Trucks (30 t)	50 km	CLCD[a]
	Dongguan to Hong Kong Port	Inland barge	128 km	CLCD[b]
	Hong Kong Port to utilization sites	Trucks (30 t)	30 km	CLCD[a]
Recycled C&D waste	C&D waste generation sites to recycled aggregate production plant	Trucks (30 t)	45 km	CLCD[a]
Recycled aggregates (C&D waste)	Recycled aggregate production plant to utilization sites	Trucks (30 t)	40 km	CLCD[a]
C&D waste (to fill sites)	C&D waste generation sites to temporary storage (fill banks), and then to fill sites in mainland China	Trucks (30 t) Inland barge	35 km 200 km	CLCD[ab]
Recycled waste glass	Waste glass collection sites to recycled aggregate production plant	Trucks (18 t)	50 km	CLCD[a]
Recycled aggregates (waste glass)	Recycled aggregate production plant to utilization sites	Trucks (30 t)	40 km	CLCD[a]
Waste glass (landfill)	Waste glass collection sites to landfill sites	Trucks (6 t) Trucks (30 t)	6 km 35 km	CLCD[a]

[a]CLCD: Heavy-duty diesel truck transportation (6 t, 18 t, 30 t)–CN–AP. Chinese Life Cycle Database Version 0.8, Sichuan University; IKE Environmental Technology Co. Ltd, 2010.
[b]CLCD: Bulk cargo shipping (2500 t)–CN–AP.

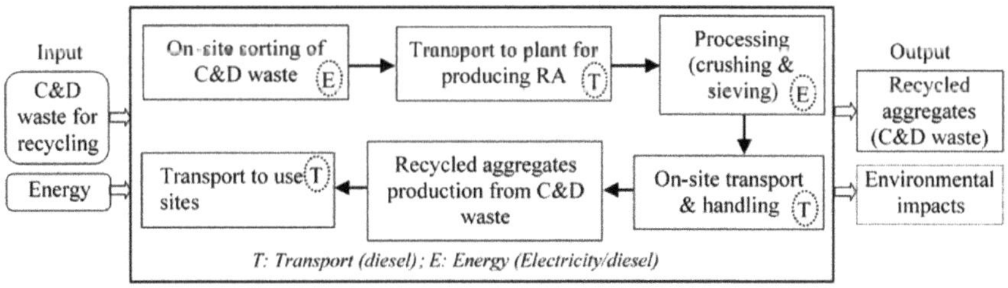

Figure 5.5 System boundary and processes of recycled aggregates produced from C&D waste.[50]

waste glass bottles can also be a good source of alternate aggregate for construction. The details of waste glass collection and management has already been described at Section 5.4.1.1. After being received at the recycling plant, the collected waste glass bottles are crushed by a hammer mill to produce glass sand (fine aggregates), after passing through sieves of different designated sizes. Mechanical forklifts are used for on-site transport, and mechanical conveyers are used to transfer the recycled glass cullet to the storage bunker.

5.4.2.2 System Boundary and Impact Assessment

A case-specific LCA study was conducted to assess and compare the environmental impacts associated with the production of different types of aggregate with a "cradle-to-site" system boundary in Hong Kong. For RA production, the environmental impacts of C&D waste and waste glass landfilling were considered as avoided impacts. The LCI data for this comparative assessment of aggregates production are shown in Tables 5.6 and 5.7.

5.4.2.3 Comparative Results of Aggregate Production

The LCA results for producing 1 tonne of fine aggregates is shown in Figure 5.6. It can be seen that about 64% and 73% lower GHG emissions are

Table 5.7 Energy requirements of the production of aggregates.[50]

Materials	Activity	Energy requirements	Sources of data	Upstream database
Natural fine (RS)	Extracting, on-site handling	69 MJ t^{-1} (diesel)	Measured	CLCD[a]
Natural fine (CS)	Extracting, screening, crushing and sieving	27 MJ t^{-1} (electricity) & 50 MJ t^{-1} (diesel)	Measured	CLP[b]; CLCD[c]
Natural coarse (CS)	Extracting, screening, crushing and sieving	22 MJ t^{-1} (electricity) & 51 MJ t^{-1} (diesel)	Measured	CLP[b]; CLCD[c]
Recycled fine (C&D waste)	Sorting (on-sites)	35 MJ t^{-1} (diesel)	Measured	CLCD[a]
	Processing (crushing, sieving, on-site transport and handling)	11 MJ t^{-1} (electricity) & 21 MJ t^{-1} (diesel)	Measured	CLP[b]; CLCD[c]
Recycled coarse (C&D waste)	Sorting (on-sites)	35 MJ t^{-1} (diesel)	Measured	CLCD[a]
	Processing (crushing, sieving, on-site transport and handling)	4 MJ t^{-1} (electricity) & 21 MJ t^{-1} (diesel)	Measured	CLP[b]; CLCD[c]
Recycled fine (waste glass)	On-site transport and handling, milling and sieving	47 MJ t^{-1} (electricity) & 21 MJ t^{-1} (diesel)	Measured	CLP[b]; CLCD[c]
Waste glass landfill	Landfill of glass/C&D waste	Referring to the ELCD database	—	ELCD[d]

[a]CLCD: Diesel consumption, China Southern Grid. Chinese Life Cycle Database.
[b]CLP: Sustainability Report 2014, the China Light and Power (CLP), Hong Kong, 2014.
[c]CLCD: Electricity generation, China Southern Grid. Chinese Life Cycle Database.
[d]ELCD: Landfill of glas/inert waste EU-27. ELCD (European reference Life Cycle Database), Joint Research Centre (JRC) of the European Commission, ELCDDATA 13292600296.

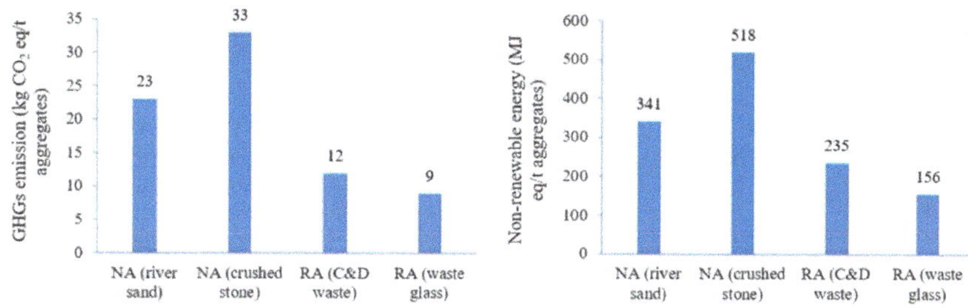

Figure 5.6 Comparative impacts of fine aggregates production.

associated with the fine aggregates production from C&D waste and waste glass, respectively, compared with the crushed stone; while the difference is about 48% and 61%, when compared with river sand. For the category of non-renewable energy consumption, RA produced with C&D waste and waste glass is, respectively, about 31% and 54% lower; and 55% and 70% lower, compared with NA from river sand and crushed stone. This is because the transportation (*e.g.* extraction sites in mainland China to use sites in Hong Kong) of the natural fine aggregates is associated with the highest energy consumption (about 71% for river sand and 66% for crushed stone), followed by extracting and handling (29% for river sand and 34% for crushed stone). Similarly, the transportation of C&D waste and waste glass to the recycled aggregate production plant and then the produced recycled aggregates to use sites is responsible for 67–69% of the total energy consumption. The other 31–33% is due to the on-site sorting and handling, and processing of the C&D waste and waste glass.[45] However, considerable environmental impacts, associated with the transport and disposal of the materials to public fills and landfills, can be avoided.

Due to the scarcity of quarry sites and the shortage of space for disposing the inert C&D waste and waste glass in Hong Kong, there is a huge potential for the production and use of recycled aggregates from locally available waste resources. However, the uptake is hampered by the current rate of recycled aggregate production, which is very low, due to the lack of appropriate sorting and recycling systems of C&D waste, and recycling and collection systems of waste glass. In addition, there is a lack of appropriate technology and downstream users, and a lack of willingness to reutilize recycled materials by the construction industry, as well as a lack of relevant policy, which hinders recycling and reuse.

5.4.3 LCA of Concrete Products Production

The use of alternative materials, including recycled aggregates in concrete product production, lowers the environmental impact, as well as saving resources. An example of different types of concrete paving blocks produced with recycled aggregates and fly ash (FA) is provided in this section.

The impacts of the production of the following paving blocks are assessed and compared:

- Normal block (NB): produced with conventional cement and natural aggregates, with other materials such as chemical admixtures and pigment
- First-generation eco-blocks (1G eco-blocks): manufactured by replacing 50% natural aggregates with recycled aggregates from C&D waste, and substituting 25% ordinary cement with FA, with other materials such as chemical admixtures and pigment
- Second-generation eco-blocks (2G eco-blocks): produced by replacing 80% natural aggregates by recycled aggregates (in which 55% RA from C&D waste and 25% recycled fines from waste glass), and substituting 25% ordinary cement by FA, with other materials including chemical admixtures and pigment
- Third-generation eco-blocks (3G eco-blocks): similar to 2G eco-blocks, but with a small percentage of nano TiO_2 as photocatalyst for the purpose of removing air pollutants such as VOC, NO_x, *etc.*

The LCI data regarding the mix-designs, and transportation and energy consumption for the concrete paving blocks production are shown in Tables 5.8–5.10 (based on our previous study.)[53]

A comparative GHG emission and non-renewable energy consumption for concrete paving blocks production with a cradle-to-site system boundary is shown in Figure 5.7. The results showed that about 19–23% lower GHG emissions and 21–27% lower non-renewable energy consumption is associated with the different types of paving eco-blocks production, compared to the normal paving blocks. Among the studied blocks, 2G had the lowest impacts, while slightly higher impacts are observed for 3G eco-blocks, due to the use of nano-TiO_2. However, the use of nano-TiO_2 in the eco-blocks had additional advantages, including its anti-microbial properties, and its ability to remove NO_x and other air pollutants.[54] The use of cement contributed to 77–86% of the total impacts for the concrete paving block production.[53] Thus, the savings are mainly achieved by using RA and FA in block production.

5.4.4 LCA of Concrete Production

5.4.4.1 The Production Process

Concrete is a composite material consisting of cement, aggregates, supplementary materials, water and admixtures, and is one of the most used construction materials globally. But the production of concrete is associated with significant environmental burdens including energy consumption, GHG emissions, resource consumption and other environmental impacts.[13] Thus, the environmental impacts have been assessed concurrently with

Table 5.8 Mix-proportions for paving blocks production in Hong Kong (by weight).[53]

Name of block	Processes	Natural fine (river sand)	Natural coarse (crushed stone)	Natural fine (crushed stone)	Recycled fine (waste glass)	Recycled coarse (C&D waste)	Recycled fine (C&D waste)	*Amount of aggregates*	Cement	Fly ash	Pigment	Admixture	TiO_2	*Amount of other materials*
NB	MP 1	50%	30%	20%	—	—	—	***80%***	100%	—	2–7%[a]	0.5–1%[a]	—	***20%***
	MP 2	45%	30%	25%	—	—	—	***80%***	100%	—	2–7%[a]	0.5–1%[a]	—	***20%***
1G EB	MP 1	50%	15%	—	—	15%	20%	***75%***	75%	25%	2–6%[a]	—	—	***25%***
	MP 2	35%	15%	—	—	15%	35%	***75%***	75%	25%	2–6%[a]	—		***25%***
2G EB	MP 1	25%	15%	—	25%	15%	20%	***75%***	75%	25%	2–6%[a]	—	—	***25%***
	MP 2	10%	10%	—	25%	10%	45%	***75%***	75%	25%	2–6%[a]	—	—	***25%***
3G EB	MP 1	25%	15%	—	25%	15%	20%	***75%***	75%	25%	2–6%[a]	—	0.5–1%[a]	***25%***
	MP 2	10%	10%	—	25%	10%	45%	***75%***	75%	25%	2–6%[a]	—	0.5–1%[a]	***25%***

[a]By the weight of cement; NB, Normal blocks; 1G EB, 1st generation eco-blocks; 2G EB, 2nd generation eco-blocks; 3G EB, 3rd generation eco-blocks; MP 1, Manufacturing process 1; MP 2, manufacturing process 2.

Table 5.9 Materials and transport for paving blocks production in Hong Kong.[53]

Materials	Transported from	Transport type	Distance (km)
Natural aggregates (RS and CS)	Previous section (Table 4.6)		
Cement	Green Island Cement (local)	Trucks (30 t)	5 km
TiO_2	Germany	Ocean Ship; Trucks (18 t)	18 675 km; 6 km
Admixture	Guangzhou, China	Trucks (18 t)	170 km
Pigment	Guangzhou, China	Trucks (18 t)	170 km
Fly ash (recycled)	CLP power plant (local)	Trucks (30 t)	5 km
Fly ash (landfill)	CLP power plant (local)	Trucks (30 t)	7 km
RA (from C&D waste)	Previous section (Table 4.6)		
RA (Waste glass)	Previous section (Table 4.6)		
Packaging wood	Guangdong Province, China	Trucks (30 t)	180 km
Paving blocks transport to user site	Manufacturing site (local)	Trucks (30 t)	41 km

material development to understand the potential benefits of using different cementitious materials in concrete production.[55]

The typical production process of concrete is shown in Figure 5.8. The production process involves the production of cement, aggregates from quarries, pozzolans [used as supplementary cementitious materials (SCM)] and chemical admixtures. All the materials are then transported to concrete batching plants, at which batching and mixing with water are done to produce ready-mix concrete.

In this LCA study, a cradle-to-gate system boundary is considered for ready-mix concrete production in Hong Kong. The functional unit of concrete is 1 m^3. The LCI data, raw materials and production processes of different types of concrete are shown in Table 5.11.[13] It is well established that the use of different industrial by-products [such as fly ash (FA), granulated blast furnace slag (GBFS), silica fume (SF), *etc.*] as SCMs in concrete production can reduce the environmental impacts significantly. Therefore, the use of SCMs in concrete can be an effective open-loop recycling system that enhances the opportunity to upcycle SCMs, as well as to promote CE in concrete production. The mix-designs of the studied OPC concrete and concretes using SCMs are shown in Table 5.12.[13]

5.4.4.2 Case-specific Results of Concrete Production

According to the mix-designs of the concrete given in Table 5.12, the LCA results of different concretes adopting the system expansion approach for SCMs are shown in Table 5.13 and Figure 5.9. The system expansion approach is an environmental impacts distribution method of by-products,

Table 5.10 Energy requirements for materials/processes and used databases.[a, 53]

Materials/processes	Energy requirement	Upstream data source
Natural aggregates (RS and CS)	Previous section (Table 4.7)	
RA (from C&D waste)	Previous section (Table 4.7)	
RA (waste glass)	Previous section (Table 4.7)	
Landfill of waste materials (C&D waste and waste glass)	R*	ELCD[b]
Cement production	R*	Zhang *et al.*[24]; CLCD[c]
TiO_2 production	R*	Ecoinvent[d]
Admixture production	R*	EFCA[e]
Pigment production	R*	CPM LCA[f]
Packaging wood production	R*	USLCI[g]
Block making process	108.94 MJ t^{-1} blocks (electricity)	CLP[h]; CLCD[i]
On-site handling	7.8 MJ t^{-1} blocks (diesel)	CLCD[j]
Running pump for water	0.68 MJ t^{-1} blocks (electricity)	CLP[h]; CLCD[i]
Transport, road	R*	CLCD[k]
Transport, sea (inland barge)	R*	CLCD[l]
Transport, sea (ocean ship)	R*	Ecoinvent[m]

[a]R*, Referring to the databases/references.
[b]ELCD, European reference Life Cycle Databas.
[c]CLCD, Cement Production, Chinese Life Cycle Database Version 0.8, Sichuan University; IKE Environmental Technology Co. Ltd, 2010.
[d]Ecoinvent 3.0: Titanium dioxide (RER). Ecoinvent System Processes, Swiss Centre for Life Cycle Inventories.
[e]EFCA, European Federation of Concrete Admixture Associations.
[f]CPM LCA, Center for Environmental Assessment of Product and Material Systems Lifecycle Assessment Database.
[g]USLCI, U.S. Life Cycle Inventory Database.
[h]CLP: Sustainability Report 2014, the China Light and Power (CLP), Hong Kong, 2014.
[i]CLCD: Electricity generation, China Southern Grid. Chinese Life Cycle Database.
[j]CLCD: Fuel combustion-Diesel–CN–AP
[k]CLCD: Heavy-duty diesel truck transportation (6t, 18t, 30t)–CN–AP. Chinese Life Cycle Database.
[l]CLCD: Bulk cargo shipping (2500 t)–CN–AP.
[m]Ecoinvent 3.0 (2013b) Transport, freight, sea, transoceanic ship (GLO). Ecoinvent System Processes, Swiss Centre for Life Cycle Inventories.

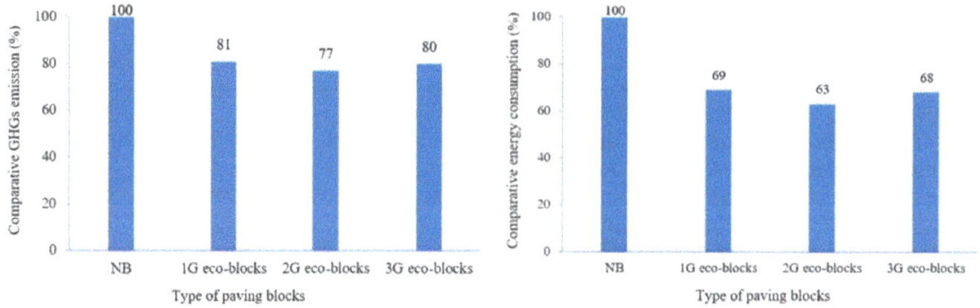

Figure 5.7 Comparative impacts of different types of paving blocks production.

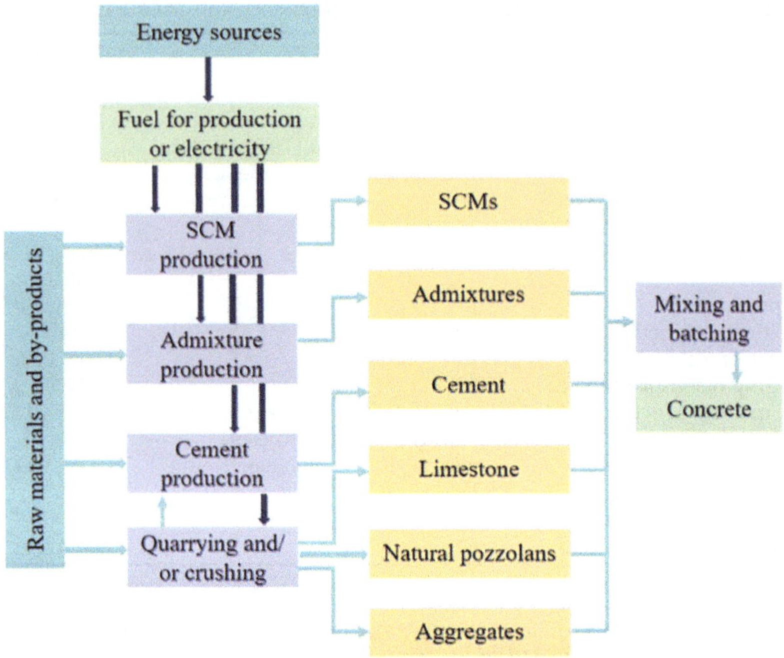

Figure 5.8 Typical process flow of concrete production.
Adapted from ref. 58 with permission from Macmillan Publishers Ltd, Copyright 2018.

Table 5.11 Sources of energy for materials/processes for concrete manufacturing.[a, 13]

Materials/processes	Energy requirement	Data sources of upstream data
Natural aggregates (RS and CS)	Previous section (Table 4.7)	
Cement production	Previous section (Table 4.3)	
FA	9.3 kWh t^{-1} [b]	CLP[d]; CLCD[e]
GBFS	72.15 kWh[c]	CLP[d]; CLCD[e]
GBFS landfill	R* (as inert waste)	Ecoinvent[f]
SF	10 kWh t^{-1} [b]	CLP[d]; CLCD[e]
Concrete (batching) production	2.5 kWh t^{-1} of concrete[24]	CLP[d]; CLCD[e]
Transport, road	R*	CLCD[g]
Transport, sea (inland barge)	R*	CLCD[h]

[a]R*, Referring to the database/references.
[b]MPA: The concrete industry sustainability performance report 1st Report, 2009. Mineral Products Association (MPA), The Concrete Centre.
[c]Dunlap R. Life cycle inventory of slag cement manufacturing process: project CTL. No. 312012. Illinois: Construction Technology Laboratories.
[d]CLP: Sustainability Report 2014, the China Light and Power (CLP), Hong Kong, 2014.
[e]CLCD: Electricity generation, China Southern Grid. Chinese Life Cycle Database Version 0.8, Sichuan University; IKE Environmental Technology Co. Ltd, 2010.
[f]Ecoinvent. Treatment of inert waste, inert material landfil (waste treatment) RoW. Ecoinvent System Processes, Swiss Centre for Life Cycle Inventories, 2013.
[g]CLCD: Heavy-duty diesel truck transportation (6 t, 18 t, 30 t)–CN–AP.
[h]CLCD: Bulk cargo shipping (2500 t)–CN–AP.

Table 5.12 Mix-design (M) of different concrete used in this study.[13]

Materials ($kg\,m^{-3}$)	M-1	M-2	M-3	M-4
Ordinary Portland cement (OPC)	445	333	238	315
Granulated blast furnace slag (GBFS)	0	142	237	0
Fly ash (FA)	0	0	0	105
Silica fume (SF)	0	0	25	0
Coarse aggregates	905	935	935	1020
Fine aggregates	745	680	605	718
Water	208	221	221	172
Admixture	1.69	1.81	2.1	1.80
28 days compressive strength (MPa)	58.7	60.8	66	53.3

Table 5.13 Impacts of different types of concrete using the system expansion approach.

	Concrete mix-designs			
Impact category	M-1	M-2	M-3	M-4
Global warming (kg CO_2 eq)	494	391	301	371
Non-renewable energy (MJ primary)	3464	2976	2499	2826

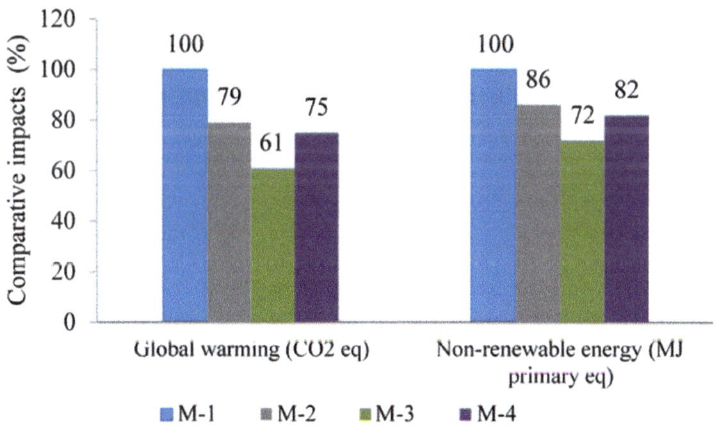

Figure 5.9 Impact comparisons for different types of concrete using the system expansion approach.

which is advantageous over traditional allocation approaches,[13] as this approach can ensure mass and energy balance, whereas the allocation approaches nearly always fail to do so.[56] In this approach, an assessment is required of the systems which are likely to be affected by the use of SCMs: namely, OPC production, when SCMs are considered as by-products.[57] The results showed that concrete produced with SCMs are associated with much lower environmental impacts compared with concrete produced with only

OPC (Table 5.13). For example, about 21%, 39% and 25% lower global warming impact is associated with M-2, M-3 and M-4 concrete (mix-design of concrete given in Table 5.12), respectively, than for the pure OPC concrete (M-1). The corresponding reduction of non-renewable energy consumption is 14%, 28% and 18%, respectively. The results indicate that significant reduction of environmental loads is observed for assessing concrete made with SCMs. OPC is the principal material for concrete, and contributes to more than 80–88% of the total impacts (depending on the types of concrete). In addition, aggregates contribute to 10–15% of the total impacts of concrete production, while other materials and production processes contribute to only 1% of the total impacts. The results may vary when considering different assumptions, as well as the type and percentage of SCMs used in different studies. For example, Zhang *et al.*[24] reported 0.181 kg CO_2 eq. GHG emissions per kg concrete produced in Hong Kong, with OPC and FA in a cradle-to-gate system boundary, which did not take an allocation approach to the FA (*i.e.* FA was considered as waste), while 0.159 kg CO_2 eq. GHG emissions was found by the authors of the chapter for the same system boundary adopting the system expansion approach.

The results of the case study for concrete production presented in this chapter showed that the use of SCMs reduces the environmental impacts significantly. However, there is no industry in Hong Kong producing GBFS and SF. In addition, FA produced in local power plants can only meet half of the demand in Hong Kong. Thus, Hong Kong mostly relies on importing SCMs to produce concrete. Therefore, more detailed studies on assessing the importing materials (with resource constraints) and cost–benefit analyses are needed in future LCA studies.

The use of mixed waste glass (post-consumer) as alternative cementitious materials in an open-loop circular approach can reduce the overall environmental impact in concrete production significantly.[59] A few studies have been conducted on the reuse of glass powder in concrete production, with their associated environmental impacts. Jiang *et al.*[60] assessed the environmental impacts of conventional concrete produced with OPC compared with concrete produced with partially replaced OPC by glass powder (Figure 5.10). It can be seen that about 4%, 10% and 18% lower GHGs emissions were associated with concrete produced with waste glass powder compared to OPC concrete, with different strength class of concrete such as 20 MPa, 30 MPa and 35 MPa, respectively. In addition, about 20% lower GHGs emission for glass powder concrete (partially replaced OPC) compared with OPC concrete was found by Deschamps *et al.*[59]

However, a case study conducted in the USA revealed that recovery of glass cullet to be used as SCM might not be economically feasible compared with the conventional SCM (*e.g.* FA) if the processing plant only handles small quantities of waste glass.[61] In Hong Kong, more than 80% of the post-consumer waste glass is still disposed of in landfill. Among the vast amount of waste glass generated in Hong Kong, there is very high potential to reuse them as SCM in cement or concrete production. Most importantly, there is

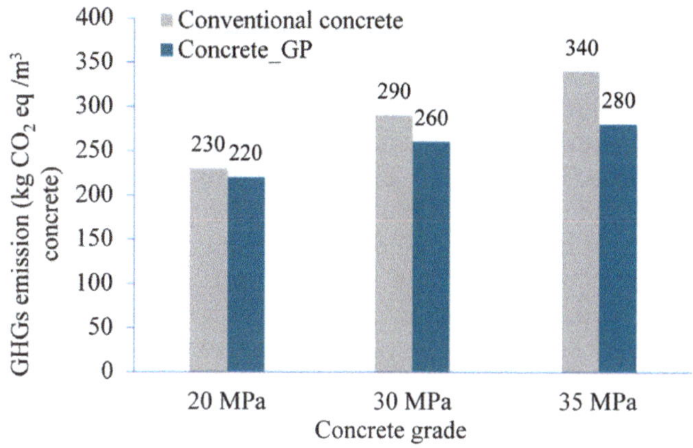

Figure 5.10 GHGs emission for concrete produced with glass powder.

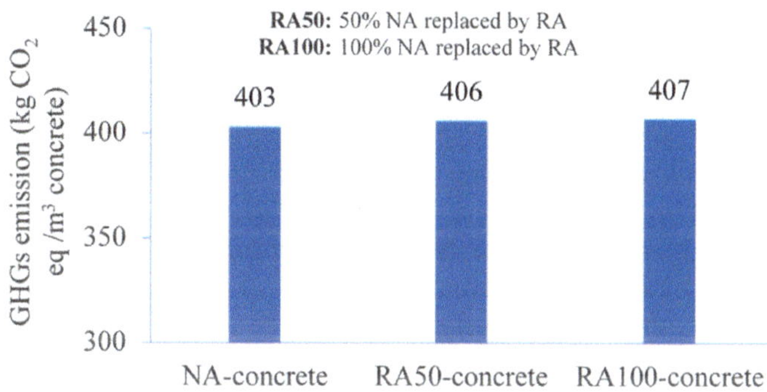

Figure 5.11 GHGs emission for concrete produced RA.

no alternate downstream user of the waste glass (*e.g.* glass manufacturer) in Hong Kong, and thus this would be a potential solution for the waste glass management problem.

RA produced from C&D waste can be reused in RA-concrete production to partially replace NA, under a closed-loop recycling system. Studies revealed that RA-concrete production is associated with similar environmental impacts to NA-concrete.[60] Although the use of RA has significant environmental benefits,[31] the use of RA in concrete requires additional cement for attaining the equivalent concrete strength that would offset the potential environmental benefits (Figure 5.11).[62] However, the transport distance of C&D waste materials can significantly influence the environmental impacts of RA. It can be seen from Figure 4.5 that RA production in Hong Kong poses much lower environmental impacts than importing NA, as much shorter transport distances are needed. Hence, the production of RA, as well as RA products, *e.g.* concrete paving eco-blocks could result in higher environmental savings (Figure 5.7).

5.5 Conclusions and Future Outlooks

With increasing concern about sustainability in the construction industry, the use of different alternative materials in the production of durable construction materials is increasing. This chapter presents the case-specific findings of production of several commonly used construction materials with different waste materials in Hong Kong. By providing examples using LCA as a methodology, our results showed that the use of alternative clinker materials and biofuel from wood waste reduces the environmental impacts for cement production significantly. In Hong Kong, the production of RA from C&D waste and waste glass reduces 48–73% of the total GHG emissions compared to using NA from river sand and crushed stones. Similarly, reusing RA in concrete blocks production reduces a considerable amount of environmental impacts, and saves substantial amounts of natural aggregates, as well as minimizing waste disposal problems. In addition, about 4–20% of the total GHG emissions can be reduced by using glass powder as SCM in producing different grades of concrete. However, the CE principle in the construction sector, especially in cement and concrete is still at the early stage of development. This is because proper methodologies, guidelines, and indicators are still yet to be developed and harmonized. As CE is now interlinked with LCA, not only for reducing natural resources consumption and environmental impacts, but also for reutilizing waste materials as secondary resources, the future research trend should focus on how to appropriately implement the CE concept by addressing the above-mentioned challenges for case-specific studies in the construction sector.

Acknowledgements

The authors wish to thank The Hong Kong Polytechnic University for funding support.

References

1. S. P. Holt and N. D. Berge, *J. Clean. Prod.*, 2018, **195**, 1057.
2. R. M. Andrew, *Earth Syst. Sci. Data*, 2018, **10**, 195.
3. https://www.statista.com/statistics/373845/global-cement-production-forecast/ (accessed July 2018).
4. A. P. Gursel, E. Masanet, A. Horvath and A. Stadel, *Cem. Concr. Compos.*, 2014, **51**, 38.
5. http://www.rockproducts.com/features/13045-world-aggregates-market.html#.XKKPPKQpCUk (accessed April 2019).
6. S. Pauliuk, *Resour., Conserv. Recycl.*, 2018, **129**, 81.
7. T. Domenech, R. Bleischwitz, A. Doranova, D. Panayotopoulos and L. Roman, *Resour., Conserv. Recycl.*, 2019, **141**, 76.
8. Y. Tao, S. Evans, Z. Wen and M. Ma, *J. Clean. Prod.*, 2019, **213**, 1172–1187.

9. R. van Berkel, *J. Ind. Ecol.*, 2010, 371–373.
10. B. Baldassarre, M. Schepers, N. Bocken, E. Cuppen, G. Korevaar and G. Calabretta, *J. Clean. Prod.*, 2019, **216**, 446.
11. M. Smol, J. Kulczycka, A. Henclik, K. Gorazda and Z. Wzorek, *J. Clean. Prod.*, 2015, **95**, 45.
12. S. Marinković, J. Dragaš, I. Ignjatović and N. Tošić, *J. Clean. Prod.*, 2017, **154**, 633.
13. M. U. Hossain, C. S. Poon, Y. H. Dong and D. X. Xuan, *Renew. Sustain. Ener. Rev.*, 2018, **82**(1), 597.
14. ISO 14040 International Standard, 2006.
15. D. Garcia-Gusano, D. Garrain, I. Herrera, H. Cabal and Y. Lechon, *J. Clean. Prod.*, 2015, **104**, 328.
16. T. Y. Huang, P. T. Chiueh and S. L. Lo, *Resour., Conserv. Recycl.*, 2017, **123**, 255.
17. H. Mikulcic, J. J. Klemes, M. Vujanovic, K. Urbaniec and N. Duic, *J. Clean. Prod.*, 2016, **136**, 119.
18. Y. Li, Y. Liu, X. Gong, Z. Nie, S. Cui, Z. Wang and W. Chen, *J. Clean. Prod.*, 2016, **120**, 221.
19. D. Yang, L. Fan, F. Shi, Q. Liu and Y. Wang, *Resour., Conserv. Recycl.*, 2017, **119**, 60.
20. A. C. Bourtsalas, J. Zhang, M. J. Castaldi and N. J. Themelis, *J. Clean. Prod.*, 2018, **181**, 8.
21. N. W. Ayer and G. Dias, *J. Clean. Prod.*, 2018, **175**, 237.
22. M. Georgiopoulou and G. Lyberatos, *J. Environ. Manage.*, 2018, **216**, 224.
23. J. R. Fyffe, A. C. Breckel, A. K. Townsend and M. E. Webber, *Waste Manage.*, 2016, **47**, 276.
24. J. Zhang, J. C. P. Cheng and I. M. C. Lo, *Int. J. Life Cycle Assess.*, 2014, **19**, 745.
25. C. Dossche, V. Boel, W. De Corte, P. Van den Heede and N. De Belie, *Cem. Concr. Compos.*, 2016, **73**, 192.
26. P. Hajek, C. Fiala and M. Kynclova, *Struct. Concr.*, 2011, **12**(1), 13.
27. E. Crossin, *J. Clean. Prod.*, 2015, **95**, 101.
28. E. L. Tucker, C. C. Ferraro, S. J. Laux and T. G. Townsend, *Resour., Conserv. Recycl.*, 2018, **129**, 240.
29. ISO 14040-14044 International Standard, 2006.
30. H. Dong, *Life Cycle Sustainability Assessment Modeling of Building Construction*, Ph.D. Thesis, The University of Hong Kong, 2014.
31. M. U. Hossain, *Life Cycle Assessment of Recycled Construction Materials: Methodology Framework Development and Results Evaluation*, Ph.D Thesis, The Hong Kong Polytechnic University, 2017.
32. O. Jolliet, M. Margni, R. Charles, S. Humbert, J. Payet and G. Rebitzer, *Int. J. Life Cycle Assess.*, 2003, **8**(6), 324.
33. J. B. Guinee, *Int. J. Life Cycle Asses.*, 2002, **7**, 311.
34. M. Goedkoop, R. Heijungs, M. Huijbregts, A. De Schryver, J. Struijs and R. Van Zelm, *A Life Cycle Impact Assessment Method which Comprises*

Harmonised Category Indicators at the Midpoint and the Endpoint Level. Report I: Characterisation, 1st edn, 2009.
35. J. Potting and M. Hauschild, Danish Environmental Protection Agency, Copenhagen, 2004.
36. N. Itsubo, M. Sakagami, T. Washida, K. Kokubu and A. Inaba, *Int. J. Life Cycle Assess.*, 2004, **9**(3), 196.
37. M. Goedkoop, R. Spriensma, M. Van Volkshuisvesting, R. O. Milieubeheer and C. D. Communicatie, *The Ecoindicator 99: A Damage Oriented Method for Life Cycle Impact Assessment*, Ministerie van Volkshuisvesting, Ruimtleijke Ordening en Milieubeheer, The Netherlands, 1999.
38. J. C. Bare, G. A. Norris, D. W. Pennington and T. McKone, *J. Ind. Ecol.*, 2003, **6**(3–4), 49.
39. B. R. Speck, S. Selke, R. Auras and J. Fitzsimmons, *Pack. Tech. Sci.*, 2015, **28**, 579.
40. A. Pinsonnault, P. Lesage, A. Levasseur and R. Samson, *Int. J. Life Cycle Assess.*, 2014, **19**, 1843.
41. C. Rodríguez, A. Ciroth and M. Srocka, *9th International Conference LCA of Food San Francisco*, USA, 2014.
42. BSI, BS 8001:2017, *Framework for Implementing the Principles of the Circular Economy in Organizations – Guide*, The British Standards Institution, London, 2017.
43. R. Bradley, I. S. Jawahir, F. Badurdeen and K. Rouch, *Resour., Conserv. Recycl.*, 2018, **135**, 141.
44. M. Dieterle, P. Schäfer and T. Viere, *Proc. CIRP*, 2018, **69**, 764.
45. M. Haupt, C. Vadenbo and S. Hellweg, *J. Ind. Ecol.*, 2017, **21**, 615.
46. M. Niero and S. I. Olsen, *Resour., Conserv. Recycl.*, 2016, **114**, 18.
47. J. Deschamps, B. Simon, A. Tagnit-Hamou and B. Amor, *J. Clean. Prod.*, 2018, **185**, 14.
48. C. Horsley, M. H. Emmert and A. Sakulich, *Fuel*, 2016, **184**, 481.
49. M. U. Hossain, C. S. Poon, I. M. C. Lo and J. C. P. Cheng, *Resour., Conserv. Recycl.*, 2017, **120**, 199.
50. M. U. Hossain, C. S. Poon, I. M. C. Lo and J. C. P. Cheng, *Resour., Conserv. Recycl.*, 2016, **109**, 67.
51. Y. Zhang, J. Mckechnie, D. Cormier, R. Lyng, W. Mabee, A. Ogino and H. L. Maclean, *Environ. Sci. Technol.*, 2010, **44**, 538.
52. M. U. Hossain, S. Y. Leu and C. S. Poon, *J. Clean. Prod.*, 2016, **113**, 400.
53. M. U. Hossain, C. S. Poon, I. M. C. Lo and J. C. P. Cheng, *Int. J. Life Cycle Assess.*, 2016, **21**(1), 70.
54. M. Z. Guo, T. C. Ling and C. S. Poon, *Cem. Concr. Compos.*, 2017, **83**, 279.
55. S. A. Miller, A. Horvath and P. J. M. Monteiro, *Environ. Res. Lett.*, 2016, **11**, 074029.
56. B. P. Weidema and J. H. Schmidt, *J. Ind. Ecol.*, 2010, **14**, 192.
57. E. Crossin, *J. Clean. Prod.*, 2015, **95**, 101.
58. S. A. Miller, A. Horvath and P. J. M. Monteiro, *Nat. Sustain.*, 2018, **1**, 69.

59. J. Deschamps, B. Simon, A. Tagnit-Hamou and B. Amor, *J. Clean. Prod.*, 2018, **185**, 14.
60. M. Jiang, X. Chen, F. Rajabipour and C. T. Hendrickson, *J. Infrastruct. Syst.*, 2014, **20**(4), 04014020.
61. E. L. Tucker, C. C. Ferraro, S. J. Laux and T. G. Townsend, *Resour., Conserv. Recycl.*, 2018, **129**, 240.
62. T. Ding, J. Xiao and V. W. Y. Tam, *Waste Manage.*, 2016, **56**, 367.

CHAPTER 6

Assessing the Environmental Benefits of Using Glass Powder as a Supplementary Cementitious Material in a Context of Open-loop Recycling

J. DESCHAMPS,[a,b] A. TAGNIT-HAMOU,[b] B. FOURNIER[c] AND B. AMOR*[a]

[a] Interdisciplinary Research Laboratory on Sustainable Engineering and Ecodesign (LIRIDE), Sherbrooke University, Department of Building and Civil Engineering, 2500, Boul. de l'université, Sherbrooke J1K 2R1, Canada; [b] SAQ Industrial Chair on the Valorization of Glass in Materials, Sherbrooke University, Department of Building and Civil Engineering, 2500, Boul. de l'université, Sherbrooke J1K 2R1, Canada; [c] Laval University, Department of Geology and Geological Engineering, 2325 Rue de l'université, G1V 0A6, Quebec City, Canada
*Email: ben.amor@usherbrooke.ca

6.1 Introduction

Circular economy (CE) is an increasingly cited concept due to the continuous debate and misconceptions about waste and resources management,[1] in addition to the growing interests from the life cycle assessment (LCA) community.[2] In contrast to the "make, use, and dispose" approach, CE advocates the minimal extraction of raw materials and the maximization of

Life Cycle Assessment: A Metric for the Circular Economy
Edited by Aiduan Borrion, Mairi J. Black and Onesmus Mwabonje

Published by the Royal Society of Chemistry, www.rsc.org

service life by promoting reuse, recycling, and recovery. In other words, the concept is to transform waste into a valuable resource by designing the product to be easily reparable, recoverable, and used as primary materials for the same industry (closed-loop) or another industry (open-loop). For example, closed-loop systems help in rethinking the way in which products are owned. By transforming the ownership of a product into a "renting service", the latter can be redirected to the manufacturer. Indeed, in functionality economy, which is one of the concepts within CE, the manufacturer of a product keeps the ownership of the product and sells only the service (*i.e.* the function of the product). Because the product was designed under CE principles, it can be easily repaired or disassembled, to be used as primary materials to create other products, or the same product again. Open-loop recycling reflects a biological-inspired concept. In the natural living world, there is no waste, as the remains of one species become reused in the breeding ground for another. The goal of open-loop recycling is to replicate this symbiosis for manufactured products at the industrial scale.[3] One great application of such a model is the mixed waste glass used as a supplementary cementitious material (SCM).

Finding industrial solutions for recycled materials is among the greatest challenges of sustainable development. Mixed waste glass from selective sorting illustrates this situation. Despite being an abundant resource, to some extent endlessly recyclable, researchers have struggled to find viable outcomes for mixed waste glass. In Quebec (Canada), where the recovery rate of glass is near 85%,[4] most of it is sent to landfill sites, and some of it is used as a low-cost covering material for access roads.[5] Because mixed waste glass contains different types of glass, as well as small impurities, such as metallic or plastic residues, it does not meet the quality standards to be used in existing industrial production chains.[6,7] However, given the pozzolanic properties of finely ground glass,[8,9] a new type of recycling solution (open-loop) as an alternative SCM is now possible. Such a recycling approach will make the development of concrete with a lower Portland cement content a promising alternative. In comparison with landfill disposal, this initiative was highlighted for its environmental benefits, as is the case for other commercial solutions such as glass wool or glass aggregate production.[7,10] It could also meet the ongoing efforts of the concrete industry to lower their environmental footprint.[11–13]

In order to assess the potential environmental impacts of concrete, life cycle assessment (LCA) is the most commonly used tool for that purpose.[14–16] This tool uses a holistic methodology by considering the inventory of all elementary flows emitted to and extracted from the environment during the entire life cycle of the studied concrete. By using an impact assessment method, these elementary flows are translated into different environmental impact indicators, such Global Warming Potential (GWP).[17] The use of LCA has resulted in multiple studies investigating the potential environmental benefits of "greener" concrete.[14–16] Indeed, the significant contribution of cement production to the environmental footprint of concrete products has driven a push to partially replace it with SCMs, such as fly ash from coal combustion, blast furnace slag from the iron and steel-making process, or silica fumes from silicon and ferrosilicon alloy production.

The LCA results of these concretes and their environmental benefit greatly depend on the allocation rules that were used to determine the environmental contribution of SCM.[18,19] Few studies have investigated the use of glass powder (GP) as an SCM in concrete. Hossain *et al.* compared mortars of similar strength (50 MPa) on a volume basis.[20] As a result, they found that, in comparison with a conventional mortar, using waste glass as an SCM to substitute 20% of the clinker can reduce energy consumption by 16%, and greenhouse gas emissions by 17%. Jiang *et al.* analysed the environmental impacts of concrete containing glass powder and compared the results to those of a conventional concrete, using the volume (1 m^3) as a functional unit.[21] Despite the high uncertainty of the results, the authors found a reduction in every impact category, but highlighted the fact that comparing the environmental impacts of concrete mixtures on a volume basis does not consider the gain in strength when using GP as an SCM. This strength issue was addressed by Habert *et al.* when investigating the environmental footprint of ultra-high-performance concrete (UHPC).[22] Despite an increase in CO_2 emissions in comparison to traditional concrete (for the same volume of concrete), the environmental impact of UHPC is lower due to an improvement in strength and durability.[23] The other studies investigated the potential environmental impacts of using recycled glass in concrete as a replacement of natural aggregate.[24] Such use is less desirable in a circular economy context, being an example of downcycling. Moreover, in the literature, regarding the potential environmental impact of concrete, the use of SCMs is viewed as one of the best avenues to reduce its environmental burden.[11,25]

In a circular economy perspective, LCA is becoming the relevant tool to address environmental impacts of such glass waste open-loop recycling in concrete. In that context, Gursel *et al.*[16] suggested that important points need to be carefully treated when assessing the environmental impact of concrete using LCA to avoid misleading interpretations. These suggestions are as follows: (1) the assessment of environmental impacts needs to be holistic. This means that the panel of environmental indicators need to be extended to not solely focusing on greenhouse gas emissions, but also to avoid burdens shifting; (2) the case study should be relevant to regional and technological variations. This is particularly relevant in locations such as Quebec, where most of the electricity production comes from hydropower; (3) the justification of allocation used for SCMs must be relevant, as the use of recycled materials in concrete becomes more and more significant; and (4) the functional unit should include all relevant concrete aspects to compare concrete on the basis of the functionality that it can fulfil, and not only its physical properties (such as volume or strength).

Considering these limitations and previous research recommendations, the main objective of this study is to estimate the potential environmental burdens of a UHPC using waste GP, in a context of open-loop circular economy, as an alternative SCM, and compared to two other concretes in a real case study: a pedestrian bridge in Sherbrooke, Quebec, Canada.

6.2 Materials and Method

This section follows the ISO 14044 structure, which is defined by four main methodological steps: the goal and scope definition of the LCA; inventory analysis; the impact assessment phase; and interpretation.[26] The technical properties of the concretes under study has been covered in a previous research work.[27] Briefly, their research presented laboratory characterization of various UHPCs, as well as the real case field validation of the concrete that is used in this LCA. This study found that the UHPC using waste GP provides high workability, and enhanced rheological properties, given the very low absorption of glass particles and optimized packing density of the entire material matrix. These technical properties present a higher mechanical and durability characteristic of the UHPC.

6.2.1 Goal and Scope Definition

The main purpose of this study was to evaluate and compare the potential environmental impacts of three possible concrete decks for a pedestrian bridge in Sherbrooke, Quebec, Canada. The first one was made using ultra-high-performance glass concrete (Glass-UHPC), using GP and silica fumes as partial cement replacements. The second was a more conventional ultra-high-performance concrete (Conv-UHPC) using only silica fumes as an SCM. Finally, a 35 MPa conventional concrete was used. The product system under study was a whole pedestrian bridge, as shown in Figure 6.1. It is 4.9 m (16.1 ft) in length and 2.5 m (8.2 ft) in width.[28] In accordance with the goal

Figure 6.1 Pedestrian bridge under study.

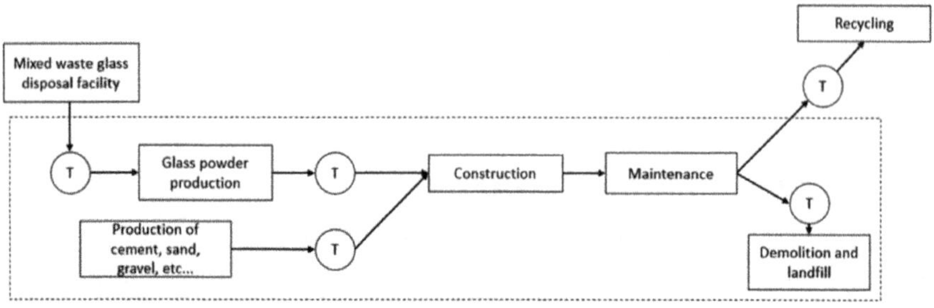

Figure 6.2 System boundaries of concrete decks under study (T refers to transport).

and scope, an attributional LCA (descriptive approach instead of a change-oriented consequential approach) is taken.

As shown in Figure 6.2, the system boundaries of this study are defined as cradle to grave (*i.e.* every step of the life cycle of each concrete deck is considered from material extraction to its end-of-life). To compare the three concretes under study on the same basis, the functional unit is defined as follows: allowing pedestrians to cross the ditch for 100 years.

6.2.2 Life Cycle Inventory

The mix design of each deck is detailed in Table 6.1. The indicated quantities represent the amount of materials required to build one cubic meter of each concrete. The mix designs of the two UHPCs were taken from the literature related to the construction of the pedestrian bridge.[27,28] The production phase includes the production of all the materials listed in Table 6.1. In the case where transport distance data were not available, market information from the ecoinvent 3.2 database was used.[29]

As can be seen in Table 6.1, UHPCs are characterized for their high cement content, low water/binder ratio, and use of a reactive pozzolanic powder such as silica fumes. Thus, traditional UHPCs have a very dense internal structure, minimizing the capillary pores, and a high-density packing, allowing for a high compressive strength.[30] Fibers are also used to improve the material's ductility and tensile capacity.[31] Both UHPCs used in this study showed a compressive strength greater than 150 MPa.[27] To be consistent with the life span defined in the functional unit (100 years), the bill of the material of each pedestrian deck, and consequently all of the reference flows, were scaled up or down. Scaling factors are shown in Table 6.2, as well as the volume and the service life of each deck. The service life of concrete construction is a complex parameter to estimate, mostly because of its prospective nature. In the case of conventional concrete, the life span was estimated using the life-365 software.[32] It is a service life prediction model, based on the simplification of Flick's second law on chloride ion diffusion in concrete. However, the software cannot predict the service life of specific concrete using special additives, such as UHPCs. Therefore, some physical

Table 6.1 Mix design for the three concretes under study.

Materials	Conventional concrete[a] (kg m[3)]	Conv-UHPC (kg m^3)	Glass-UHPC (kg m^3)	Ecoinvent data V3.2	Transport distance in km
Cement	400	825	555	Cement, Portland CA-QC	227
Plasticizer	3	46	17	Market for plasticizer for concrete GLO	—
Glass Powder	—	—	410	Created data[b]	230
Reinforcing steel	58,5	58,5	58,5	Market for reinforcing steel GLO	—
Water	178	225	226	Market for tap water CA-QC	—
Sand	801	—	—	Sand production GLO	116
Silica Sand	—	990	888	Silica sand production GLO	116
Quartz Powder	—	232	0	Silica sand production GLO (used as a proxy)	116
Silica Fume	—	229	205	Mass allocation on silicon, metallurgical grade CA-QC[c]	155
PVA Fiber	—	32,5	32,5	Created data[d]	755
Gravel	1032	—	—	Market for gravel, round GLO	—
Foam block	—	8,35	8,35	Market for polystyrene, expandable GLO	—

[a]From Mehta and Monteiro (2006).[52]
[b]From industrial source.
[c]Global data adapted with Quebec energy mix.
[d]From Life Cycle Inventories of Photovoltaics (Jungbluth *et al.*, 2012).[53]

Table 6.2 Characteristics of the three pedestrian deck types.

	Conventional concrete	Conv-UHPC	Glass- UHPC
Volume (m^3)	4	2	2
Service life (year)	50	120	120
Scaling factor	100/50	100/120	100/120

properties were examined to provide some good indications of concrete durability. UHPCs are usually recognized for their high performance in those indicators,[31] leading to a long service life estimation. For example, Gu *et al.* estimated the service life of UHPC structures in a marine environment to be up to 200 years, whereas that of normal concrete would be fewer than 20 years.[33] Yi *et al.* highlighted the advantage of UHPC in improving the service life of structures to be longer than 100 years, with minimal maintenance requirements.[34]

Regarding the Glass-UHPC used in this study, its physical properties were investigated by Soliman *et al.* in a previous paper.[27] The results of the characterization of Glass-UHPC showed no evidence of deterioration nor cracking after 1000 freezing-and-thawing cycles. Moreover, a very high resistance to chloride-ion penetration and electric resistivity (indicating a very low risk of corrosion occurring in the reinforcing steel bars) was found, in addition to high results in terms of resistance to mechanical abrasion and deicing salt scaling (particularly important as the footbridge will be subjected to the action of a snow removal and deicing salt in winter). Regarding these results, an estimation of 120 years was determined for the life spans of the Glass-UHPC and Conv-UHPC. A sensitivity analysis of these parameters was also conducted to assess to what extent the conclusion of our results remained valid.

At the time of the study, waste mixed color glass does not have a positive price value,[35] and it is available as a waste product. Therefore, following the cut-off recycled content approach,[36] the considered impacts resulting from the production of GP are limited to transportation from the recycling site to the GP factory, the energy required by machines, and the infrastructure to produce the powder, but exclude the production of the mixed color glass itself. The datasets used to model the GP production are available in the electronic supplementary materials (Table S1).

Regarding the silica fumes, the latter is considered as waste from semiconductor production.[18,37,38] However, being of a great commercial interest, particularly in the concrete sector, it is considered as a co-product.[39] In that particular situation, the cut-off rules cannot be applied. In this study, mass allocation was used by considering the mass ratio based on Chen's work:[40] a production of 0.15 kg of silica fumes for every 1 kg of silicon metal (13% mass allocation). The construction phase was adapted from ecoinvent data to match the consumption of machinery and material production used during the real construction process (*e.g.* oil, plywood formwork, and machinery). The wooden posts, galvanized steel handrail, screws, and concrete base blocks were identical for every product system under study and their service life was independent from the deck's life of service. Therefore, they can be excluded from the system boundaries.[17] In addition, the maintenance phase was considered the same for each concrete mixture, as each deck was designed to remain without requiring major changes during its life span. As a result, the maintenance phase will not appear in the results. Moreover, a small quantity of CO_2 was taken up by the concrete during a process called carbonation,[41] but the amount that it represents is so negligible that it was not considered in this study.[21,41,42] Finally, the concrete's end of life was modeled using ecoinvent data for reinforced concrete treatment, considering that 40% of the deck volume is sent to a landfill (located 29 km from the site). The other 60% will be used as recycled aggregate or pavement material.[43]

6.2.3 Life Cycle Impact Method

The impact method used to calculate the potential environmental burdens of the product system was IMPACT 2002+.[44] It is composed of 4 endpoint indicators and 15 midpoint indicators. It covers a large number of substances compared with that of other American methods.[45] The list of these indicators, and the contribution of midpoint categories to endpoint categories, are shown in Table S2 in the supplementary material. It is important to note that the damage factors characterizing impacts on Ecosystem Quality (EQ) caused by Aquatic Acidification (A-ACI) and Aquatic Eutrophication (A-EUT) are still under development. Thus, they do not contribute to the endpoint ecosystem quality. Therefore, all presented results in the following subsections follow the midpoint and endpoint indicators.

6.2.4 Interpretation

6.2.4.1 Sensitivity Analysis – Impact Assessment Method

Sensitivity analysis is highly recommended to test the robustness of results and their sensitivity to data, assumptions, and models used.[17] In our case, two different analyses were conducted. The first was on the calculation method, by changing IMPACT 2002+ to TRACI 2.1.[46] TRACI is a North American problem-oriented method that covers fewer environmental indicators. It contains the following midpoint categories: global warming, acidification, eco-toxicity, eutrophication, ozone depletion, smog (photochemical oxidation), carcinogenic, non-carcinogenic, respiratory effects, and fossil fuel depletion. In this context, we compared completeness (IMPACT 2002+) to adequacy (TRACI 2.1).

6.2.4.2 Sensitivity Analysis – Life Span

As previously mentioned, the second sensitivity analysis focused on the life span of the concrete decks. This parameter changed the quantities of materials needed to fulfil the functional unit (for 100 years). The tested life span scenarios for each deck are listed in Table 6.3.

6.2.4.3 Uncertainty Analysis

Using a large amount of data to model, as extensively as possible, the system under study requires estimations and approximations of input values. Monte Carlo analysis is a robust approach to address the propagation of data uncertainty through final results.[47] To do so, the Monte Carlo analysis, in combination with the pedigree matrix approach was used to quantify the uncertainty of the input values.[48]

Table 6.3 Life span scenarios of the three concrete mix designs under study (in years).

Scenario	Conventional Concrete	Conv-UHPC	Glass- UHPC
Base case	50	120	120
Optimistic	75	200	200
Pessimistic	35	100	100

6.3 Results and Discussion

This section is divided into two main sections. The first one presents and compares the LCA results of the three different bridges. Then, the sensitivity of key parameters, life span, and the impact assessment method were tested to investigate the robustness of our conclusions.

6.3.1 Comparative LCA Results

The midpoint and endpoint results are, respectively, shown in Figures 6.3 and 6.4. In addition, detailed numerical results are presented in the supplementary material (Table S3). It is clear that the Conv-UHPC and Glass-UHPC have lower environmental impacts than those of conventional concrete for all the assessed categories. Indeed, in comparison to conventional concrete, Conv-UHPC decreased the impact on non-renewable energy (NRE) by 11% and on mineral extraction (ME) by up to 74%. The reduction was 42% for global warming (GW). Glass-UHPC allowed for some even greater reductions in impacts compared with conventional concrete: from 20% for NRE up to 75% for ME, with a 52% reduction in GW. These results are mainly explained by the life span of each deck. Indeed, the shorter it lasts, the more materials are needed to fulfil the 100-year functional unit.

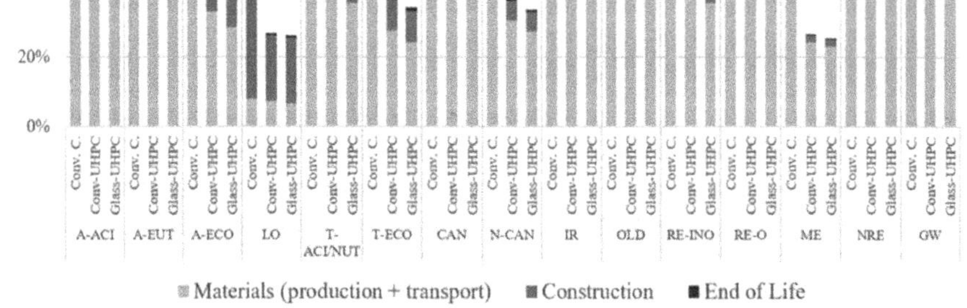

Figure 6.3 Midpoint results of the conventional concrete (Conv. C.), Conv-UHPC, and Glass-UHPC decks (IMPACT 2002+) [Aquatic acidification (A-ACI), aquatic eutrophication (A-EUT), aquatic ecotoxicity (A-ECO), land occupation (LO), terrestrial acidification/nitrification (T-ACI/NUT), terrestrial ecotoxicity (T-ECO), carcinogens (CA), non-carcinogens (N-CA), ionizing radiation (IR), ozone layer depletion (OLD), respiratory inorganics (RE-INO), respiratory organics (RE-O), mineral extraction (ME), and non-renewable energy (NRE)].

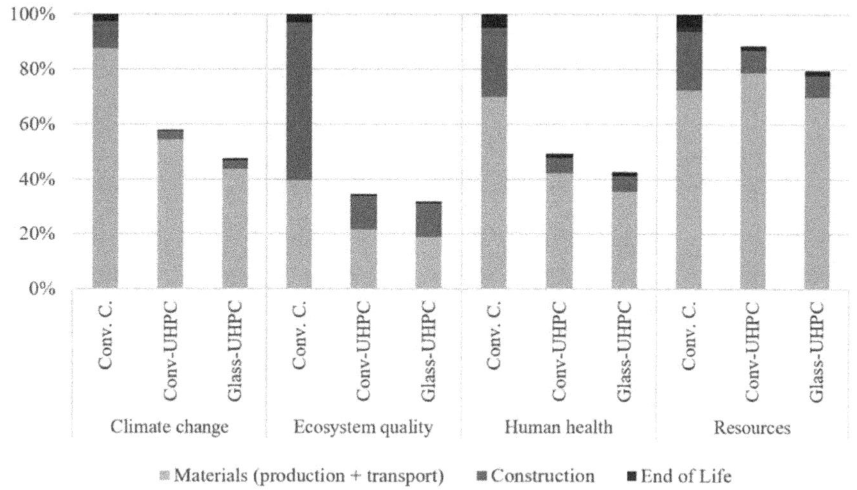

Figure 6.4 Endpoint results for the conventional concrete (Conv. C), Conv-UHPC, and Glass-UHPC decks (IMPACT 2002+).

Figures 6.3 and 6.4 also highlight that materials production is the hotspot from a life cycle perspective, while the end-of-life phase has a negligible contribution on environmental impacts. It is important to note that the construction phase is the main contributor for a few environmental indicators, mostly related to EQ, such as land occupation (LO) and terrestrial and aquatic eco-toxicity (T-ECO and A-ECO), particularly for the conventional concrete. In the case of LO, this was caused by the plywood formwork production, because the logging industry requires large portions of land for growing trees. Regarding the T-ECO and A-ECO indicators, the production of plywood requires heat that is produced on site *via* hardwood chips. The treatment of wood ash resulting from this heat production is then directly treated and spread on agricultural land, resulting in aluminum and zinc pollution in the soil, contributing to T-ECO and A-ECO. This result is also clearly visible for the Glass-UHPC and Conv-UHPC for LO, but to a lesser extent for T-ECO and A-ECO. These indicators are driven by other materials, such as PVA fibers, having a significant contribution to these impacts, mitigating the plywood production contribution.

Figure 6.5 shows the contribution analysis of the materials production phase (construction and end-of-life are excluded) to help in identifying hotspots. As expected, when referring to GW indicator, cement plays a major role. It is essentially explained by the CO_2 emissions that occur during clinker production. However, CO_2 emissions are not the only pollution that is released during clinker production. In practice, nitrogen and sulfur dioxides are also emitted and contribute to respiratory inorganic (RE-INO) effects, terrestrial acidification/nitrification (T-ACI/NUT), and aquatic acidification (A-ACI). This is why cement appears as a major hotspot in these categories. Cement also appears as a high contributor in terms of T-ECO

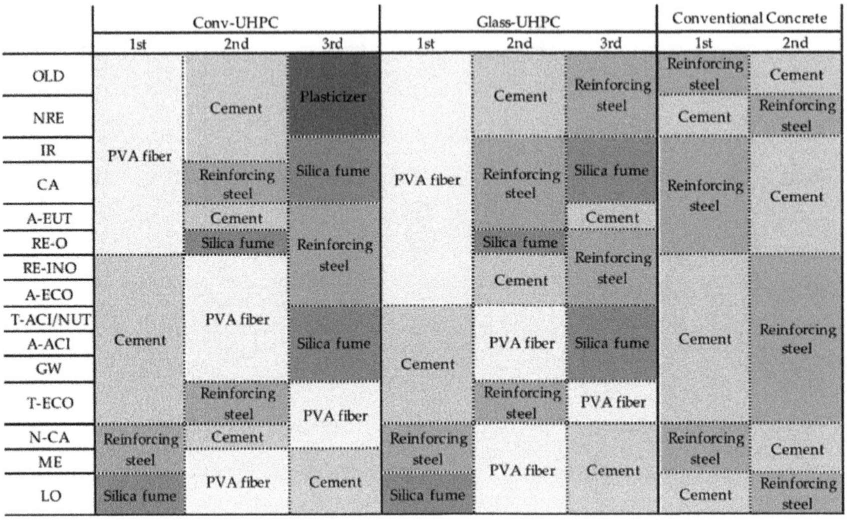

Figure 6.5 Main contributors of midpoint potential environmental categories (IMPACT 2002+) of Conv-UHPC, Glass-UHPC and Conventional Concrete for the material phase [Aquatic acidification (A-ACI), aquatic eutrophication (A-EUT), aquatic ecotoxicity (A-ECO), land occupation (LO), terrestrial acidification/nitrification (T-ACI/NUT), terrestrial ecotoxicity (T-ECO), carcinogens (CA), non-carcinogens (N-CA), ionizing radiation (IR), ozone layer depletion (OLD), respiratory inorganics (RE-INO), respiratory organics (RE-O), mineral extraction (ME), non-renewable energy (NRE)].

and A-ECO, because of blasting operations in hard coal and limestone mines that emit aluminum into the air.

Regarding conventional concrete, the two main contributors are the reinforcing steel and the cement. But as we reach the strength of the UHPC mix, the need for reinforcing steel is reduced, as well as the amount of cement needed to fulfil the service. Indeed, the decrease in the reinforcing steel and cement footprint leaves room for PVA fibers, silica fumes, and plasticizer to appear as the new hot spots for Conv-UHPC and Glass-UHPC. Silica fumes, a co-product of silicon metal production, also tend to be an important contributor. Indeed, nitrogen and sulfur dioxides emitted during the production process of silicon metal make an important contribution to the T-ACI/NUT and A-ACI indicators. However, in this case, it has been produced in Quebec, where the electricity mix is 99% hydropower energy.[49] This maintains silica fumes' contribution to a reasonable level, particularly for GW. In the case of PVA fiber, which is a major contributor in most midpoint categories for UHPCs, a deeper analysis of its environmental burden shows that it is the production of raw chemicals such as ethylene or polyvinyl chloride that dominates the impacts. The same applies for plasticizers, despite the small quantities of involved chemicals (less than 2% of the total mass for Conv-UHPC). Regarding the impact of GP production, it is only responsible for from 0.5% up to 7.2% of the overall environmental impacts of Glass-UHPC. The greatest impact of GP is captured when analyzing terrestrial eco-toxicity (7.2%) and ozone layer depletion (5.0%) indicators. This is mainly explained by all required transportation stages. Moreover, the contribution of the GP production environmental impact is mainly divided between the transportation and the energy required to crush the glass into a fine powder. However, it is important to remember that the production process occurs in Quebec, where most of the electrical energy originates from a hydroelectric power plant, hence reducing its contribution to the environmental impact.

It is worth mentioning that when the three decks are compared on the basis of their respective volume (without considering the life span and hence the functional unit), the results are completely different. Indeed, Conv-UHPC would be the solution with the greater impact on all the indicators, followed closely by Glass-UHPC. Given this, conventional concrete would decrease the impact for every indicator by at least 30%. The high Portland cement content in UHPCs and the alternative materials used, which originate from highly energetic processes (such as those of silica fumes or PVA fibers) explain this inversion of the results. The results of the comparison on midpoint and endpoint indicators of the same volume for each concrete is shown in Figures S1 and S2, respectively, in the supplementary material.

6.3.2 Interpretation

6.3.2.1 Sensitivity Analysis – Impact Assessment Method

To test the robustness of our results, a first sensitivity analysis was performed by changing the calculation method by using TRACI 2.1. Comparing similar indicators from Figures 6.3 and 6.6, very similar results for the corresponding midpoint indicators were found and thus, did not change our primary conclusions. One can argue that an inversion of the results can be found when comparing non-renewable energy (Impact 2002+) and fossil fuel

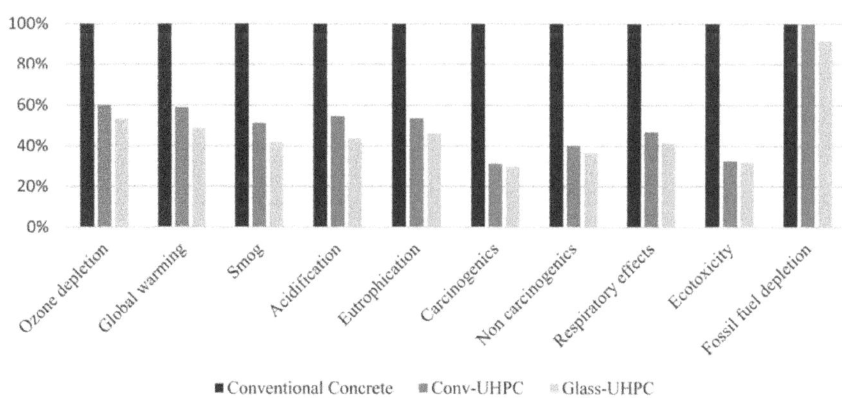

Figure 6.6 Midpoint results of the conventional concrete, Conv-UHPC, and Glass-UHPC decks (TRACI 2.1).

depletion (TRACI). It is important to realize that these two indicators do not represent the same thing. TRACI illustrates the rarefaction of fossil fuels (and the extra energy that required to extract more of these fuels), whereas IMPACT 2002+ focuses only on the consumption of non-renewable energy. Crude oil and natural gas are both rarer than hard coal. Consequently, the hard coal consumption in TRACI contribute less to the fossil fuel depletion indicator than crude oil or natural gas consumption for the same quantity (this can be noticed in supplementary material Figure S3). Both are more used in UHPCs, mainly because of the raw chemical production of PVA fibers.

6.3.2.2 Sensitivity Analysis – Life Span

Because we are considering the service that must fulfil the concrete during a certain period of time in the functional unit, the outcome of the results is highly dependent on the life span of each concrete mix. Figure 6.7 shows the

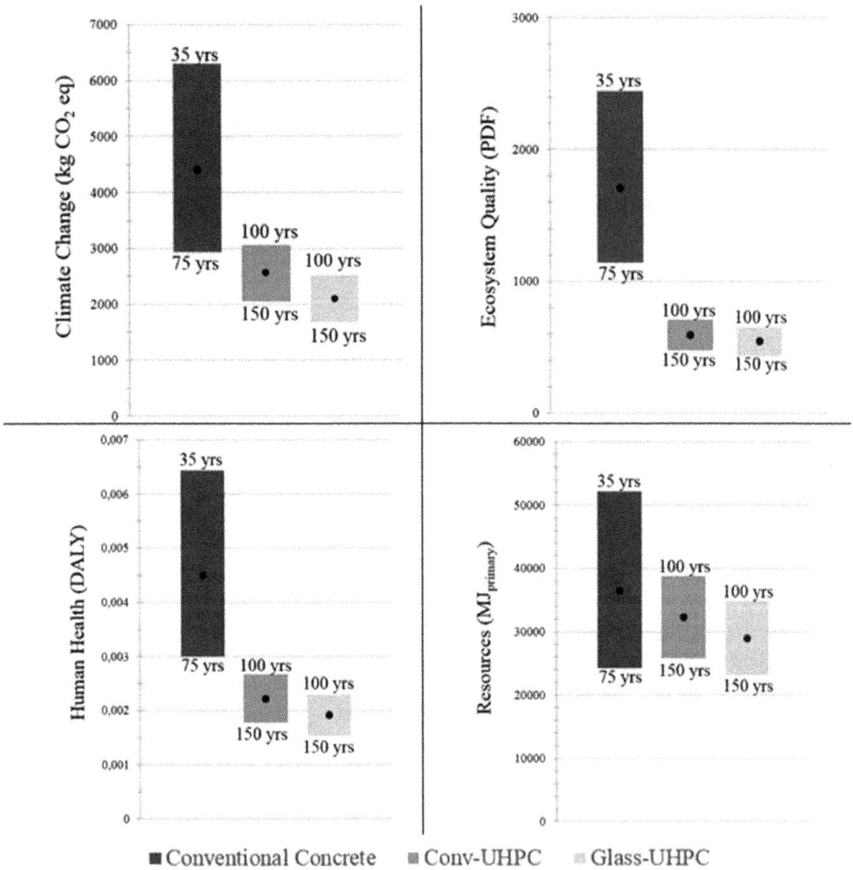

Figure 6.7 Results of the sensitivity analysis on life span of conventional concrete on endpoint indicators: climate change, ecosystem quality, human health, and resources (black points represent the original scenarios).

change in the environmental impact of the three-concrete mix, according to different life span scenarios. It appears that for ecosystem quality and human health, the Conv-UHPC and Glass-UHPC environmental impacts never overlap with those of conventional concrete. For the climate change indicator, the conventional concrete can be the more eco-efficient compared to Conv-UHPC. This is the case when the latter reaches its optimistic scenario of 75 years, and if the Conv-UHPC life span decreases to 100 years or less. For the resource indicator when the life span of conventional concrete is higher, its impact can be less than both Conv-UHPC and Glass-UHPC. Regarding Conv-UHPC and Glass-UHPC, their impact overlaps considerably because their environmental impact is already relatively similar. Nevertheless, when considering the same service life, Glass-UHPC will always result in less impact than Conv-UHPC.

6.3.2.3 Uncertainty Analysis

As LCA aims to model the inventory of the whole life cycle of a given product, it relies on many uncertainties of the background data. Monte Carlo analysis allows us to investigate the propagation of this uncertainty on the results and provides an indication of the robustness of our conclusions. Assessed independently (absolute results of each system), this high uncertainty is clearly visible for each system under study (see supplementary material Figure S4). However, the systems under study are quite similar and use the same background processes, *e.g.* they are made from the same materials or they have many common suppliers. Therefore, most of the uncertainty linked to these processes can be considered equal. As the objective of this study was a comparative LCA, this interrelation was removed from the comparison. Figure 6.8 shows the propagation of uncertainty in the midpoint results when comparing conventional concrete to Glass-UHPC. It shows that despite a great absolute uncertainty in both systems, our conclusion remains solid for most midpoints. Regarding ionizing radiation, respiratory organics, and non-renewable energy, the confidence in our conclusion is maintained (greater than 80% for the preferred scenario). A more detailed and refined model of the production of the main contributor in these midpoints, such as PVA fiber, to improve the data quality could reduce this variability.

6.3.3 Discussion

The reduction in the environmental impact resulting from the use of concrete with a superior strength grade on the life cycle of concrete structures shown in this study accords with the conclusions reached by several authors on LCA of UHPC and high-performance concrete. Habert *et al.* showed a reduction in every environmental indicator when comparing the life cycle of road bridges to high-performance or traditional concrete. They also highlighted the high uncertainties in their results.[23] A study regarding the rehabilitation of road bridges also found a potential reduction of 40% in

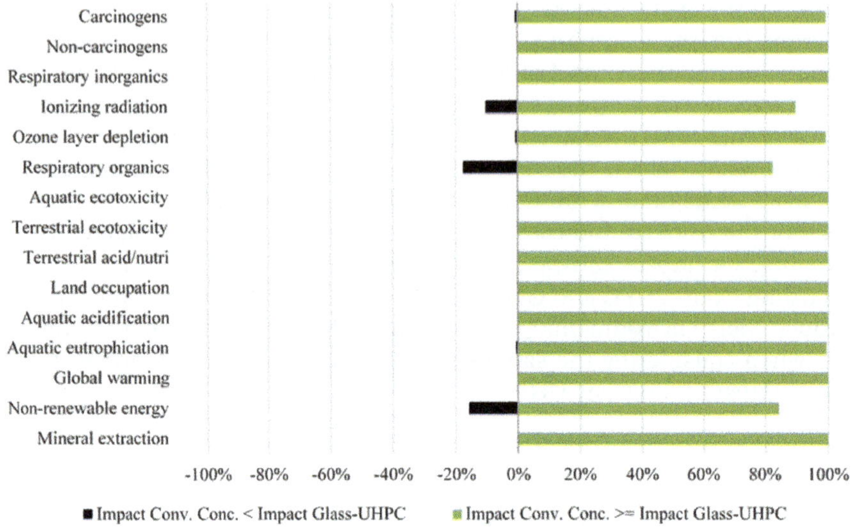

Figure 6.8 Results for Impact 2002+ midpoint indicators of the Monte Carlo analysis between conventional concrete and Glass-UHPC (confidence interval: 0.95; 5000 iterations).

greenhouse gas emissions using UHPC in comparison to standard concrete when considering the whole life cycle, and no environmental benefits were found when comparing only the production of the base materials.[50] On a UHPC bridge built in China, Dong showed that UHPC has a higher initial environmental and economic cost than normal concrete, but considering the long-term and maintenance costs, normal concrete becomes the least desirable option.[51] The results of the present study are aligned with the trends observed in the literature concerning the environmental impact of UHPC. In this study, we showed that the environmental impact of UHPC could be reduced even further by using finely ground GP from mixed waste glass. This clearly demonstrates the environmental benefits of glass powder in a circular economy context by using an open-loop recycling approach. Moreover, the high uncertainty, inherent to the LCA method, can be overcome with the use of a compared Monte Carlo analysis to clarify the confidence in the conclusions.

6.4 Conclusion

The use of GP as an SCM combines the benefits of being a low-environmental-impact material and improving the strength and durability of concrete. Our results confirm that comparing concretes only based on volume can mislead the analytical conclusions. In our case study, the great majority of impacts were decreased by at least 11% in comparison to standard Conv-UHPC and conventional concrete, and by at least 20% when comparing those of Glass-UHPC to those of conventional concrete. Nevertheless, as the sensitivity analysis highlights, these results are highly dependent on the service life assumed for each concrete and the functional unit used. Despite a high intrinsic uncertainty among the system under study, the Monte Carlo simulation showed a clear confidence in the conclusion that Glass-UHPC has a lesser environmental impact than that of conventional concrete.

The development of a concrete mix design with high-strength capacity and high durability seems to shift the usual main contributors such as cement and steel to new materials used to improve performance (fibers, silica fumes, GP, *etc.*). However, the environmental footprint of these new materials is not fully understood and needs to be investigated further in future studies to improve the accuracy of the concrete's LCA.

A part of those improvements can be made by having a better characterization of the processes producing alternative SCM, particularly if these processes are mature enough to be scaled up at the macro level, or if they represent a small volume on a lab scale. Another part of this improvement should be to revisit the allocation methodology regarding the recycled material. The shift toward the circular economy is bringing an unprecedented interest for waste material, at such a point that it is unclear whether they should be considered as such (*i.e.* waste), any longer. This fundamental question can be translated into LCA language when dealing with allocation. The question of how to allocate the burden of co-product (previously considered a waste product), or of a product reaching the end of their first life cycle and certainly not their last one) will be crucial to investigate.

Lastly, the growing interest in GP could raise the economic value of mixed glass waste and make it a valuable material. Moreover, massive use of GP as an SCM (as well as any other SCM) raises the question of its impact on the economic market of cement or SCMs. This question should be addressed in a subsequent LCA investigating the consequences of recycled GP commercialization at a larger scale. Finally, in a circular economy perspective, this work has demonstrated how LCA is becoming a relevant tool that addresses the environmental impacts of open-loop recycling for concrete. More examples are still needed to document the extent at which circular economy brings environmental benefits.

Acknowledgements

The authors would like to acknowledge the Centre de Recherche sur les Infrastructures du Béton Research Centre and the Industrial Research Chair (SAQ de valorization du verre dans les matériaux) at the University of Sherbrooke for funding this project.

References

1. F. Blomsma and G. Brennan, *J. Ind. Ecol.*, 2017, **21**, 603–614.
2. M. Geissdoerfer, P. Savaget, N. M. P. Bocken and E. J. Hultink, *J. Cleaner Prod.*, 2017, **143**, 757–768.

3. J. S. Mulrow, S. Derrible, W. S. Ashton and S. S. Chopra, *J. Ind. Ecol.*, 2017, **21**, 559–571.
4. Recyc-Québec, *Caractérisation des matières résiduelles du secteur résidentiel*, 2015.
5. S. Charron, *Bilan 2012 Recyc-Québec*, 2012.
6. B. Simon, M. Ben Amor and R. Földényi, *J. Cleaner Prod.*, 2015, **112**, 238–248.
7. E. L. Tucker, C. C. Ferraro, S. J. Laux and T. G. Townsend, *Resour., Conserv. Recycl.*, 2018, **129**, 240–247.
8. Y. Shao, T. Lefort, S. Moras and D. Rodriguez, *Cem. Concr. Res.*, 2000, **30**, 91–100.
9. R. Idir, M. Cyr and A. Tagnit-Hamou, *Cem. Concr. Compos.*, 2011, **33**, 19–29.
10. R. Saad, F. Charron-Doucet and E. Clément, *Analyse environnementale du cycle de vie de projets de commercialisation du verre mixte récupéré via des centres de tri de matières recyclables au Québec*, 2015.
11. C. Meyer, *Cem. Concr. Compos.*, 2009, **31**, 601–605.
12. J. S. Damtoft, J. Lukasik, D. Herfort, D. Sorrentino and E. M. Gartner, *Cem. Concr. Res.*, 2008, **38**, 115–127.
13. S. Lemmet, Buildings and Climate Change, Summary for Decision-Makers, UNEP SBCI, 2009, https://wedocs.unep.org/bitstream/handle/20.500.11822/32152/BCC_SDM.pdf?sequence=1&isAllowed=y (accessed Jan 2021).
14. P. Van den Heede and N. De Belie, *Cem. Concr. Compos.*, 2012, **34**, 431–442.
15. D. R. Vieira, J. L. Calmon and F. Z. Coelho, *Constr. Build. Mater.*, 2016, **124**, 656–666.
16. P. Gursel, E. Masanet, A. Horvath and A. Stadel, *Cem. Concr. Compos.*, 2014, **51**, 38–48.
17. O. Jolliet, M. Saadé-Sbeih, S. Shanna, A. Jolliet and P. Crettaz, *Environmental Life Cycle Assessment*, 2016.
18. M. U. Hossain, C. S. Poon, Y. H. Dong and D. Xuan, *Renew. Sustain. Energy Rev.*, 2018, **82**, 597–608.
19. K. E. Seto, C. J. Churchill and D. K. Panesar, *J. Cleaner Prod.*, 2017, **157**, 65–75.
20. M. U. Hossain, C. S. Poon, I. M. C. Lo and J. C. P. Cheng, *Resour., Conserv. Recycl.*, 2017, **120**, 199–208.
21. M. Jiang, X. Chen, F. Rajabipour and C. T. Hendrickson, *J. Infrastruct. Syst.*, 2014, **20**, 4014020.
22. G. Habert and N. Roussel, *Cem. Concr. Compos.*, 2009, **31**, 397–402.
23. G. Habert, D. Arribe, T. Dehove, L. Espinasse and R. Le Roy, *J. Cleaner Prod.*, 2012, **35**, 250–262.
24. M. U. Hossain, C. S. Poon, I. M. C. Lo and J. C. P. Cheng, *Resour., Conserv. Recycl.*, 2016, **109**, 67–77.
25. T. Blankendaal, P. Schuur and H. Voordijk, *J. Cleaner Prod.*, 2014, **66**, 27–36.
26. ISO 14044, *Management environnemental – Exigences et lignes directrices*, 2006.

27. N. A. Soliman, A. F. Omran and A. Tagnit-Hamou, *ACI Mater. J.*, 2016, **113**, 307–316.
28. A. Tagnit-hamou, N. Soliman, A. F. Omran, M. T. Mousa, N. Gauvreau and M. F. Provencher, *Concr. Int.*, 2015, **3**, 41–48.
29. G. Wernet, C. Bauer, B. Steubing, J. Reinhard, E. Moreno-ruiz and B. Weidema, *Int. J. Life Cycle Assess.*, 2016, **3**, 1218–1230.
30. M. Schmidt and E. Fehling, *Seventh International Symposium on Utilization of High Strength/ High Performance Concrete*, 2005, pp. 51–78.
31. T. Vande Voort, M. Suleiman and S. Sritharan, Behavior of Driven Ultrahigh-Performance Concrete H-Piles Subjected to Vertical and Lateral Loadings, https://ascelibrary.org/doi/pdf/10.1061/%28ASCE%29GT.1943-5606.0000350 (accessed Dec 2020).
32. E. C. Bentz and M. D. A. Thomas, *Life-365 User Man.*, 2013, 1–87.
33. C. P. Gu, W. Sun, L. P. Guo and Q. N. Wang, *Appl. Mech. Mater.*, 2013, **438–439**, 108–112.
34. N. H. Yi, J. H. J. Kim, T. S. Han, Y. G. Cho and J. H. Lee, *Constr. Build. Mater.*, 2012, **28**, 694–707.
35. Recyc-Québec Annual Report for 2013/2014, https://www.recyc-quebec.gouv.qc.ca/sites/default/files/documents/rapport-annuel-2013-2014.pdf (accessed Dec 2020).
36. Ecoinvent database, Allocation cut off by classification, https://www.ecoinvent.org/database/system-models-in-ecoinvent-3/cut-off-system-model/allocation-cut-off-by-classification.html (accessed Dec 2020).
37. M. L. Marceau, M. A. Nisbet and M. G. VanGeem, *Life Cycle Inventory of Portland Cement Concrete*, http://large.stanford.edu/courses/2016/ph240/pourshafeie2/docs/marceau-2007.pdf (accessed Dec 2020).
38. J. Prusinski, M. Marceau and M. VanGeem, *Eighth CANMET/ACI International Conference on Fly Ash, Silica Fume, Slag and Natural Pozzolans in Concrete*, 2004, pp. 1–26.
39. C. Chen, G. Habert, Y. Bouzidi, A. Jullien and A. Ventura, *Resour., Conserv. Recycl.*, 2010, **54**, 1231–1240.
40. P. Van den Heede, M. Maes and N. De Belie, *Constr. Build. Mater.*, 2013, **67**, 74–80.
41. D. J. M. Flower and J. G. Sanjayan, *Int. J. Life Cycle Assess.*, 2007, **12**, 282–288.
42. E. R. Teixeira, R. Mateus, A. F. Camões, L. Bragança and F. G. Branco, *J. Cleaner Prod.*, 2016, **112**, 2221–2230.
43. K. Beaulne-bélisle, B. Gariépy, J.-F. Luc Vachon, K. McGrath and J. Rosset, *Profil de la gestion des débris de construction, rénovation et démolition (CRD) au Québec*, RECYC-QUÉBEC, 2009, p. 136.
44. O. Jolliet, M. Margni, R. Charles, S. Humbert, J. Payet, G. Rebitzer and R. K. Robenbaum, *Int. J. Life Cycle Assess.*, 2003, **8**, 324–330.
45. A. Lautier, R. K. Rosenbaum, M. Margni, J. Bare, P. O. Roy and L. Deschênes, *Sci. Total Environ*, 2010, **409**, 33–42.
46. J. Bare, *US Environ. Prot. Agency*, 2012, 1–24.
47. G. P. Peters, *Int. J. Life Cycle Assess.*, 2007, **12**, 373–380.

48. B. P. Weidema and M. S. Wesnaes, *J. Cleaner Prod.*, 1996, **4**, 167–174.
49. J. Whitmore and P.-O. Pineau, État de l'énergie au Québec, 2017, Chaire de gestion du secteur de l'énergie, HEC Montréal, Décembre 2016, https://energie.hec.ca/wp-content/uploads/2016/12/EEQ2017.pdf (accessed Jan 2021).
50. G. Habert, E. Denarié, A. Šajna and P. Rossi, *Cem. Concr. Compos.*, 2013, **38**, 1–11.
51. Y. Dong, *Constr. Build. Mater.*, 2018, **167**, 414–425.
52. P. K. Povindar, K. Mehta and P. J. M. Monteiro, *Concrete: Microstructure, Properties, and Materials*, 2006.
53. N. Jungbluth, M. Stucki, K. Flury, R. Frischknecht and S. Büsser, *Life Cycle Inventories of Photovoltaics*, 2012.

CHAPTER 7

Life Cycle Assessment of Sustainable Polymer Packaging

MIAO GUO*[a,b] AND NILAY SHAH[a]

[a] Department of Chemical Engineering, Imperial College London, London SW7 2AZ, UK; [b] Department of Engineering, King's College London, WC2R 2LS, UK
*Emails: miao.guo@imperial.ac.uk; miao.guo@kcl.ac.uk

7.1 Introduction

7.1.1 Sustainable Polymers from Renewable Resources

Since the first industrial-scale production of petroleum-based polymers took place in the 1940s, global demand for plastic has increased considerably.[1] According to statistics from Plastics Europe,[2] the average annual increase in global production and consumption of plastic has reached nearly 9% since 1950; with total global production increasing from 1.5 million tonnes in 1950 to 335 million in 2016, with applications in almost every sector. Consumption on a *per capita* basis had grown to approximately 75–85 kg per year in Europe and North America by 2017.[3] Another potential area for rapid growth in plastic consumption is the Asian developing countries, where a rapidly increasing trend was observed in plastic consumption in China (almost doubled from 2009 to 2018). Amongst 20 distinct groups of plastic, five plastic families accounted for approximately 75% of European plastic consumption: polyethylene (PE); polypropylene (PP); polyvinylchloride (PVC); polystyrene (PS); and polyethylene terephthalate (PET).[2] Packaging remains the largest use for plastic, followed by the building and construction industry, and accounts for nearly 60% of Europe's plastic demands.[2] Increasing

Life Cycle Assessment: A Metric for the Circular Economy
Edited by Aiduan Borrion, Mairi J. Black and Onesmus Mwabonje

Published by the Royal Society of Chemistry, www.rsc.org

demands for petroleum-based polymers in packaging as well as other sectors, is not only a resource depletion issue (due to finite oil and gas reserves), but also causes other environmental and waste management issues *e.g.* damaging effects of plastic waste on aquatic systems.[4] Despite recycling outperforming landfill for the first time in 2016, a significant amount of post-consumer plastic waste was generated in the UK (27.1 million tonnes in 2016).[2,3]

Research efforts have been placed on renewable alternatives to fossil-derived plastics and sustainable end-of-life options. Biomass from terrestrial or aquatic plants is used as the renewable raw material and the polymers are often referred to as 'bio-based'.[5] Biodegradable polymers can be defined as 'a polymer with deterioratable physical and chemical properties which can be decomposed down to the formation of CO_2, H_2O, CH_4 and other low molecular-weight products, under the influence of microorganisms in both aerobic and anaerobic conditions'.[5] Throughout this process, all carbon should be accounted for in the carbon balance, residues produced should be non-toxic to the environment and residues along with microbial biomass should be eventually incorporated into the geochemical cycle.[6] Noticeably, some petrochemical polymers are biodegradable, and bio-based plastics are not necessarily biodegradable but can be recyclable *e.g.* bio-polyethylene (bio-PE), to replace PE. Therefore, we classify biopolymers according to origin and biodegradability (Figure 7.1).

Research on biodegradable plastics can be traced back to the 1980s.[7,8] Biodegradable polymers offer an opportunity for waste management since they can degrade under either natural environmental conditions or in municipal/industrial biological waste treatment facilities. Specifically, biodegradable polymers can be categorized as being of natural origin, synthetic origin (petroleum-origin polymers) and biodegradable polymer blends.[6,9,10]

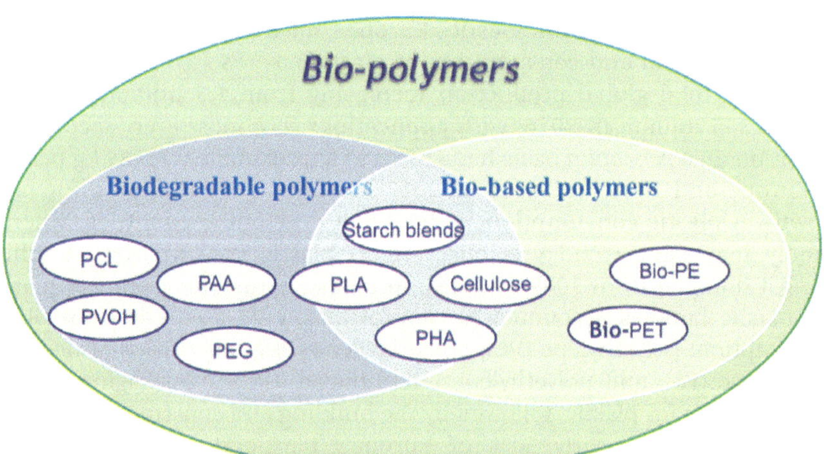

Figure 7.1 Classification of biopolymers. Notes: PEG: polyethylene glycol; PCL: polycaprolactone; PAA: polyacrylic acid; PVOH: polyvinyl alcohol; bPET: bio-based polyethylene terephthalate; bPE: bio-based polyethylene; PHA: polyhydroxyalkanoates.

Very few synthetic plastics are biodegradable. Compared with synthetic origin polymers, biopolymers based on naturally occurring long-chain molecules – *e.g.* polysaccharides, proteins, lignin and shellac – have been suggested as more environmentally 'friendly' option, as they are derived from renewable resources. The blending of bio-based polymers with synthetic polymers has also received considerable attention as this offers the capability to attain mechanical and other properties, as well as favourable degradation rates.

Annually, over 150 billion tonnes of polysaccharides are produced naturally, with humans consuming only 1% of this volume; thus, polysaccharides offer great opportunities as biopolymer feedstock. Starch, the second most abundant renewable polysaccharide in nature,[11] has been recognized as one of the promising substitutes for petrochemical plastics for a variety of applications.[12,13] Composed of repeating D-glucopyranosyl units, starch can be separated into amylose, which is linear, linked by α-(1–4) linkage, and amylopectin, which has α-(1–4)-linked backbone and α-(1–6)-linked branches.[14] Amylose, which has a molecular weight of several hundred thousand, shaped in the form of helix, is a minor component typically ranging between 20% and 30% of starch. Amylopectin, with a molecular weight in the order of several million, presenting in double helical crystalline structure, is the major component of starch.[6] Starch and starch-derived glucose is biodegradable and easily metabolized by a range of micro-organisms. Starch-based materials derived from corn, potato, wheat, tapioca and rice *etc.* have been widely developed. The various starch-based polymers can be classified into several categories:[5,10,15]

- Starch-filled plastics;
- Chemically modified starch;
- Plasticized thermoplastic starch;
- Biopolymers synthesized from starch and starch-derived glucose;
- Starch compounded with biodegradable polymers; and
- Starch-derived recyclable polymers.

In starch-filled plastics, starch is used as a biodegradable filler in synthetic polymers, such as starch-filled low-density polyethylene (LDPE) film[16] or PS sheet.[17] Generally, the starch filler component is lower than 20%.[10] However, the starch filler only enhances the disintegration of these blends in nature, not necessarily its biodegradation.[10,17,18] Quite often, the properties of native starch are inadequate for the application, and chemically modified starches have been developed to overcome performance shortcomings.[19] An example of this is the series of papers/patents describing the co-polymerization of starch and other polymers, such as polyacrylamide-starch complex[20,21] and the starch esters prepared by a non-aqueous process.[22] Thermoplastic starch (TPS) is processed in the presence of plasticizers such as water, selected amino acids, or glycerol.[22,23] There has been increasing interest in the utilization of TPS for biodegradable materials and the biodegradability of TPS has been broadly researched.[6,18] The synthesized biopolymers from starch are either produced *via* chemical synthesis or fermentation by microbes. Polylactic acid (PLA) is a leading example of a biodegradable biopolymer, synthesized from lactic acid derived from dextrose from starch, *via* fermentation. Starch polymers compounded (blended) with other polymers have also developed rapidly, such as the blends commercialized under the Mater-Bi®[24] trademark. Examples of recyclable biopolymers include bio-PET and Bio-PE (*e.g.* CocaCola PlantBottle[25] and Braskem I'm Green[26]). These renewable derived bioplastics offer chemically and physically identical properties to petroleum-based counterparts; these can be potentially produced using the same processing technologies and deliver the same recyclability as petroleum polymers.

7.1.2 Life Cycle Assessment (LCA)

The ISO 14040 series of standards provide a framework for undertaking LCAs.[27] Within ISO standards, no specific and exact LCA methodology is defined, but LCA frameworks and guiding principles are given. LCA is further extended to life cycle sustainability assessment (LCSA), encompassing three pillars, *i.e.* economic (life cycle costing), environmental (LCA) and social (social-LCA) footprints.[28] This chapter will focus only on environmental LCA. LCA consists of four interactive phases (goal and scope definition, inventory analyses, impact assessment and interpretation),[27–29] capturing every life cycle stage from raw material acquisition to final disposal. The first phase scopes the fundamental LCA elements including system boundaries and functional unit, and defines the principle and methodologies to be adopted. At the life cycle inventory stage (LCI), data are collected within the system boundary and assessed against data-quality requirements. Inventories are classified and linked with the impact categories at the life cycle impact assessment (LCIA) stage, which comprises the mandatory element of characterization and other optional ones (normalization, weighting). Inventory outputs are converted to category indictors using characterization methods, which can be further linked to category endpoints defined as 'attribute or aspect of natural environment, human health or resources identifying an environmental issue'.[29] The aggregated indicator results provide characterized profiles of the studied products, which can be normalized to give information on the relative magnitude of each indicator result. LCIA outcomes are presented in the interpretation phase, in accordance with the goal and scope, to reach conclusions and recommendations. Generally, two types of LCA have been distinguished: retrospective *i.e.* attributional LCA (ALCA); and prospective, *i.e.* consequential, LCA (CLCA). The former focuses on describing the environmentally relevant flows to and from a life cycle and its subsystems; the latter investigates how environmentally relevant flows change in response to possible decisions.[30] ALCA and CLCA can be generalized as eqn (7.1).[31]

$$EI_{kpi} = \sum_{r}\sum_{s} EIf^{\text{in}}_{r,kpi} X^{\text{in}}_{r,s} + \sum_{c}\sum_{s} EIf^{\text{out}}_{c,kpi} X^{\text{out}}_{c,s} \tag{7.1}$$

Where the variable EI_{kpi} denotes the total environmental profiles of a given process, product or system (per functional unit) expressed as key performance indicator *kpi*. EI_{kpi} is determined by the characterization impact factors for input resource *r* ($EIf^{in}_{r,kpi}$) or emitted compound *c* ($EIf^{out}_{c,kpi}$) and the input or output flows ($X^{in}_{r,s}$ or $X^{out}_{r,s}$) at each life cycle stage *s*. Negative $EIf^{out}_{c,kpi}$ represents the environmental credits allocated to the system due to the saving or substitution. Despite the simplified linear form, processes and systems often lead to non-linear functions *e.g.* closed-loop manufacturing.

For multiple-product systems, three allocation approaches are applicable to partition the material/energy flows and their associated impacts amongst the co-products *i.e.* allocation by physical relation (*e.g.* mass, volume *etc.*), allocation by economic values or system boundary expansion.[32,33] In CLCAs, allocation by mass or economic values can be avoided by applying system boundary expansion.[30,34] Although there is no universal consensus on the allocation method in ALCAs, the avoidance of allocation through system expansion (followed by physical-based allocation) has been indicated as the preferred approach for LCA in ISO LCA guidelines[33,35] and PAS 2050.[36]

7.1.3 LCAs of Sustainable Polymers

As generalized in Figure 7.2, bio-based polymer life cycles largely consist of four phases: terrestrial or aquatic plant cultivation; polysaccharides extraction; bio-based polymer production and end-of-life. Notably, end-of-life can be classified as thermochemical (*e.g.* incineration, gasification, pyrolysis), biochemical (*e.g.* anaerobic digestion, composting) and chemical/mechanical

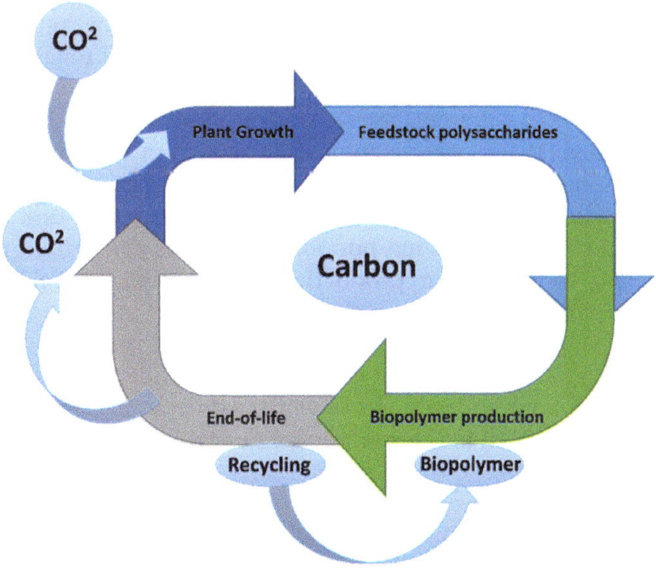

Figure 7.2 Bio-based polymer life cycle.

routes (*e.g.* recycling). As demonstrated in Figure 7.2, biogenic carbon plays a significant role in the life cycle of sustainable polymers derived from biological renewable resources. The different processes can be broadly categorized as carbon sinks and carbon emitters. Carbon sinks (*e.g.* carbon capture and utilization technologies, carbon recovery) sequester biogenic carbon from atmosphere and contribute to climate change mitigation by removing and fixing CO_2; whereas carbon emitters (*e.g.* combustion processes) cause negative impacts on climate change. Similarly, based on energy profiling, conversion processes can be classified as energy sinks, which act as consumers of CO_2, and energy sources (energy generation). Potentially, energy sinks can be shifted to energy sources, for example, through value recovery from low-grade waste. LCA offers a quantitative approach to systematically evaluate environmental sustainability and inform decision-making.

LCA has been widely applied to bio-based plastic studies, including biodegradable polymers *e.g.* PLA,[37,38] PHA, PHB,[39–41] starch-based composites[42] and recyclable biopolymers *e.g.* bio-PET,[43–45] and bio-PE.[26,46] Research efforts have been focused on comparing bio-derived polymers with petroleum counterparts and exploring waste treatment strategies. This chapter presents two LCA case studies to discuss different types of starch-based biodegradable biopolymers, including starch-based biocomposite, biosynthesized polymer. The focus of these two case studies is to highlight environmentally damaging steps and compare the bioplastic with petrochemical counterparts.

7.2 Case Study 1 – Starch-based Bio-composites

7.2.1 LCA Methodology

ALCA has been applied to evaluate the cradle-to-grave environmental profiles of starch-based bio-composites, where the functional unit was defined as 'a unit (*e.g.* 1 kg) starch-based biocomposite at factory gate'. Three composites derived from different starch sources, *i.e.*, wheat (W-Bio), potato (P-Bio) and maize (M-Bio) were compared. The system boundary definition of starch-based bio-composite is given in Figure 7.3, where the wheat starch-based biopolymer is illustrated as an example. The following unit processes are considered within the system boundary: biomass cultivation; feedstock processing (*e.g.* wheat starch production); and polymer packaging production, distribution and transportation.

The economic allocation approach was followed for production processes where multiple products occur. For end-of-life scenarios, a system boundary approach was adopted. Closed-loop recycling was assumed for the polymer residue disposal involved in the starch-based composites and the functional equivalent quantity of fossil-based counterparts were allocated as a credit, as a consequence of the recycling process. A carbon counting approach was used to track the carbon footprint during the life cycle of biopolymers specifically, to determine: (1) biogenic carbon content of the biopolymers; (2) fossil carbon contained in PVOH and other additives of the biopolymers;

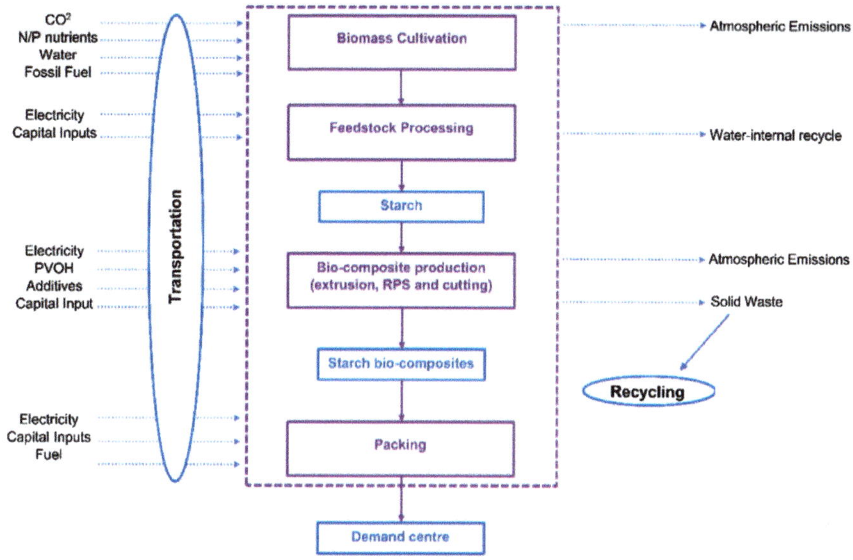

Figure 7.3 Starch-based biocomposite (RPS = Regular Packing & Stacking technology).

(3) the downstream release of biogenic and fossil carbon during subsequent processing stages over the polymer life cycles. A mid-point (problem-oriented) approach CML baseline (a characterization methodology developed by developed by the Institute of Environmental Sciences, Leiden University) has been applied to assess the impacts; analyses were performed on SimaPro LCA platform.

7.2.2 Inventory Analyses

The main input-output data for the three starch-PVOH biocomposites were provided by manufacturers and suppliers of feedstock and additives for biopolymer production.[47] The inventory for crop cultivation was based on process-based simulation and primary data from field measurements.[48] The production process was modelled for the PVOH biocomposites with a density of 1.3 g cm^{-3}, 87–89% hydrolysis degree and an average molecular weight of 2488. PVOH inventory, including baseline and uncertainty range are summarized in Table 7.1, and more details can be found in the publication by Guo, 2012.[47] Detailed inventories for crop cultivation and biocomposite production can be found in published research.[48]

7.2.3 'Cradle-to-factory-gate' LCIA for Starch-based Composites

As shown in Figure 7.4, wheat-based biocomposite (W-Bio) outperforms the other two starch biocomposites in most impact categories, except for

Table 7.1 Inventory for PVOH production.

	Input	Output	Lower range	Upper range	Assumption
Polymerization (tonne)					
Methanol[a]	8.1 E-02	—	5.5E-02	1.1E-01	• Conversion 65% (Uncertainty range50%-95%)[b]
VAC input[a]	1.5E+00	—	1.1E+00	2.0E+00	
AIBN[a]	4.1 E-04	—	2.8E-04	5.3E-04	
PVAc		1	—	—	• 90% recovery rate[c]
Hydrolysis (tonne)					
Methanol[e]	1.1E+00		5.0E-01	1.8E+00	• Conversion efficiency 90% in first reactor then reach 100% in second reactor[d]
NaOH[e]	9.8E-02		1.9E-02	1.7E-01	
PVAc	1.9E+00		—	—	
PVOH	—	1	—	—	
Methyl acetate	—	1.7 E+00	—	—	• 90% recovery rate[c]
Total energy consumption (kj per tonne PVOH produced)					
Electricity	1.1E+03	—	—	—	—
Natural gas	3.5E+04	—	7.0E+03	1.1E+05	

[a]Ref. 49.
[b]Ref. 49 and 50.
[c]Ref. 49 and 51.
[d]Ref. 51.
[e]Ref. 49 and 52.

acidification and Photochemical Oxidation (POCP), due to the higher field emissions occurring in wheat cultivation.[48] The contributional analyses results (*i.e.* hot-spot analyses) for the three starch-based biocomposites are broadly similar. The production of the two major components of starch-based biocomposites, *i.e.* PVOH and starch were dominant contributors to the environmental impacts across all impact categories. Biopolymer production process, including energy consumption and additives, as well as infrastructure inputs in total accounted for 20–40% of overall environmental burdens. This study suggests that the contribution of PVOH was significant in energy consumption and POCP impacts, which is in agreement with a previous study on maize starch-based biocomposite.[53]

The agricultural system was the major driver for acidification and eutrophication impacts over the 'cradle-to-gate' life cycle stages of starch-based biocomposites. The agricultural system, together with PVOH production dominated Global Warming Potential (GWP100) and Ozone Depletion Potential (ODP) impacts of starch-based bio-composites, whereas PVOH production appeared as the main cause of abiotic depletion and POCP scores, which was attributable to the resources consumed during production of feedstocks (C_2H_4, CH_3OH, CO) for PVOH, and atmospheric emissions from vinyl acetate (VAC) and its feedstocks (C_2H_4, $C_2H_4O_2$) manufacturing. Apart from these two contributors, the embedded emissions in infrastructure were the driving factor for impacts of starch-based biocomposites in human and eco-toxicity impact categories. Detailed analyses in each impact category are presented below.

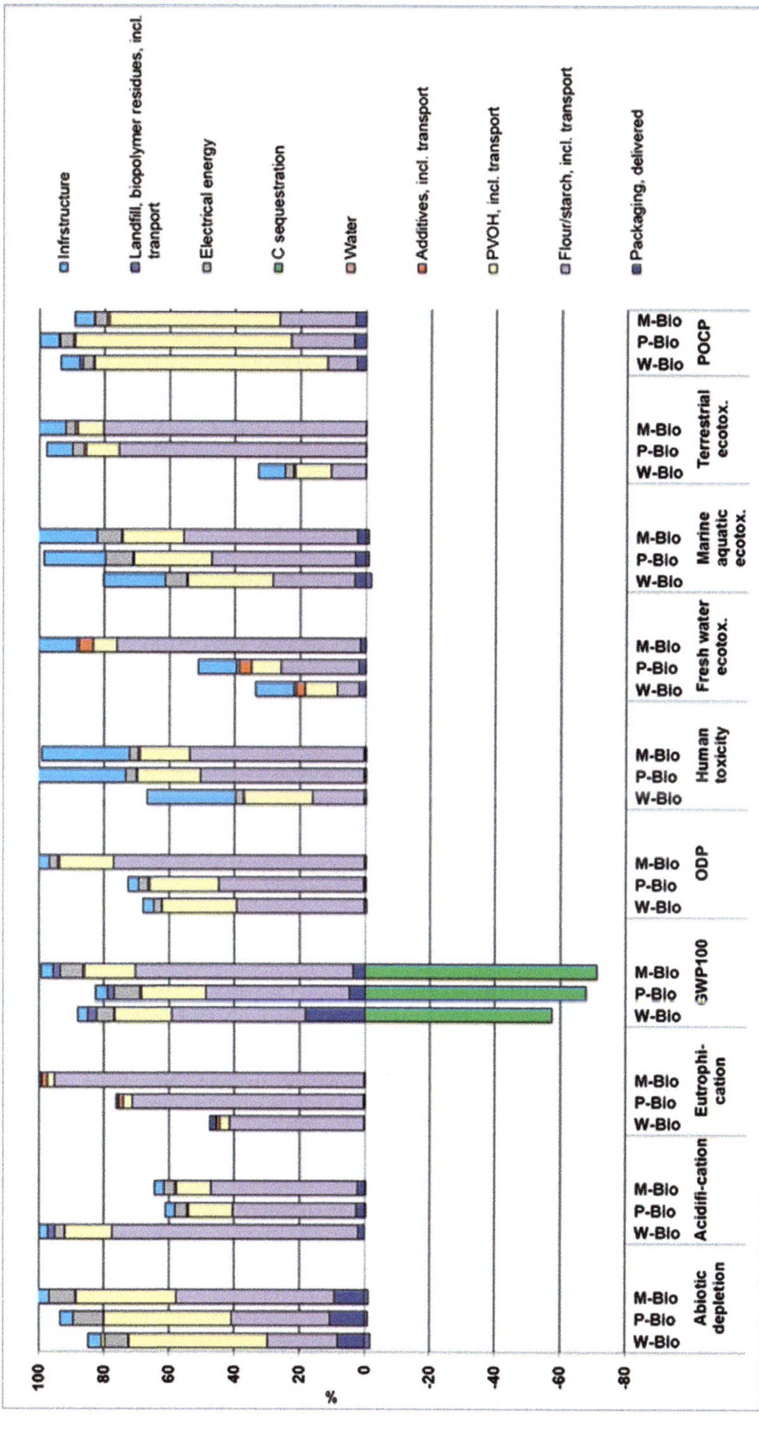

Figure 7.4 'Cradle-to-factory-gate' characterized LCIA profiles of starch-PVOH biocomposite (unit: kg biocomposite, CML baseline).

7.2.3.1 Abiotic Depletion

In abiotic depletion, recycling of wasted packaging were presented as minor negative values below the line, where the recycled PP packaging was treated as 'credits' by following a system boundary expansion approach. However, this beneficial effect was overridden by the burdens induced by other processes, amongst which PVOH is presented as a dominant contributor, causing nearly 60% of abiotic depletion. It was mainly attributable to resources (*e.g.* natural gas and crude oil) consumed during production of feedstocks (C_2H_4, CH_3OH, CO) for PVOH manufacturing. Two feedstocks, CO and CH_3OH, were highly fuel-dependent: the former was modelled as the product from combustion of heavy fuel oil, whereas the latter was derived from natural gas; CO production was also an energy (electricity) demanding process. Three other resource consumers were: wheat starch production; electricity; and transportation inputs involved in starch-based biocomposite production, in total accounting for 39% of resource depletion.

7.2.3.2 Global Warming Potential (GWP100)

The CO_2 sequestered in starch-based biocomposites represents a carbon sink (shown as negative values below the line in Figure 7.4). However, such beneficial effects were not sufficient to offset the carbon emissions caused by agricultural systems and PVOH production (80% of GWP100 positive impacts in total). The Greenhouse Gases (GHGs) generated from fuel combustion, either for energy generation or feedstock manufacturing during the life cycle of VAC, caused nearly 20% of GWP100 burdens (above-line positive value). The remaining 20% of the burdens on GWP100 were caused by other GHGs emitters, *i.e.* fuel combustion in vehicle engines for transport and for the electricity generation stage.

7.2.3.3 Acidification and Eutrophication

Both acidification and eutrophication impacts were dominated by crop cultivation and starch production processes, whereas the PVOH component only shared 6.3–14.7% of burdens, caused by the feedstock production (C_2H_4, CH_3OH, CO) for PVOH production. The major contributors to impacts of PVOH in acidification were the SO_2 and NO_x emissions from electricity generation (CO and CH_3OH production are energy-intensive processes), and the SO_2 and NO_x released from production of feedstock C_2H_4. Impacts produced by PVOH in eutrophication were primarily caused by NO_x and Chemical Oxygen Demand (COD) release (COD released from VAC production and the COD and NO_x released from fuel processing and delivery for PVOH production demand).

7.2.3.4 Human and Eco-toxicity

The infrastructure components involved in PVOH and starch-based biocomposite manufacturing were drivers for environmental profiles on all toxicity impact categories. Impacts on human toxicity were attributed to chromium, arsenic, and Polycyclic Aromatic Hydrocarbons (PAHs) released from producing construction materials. The impacts on fresh water/marine aquatic ecotoxity was driven by the water emissions from disposal of slag or ash during metal production or electricity generation, *i.e.* nickel, vanadium, cobalt, barium, beryllium. The environmental burdens on terrestrial ecotoxicity were shared by electricity power transmission, production of infrastructure materials (copper and steel), and lignite and heavy fuel oil burning process (due to the chromium emission to soil and atmospheric emissions of mercury vanadium and arsenic).

7.2.3.5 Ozone Depletion Potential (ODP)

One third of impacts on ODP were contributed by the PVOH component, whereas the remaining two thirds were dominated by wheat starch (37%) and transport (21%). The ODP burdens were primarily driven by the production and delivery of fuels (natural gas and heavy fuel oil) utilized for PVOH feedstock production and used for NH_3 and diesel production, *e.g.* $CBrF_3$ emitted from crude oil production and $CBrClF_2$ released from transport of natural gas.

7.2.3.6 Photochemical Oxidation (POCP)

The POCP scores were dominated by PVOH production (80% of the impacts) due to the C_2H_4 and $C_2H_4O_2$, CH_3OH released during production of VAC and its feedstocks (C_2H_4 and $C_2H_4O_2$). An additional 20% of POCP was shared by other inputs (infrastructure, electricity, transportation, flour/starch), their individual contribution varied between 4% and 6.5% (due to CO, SO_2, CH_4 emitted either from electricity generation, field operation, and transport).

7.2.4 Discussion

In this study, hot-spots were identified at each life cycle stage of starch-PVOH biocomposites, *e.g.* PVOH production and agriculture, which represent the main energy sinks and carbon emitters. Through developing a transparent LCA inventory for PVOH, our research fills a data gap. Our sensitivity analyses indicated uncertainty in PVOH inventory causes 10–20% change in the overall environmental impacts of biocomposites.

We compared three starch-based biocomposites with their petrochemical counterparts, *e.g.* PE, PS. Our results suggested the comparisons are highly sensitive to their packaging applications, starch sources and end-of-life scenarios,[54,55] whereas the comparison results are largely independent on the PVOH uncertainty range. Overall, starch biocomposites represent promising alternatives to petrochemicals in packaging applications. Up to the manufacturing gates, the environmental advantages of starch-based composites over petrochemical plastics can be concluded on four impact categories, *i.e.* abiotic depletion, ODP, GWP100 and POCP, however, starch biocomposites incur higher acidification and eutrophication burdens than petrochemicals, due to N emissions from the agriculture system. The cradle-to-gate LCA studies highlight the ranking of the waste treatment routes, which depend on their chemical composition, notably, anaerobic digestion represents a promising option in terms of carbon sink effects for bioplastics containing nitrogen and sulphur.[55] However, starch-based composite hardly compete with petrochemicals plastics with recycling end-of-life.

7.3 Case Study 2 – Energy Profiling of Starch-glucose Biosynthesis Polymers

7.3.1 LCA Methodology

PHBV (poly-hydroxy butyrate-*co*-valerate) is part of the poly hydroxyl alkanoate (PHA) family, PHBV can be derived, using biosynthesis technologies, from a range of carbohydrates sources, such as starch. An ALCA was carried out to evaluate the long-term quantitative and qualitative depletion of fossil resources, resulting from the fossil energy demand for producing and disposing of newly developed PHBV-based packaging, compared with conventional petrochemical-sourced counterparts. This case study aims to provide insights on the opportunities available for future development of PHBV-based bio-packaging to deliver a sustainable production system, with minimized 'damage' to fossil fuel resources from a long-term perspective.

The functional unit defined for the LCA was the production and end-of-life disposal of a single food packaging tray (with external dimensions of 175 mm×130 mm×28 mm) derived from PHBV-Ecoflex biocomposite or petrochemical polymers (PET/PP).

The product systems for PHBV-Ecoflex biocomposite and petrochemical counterparts are illustrated in Figures 7.5 and 7.6, respectively. The following unit processes were considered within the system boundary of bio-packaging: corn farming; milling and enzymatic hydrolysis for glucose production; PHBV granulate manufacturing; compounding and processing to produce PHBV-Ecoflex bio-packaging; and end-of-life (incineration with energy recovery (EfW)). For the petrochemical counterparts, cracking (*e.g.* propylene and ethylene production), intermediate products production (*e.g.* terephthalic acid), polymerization and packaging production and end-of-life phases (EfW, recycling and landfill) were included in the analysis (see Figure 7.6). Based on the functional unit definition, the use-phase of a full life cycle is excluded from the assessment, primarily because it is assumed that the fossil fuel consumption of this phase is negligible.

An economic allocation approach was applied to the PHBV production process, where multiple physical products occur (except for carbon). In order to 'track' the biogenic carbon footprint during the life cycle of PHBV, a stoichometric carbon counting approach was applied, to determine: (1) biogenic carbon sequestration into the PHBV (from atmospheric CO_2 removal during the corn crop growth); and (2) downstream release of biogenic

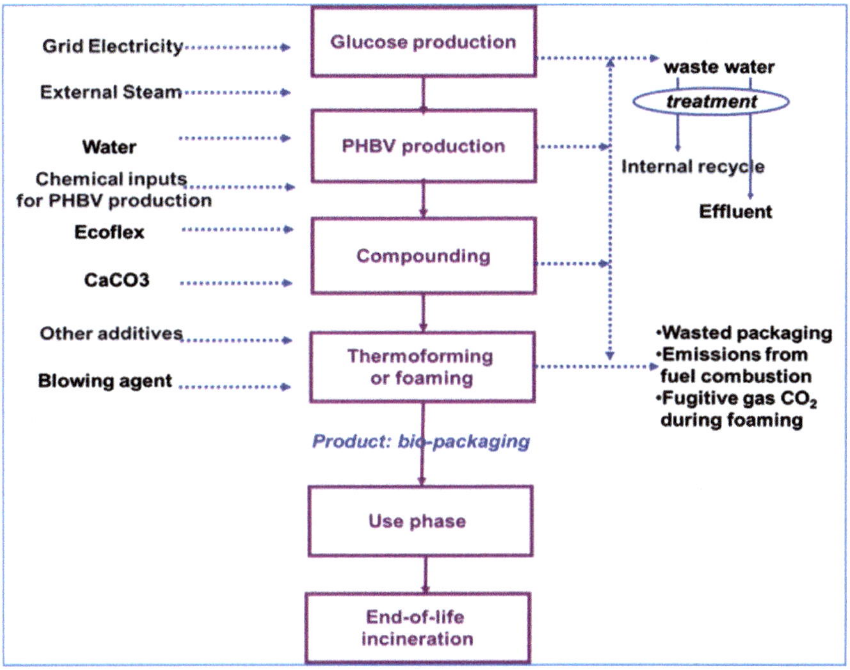

Figure 7.5 Bio-packaging product systems modelled.

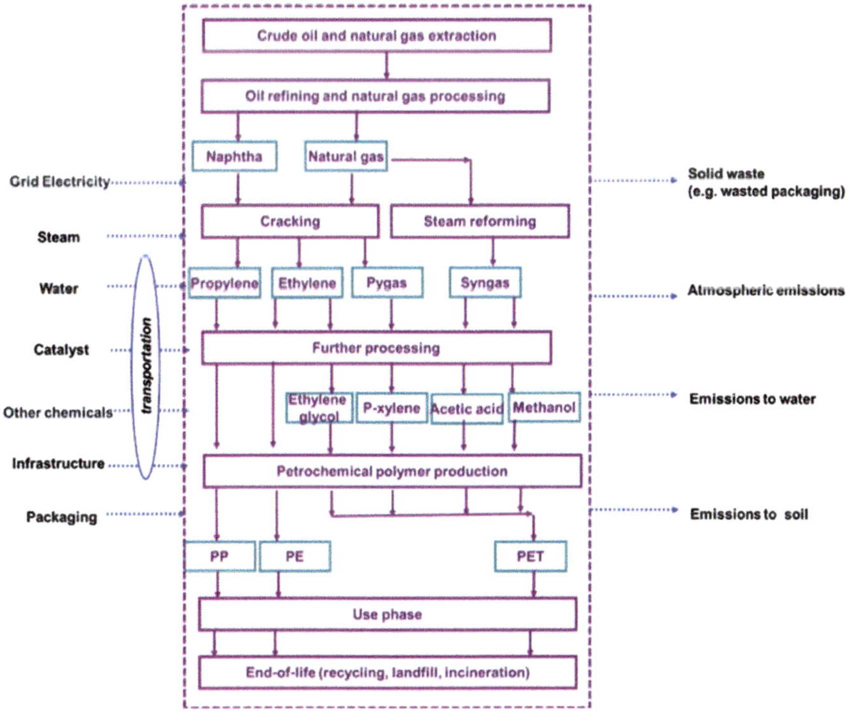

Figure 7.6 Petrochemical product system modelled.

and fossil carbon during the subsequent processing, and end-of-life phases over the life cycle.

The system boundary expansion approach, *i.e.* allocation by substitution credits for burdens avoided was applied to the end-of-life disposal for PHBV-based packaging. In the incineration processes, the inputs and outputs for the equivalent quantity of electrical and/or thermal power generated by the average electricity/heat generation supply mix for the national grid were allocated as a 'credit' to the modelled systems. At the end-of-life any PET or PP may be re-used within the same product system, or may be used in other product systems (recycled material undergoes changes in inherent properties). The former is referred to as 'closed-loop recycling' the latter as 'open-loop recycling'. In this study, a closed-loop was assumed, *i.e.* the virgin material for the life cycle is assumed to be displaced by recycled material within the same product system.

A damage-oriented (endpoint) approach [Eco-Indicator hierarchist version (EI H)[56]] was applied in this case study. The rationale for choosing EI H is that this characterization model handles fossil fuel and minerals as two separate resource-related impact categories, and takes into account the long-term trend of resource quality depletion from a balanced time-horizon perspective.[40] The surplus energy concept applied in the EI H fossil fuel impact category (surplus energy is referred to as the difference between the energy needed to extract a resource now and at a point in the future) provides a relative measure for the 'damage' that the depletion of a fossil resource causes. The LCA was performed in SimaPro.

7.3.2 Experimental Methods

The elemental analysis for carbon, hydrogen, nitrogen and sulphur contained in PHBV biopolymer were carried out by OEA Laboratories Ltd using a Thermoquest EA1110 elemental analyser. The calorific value measurement was conducted using an Oxygen Bomb Calorimeter (Parr 6100) to determine the higher heating value (HHV) of thermoformed or foamed biocomposite. The lower heating value (LHV) was calculated from the HHV and the total hydrogen content in the biocomposite according to LHV $(\mathrm{J\,g^{-1}}) = \mathrm{HHV}\ (\mathrm{J\,g^{-1}}) - 218.13 \times \mathrm{H}$ percent (wt %).[57]

7.3.3 Life Cycle Inventory Analysis

7.3.3.1 Biocomposite Production

The inventories for corn-glucose and PHBV production were provided by the glucose supplier and Tianan Biologic Material Co., Ltd., respectively, which have been detailed in ref. 40. The dataset for amylase enzyme production was provided by Novozymes A/S to represent the site-specific enzymes used for glucose production in China. The Ecoflex material was modelled as a petrochemical polymer derived from crude oil and natural gas.

The composition of Ecoflex was based on the results reported by Witt *et al.* (2001) (terephthalic acid: 37.35% w/w, adipic acid: 26.23% w/w, 1, 4-butanediol: 36.42% w/w). The inventory for engineered $CaCO_3$ was provided by a consortium under the project Farm Produced Ultrathin Lightweight Packaging (TS/G000859/1).

Two PHBV-Ecoflex biocomposite products produced *via* different technologies were modeled, *i.e.* thermoformed and foamed bio-packaging. The input-output data for blowing, thermoforming and foaming processes were based on the trials carried out by consortium under the project Farm Produced Ultrathin Lightweight Packaging (TS/G000859/1). The blowing agent involved in the foaming process was modelled based on its theoretical components (citric acid and sodium bicarbonate). The production processes of other input materials were derived from the Ecoinvent database.

7.3.3.2 LCI for Bio-packaging and Petrochemical Counterparts

The packaging modelled was a food packaging tray derived from PHBV-Ecoflex biocomposite or a petrochemical polymer (PET or PP). The key parameters and data sources for the biocomposite and petrochemical counterparts are specified in Table 7.2. During the foaming process, the active compounds contained in the blowing agent are sodium bicarbonate and citric acid. Their thermo decomposition efficiency (to release CO_2) was assumed as 90%, and 5% of CO_2 was assumed to escape from the surface as fugitive gas during the foaming process (data obtained directly from a foaming technology company). Two scenarios were modelled for foamed biocomposites. Scenario 1 represents the density reduction (36.5%) occurring in laboratory foaming experiments, scenario 2 represents an estimated achievable density reduction (60%) in commercial foaming processes (estimation from foaming technology company).

Table 7.2 Specifications for the bio-packaging *vs.* PET/PP packaging modelled.

	Thermoformed bio-composite	Foamed bio-composite	PET	PP
Density ($g\,cm^{-3}$)	1.36	Scenario 1: 0.86 Scenario 2: 0.54	1.4	0.9
Recycled content (%)	Virgin	Virgin	50%	Virgin
Input per formed tray	Compounded biocomposite, energy	Compounded biocomposite, energy	PET granulate, energy	PP granulate, energy
Output per formed tray	Waste packaging, biocomposite residues	Waste packaging, biocomposite residues	Wasted packaging, PET residues	Wasted packaging, PP residues
Data sources	BLT consortium		Sharp Interpack Ltd.	

Table 7.3 Key parameters for incineration.

Inputs per kg waste	Medium incineration	Large incineration
Operation energy and material inputs		
Electricity input (MJ)	4.63e-3	4.27e-3
Diesel input (KG)	7.42e-5	4.83e-5
Natural gas (MJ)	8.50e-4	1.15e-4
Light fuel oil (KG)	0	1.91e-4
Liquefied natural gas (KG)	0	6.35e-7
Water (KG)	1.25e-1	1.21
Activated carbon (KG)	4.71e-4	1.38e-4
Hydrochloride acid (KG)	5.80e-5	2.42e-4
Boiler feed water chemicals (KG)	1.61e-5	1.46e-4
Condenser treatment chemicals (KG)	1.43e-5	0
Lime (KG)	1.13e-2	1.09e-2
Urea powder (KG)	1.47e-3	
Sodium hydroxide (KG)	3.28e-5	2.33e-4
Cooling water treatment chemicals (KG)	0	1.92e-5
Sulphuric acid (KG)	0	1.89e-4
Sodium hydrochloride (KG)	0	1.20e-4
Outputs		
Emissions	Complete combustion assumed, *i.e.* NO_x, CO_2 *etc.* as main emissions to atmosphere	
Output energies	Recovered electrical and thermal energies exported. Assumption: total 85% energy recovery (heat: electricity energy ratio = 3/2).	

7.3.3.3 LCI of End-of-life

Two incineration (EfW) scenarios for bio-packaging disposal, to represent UK medium- and large-scale incineration plants were modelled, where the operational parameters and infrastructure data were derived from the WRATE model (see Table 7.3). The physical and chemical properties of thermoformed and foamed biocomposites characterized in the lab were used as inventory for incineration scenarios (Table 7.4). The end-of-life scenarios modelled for petrochemical counterparts (incineration, recycling, landfill) are derived from the Ecoinvent database.

7.3.4 Life Cycle Impact Assessment

7.3.4.1 Cradle-to-factory-gate Fossil Energy Profiles for Bio-packaging vs. *Petrochemical Counterparts*

The fossil energy profiles for PHBV production have been presented in our previous study,[40] where corn farming and glucose production dominated as energy sinks. As presented in Figure 7.7, the PHBV polymer accounts for 35–38% of fossil fuel surplus energy demand for bio-packaging up to the factory gate. Another important component, Ecoflex, caused 53–58% of the

Table 7.4 Composition of thermoformed and foamed biocomposites.[a]

	Thermoformed bio-composites	Foamed bio-composite
TS (% total mass)	99.86%	97.99%
VS (% total mass)	80.91%	79.15%
C (% of total mass)	50.34%	50.65%
N (% of total mass)	0.28%	0.28%
S (% of total mass)	0%	0%
H (% of total mass)	5.70%	5.84%
LHV (Mj kg^{-1} fresh sample)	20.05	20.42
HHV (Mj kg^{-1} fresh sample)	20.07	20.43

[a]LHV = low heating value; HHV = high heating value.

impact scores. Over 80% of the environmental burdens of Ecoflex were attributed to the adipic acid and 1,4-butanediol, due to the intensive energy and chemical inputs involved, *e.g.* cyclohexanol input for adipic acid production. Calcium carbonate and other additives contained in biocomposites made only minor contributions to the fossil fuel scores of the biocomposites. The electrical energy input for thermoforming is shown to be greater than the electrical energy requirement for foaming.

In general, foamed bio-packaging delivered a better product system than the thermoformed alternative. The foamed biocomposite showed superior 'cradle-to-factory-gate' fossil fuel scores to PET and PP, due to foaming technology (density reduction), whereas the thermoforming bio-packaging cannot compete with petrochemical counterparts (PET/PP) on fossil fuel surplus energy demand at factory gate. However, these LCA findings were based on the technology developed in 2013 for PHBV and Ecoflex polymers, which are still at the development stage, and are compared with the well-developed petrochemical polymers PET and PP. Moreover, the PET packaging contained 50% recycled content. The Ecoflex modelled here was based on secondary data and the fossil fuels (natural gas and crude oil) were modelled as energy source and feedstock. The potential for the further optimization in the PHBV biopolymer production technologies has been analysed by Guo *et al.*,[40] which could shift the PHBV-Ecoflex biocomposites away from energy sinks.

7.3.4.2 *Cradle-to-grave Fossil Energy Profiles for Current and Prospective Bio-packaging*

The LCIA fossil energy profiles modelled for thermoformed and foamed bio-packaging, over their whole life cycles with energy recovery (EfW) as the end-of-life option, are presented in Figure 7.8. Different incineration processes did not make a significant difference to the 'cradle-to-grave' energy demand profiles of biocomposites. This can be explained by the contributional analysis where the production stage dominated the fossil energy scores of biocomposites over their life cycles. The renewable energy recovered during waste treatment acted as energy sources, bringing environmental savings *via* energy

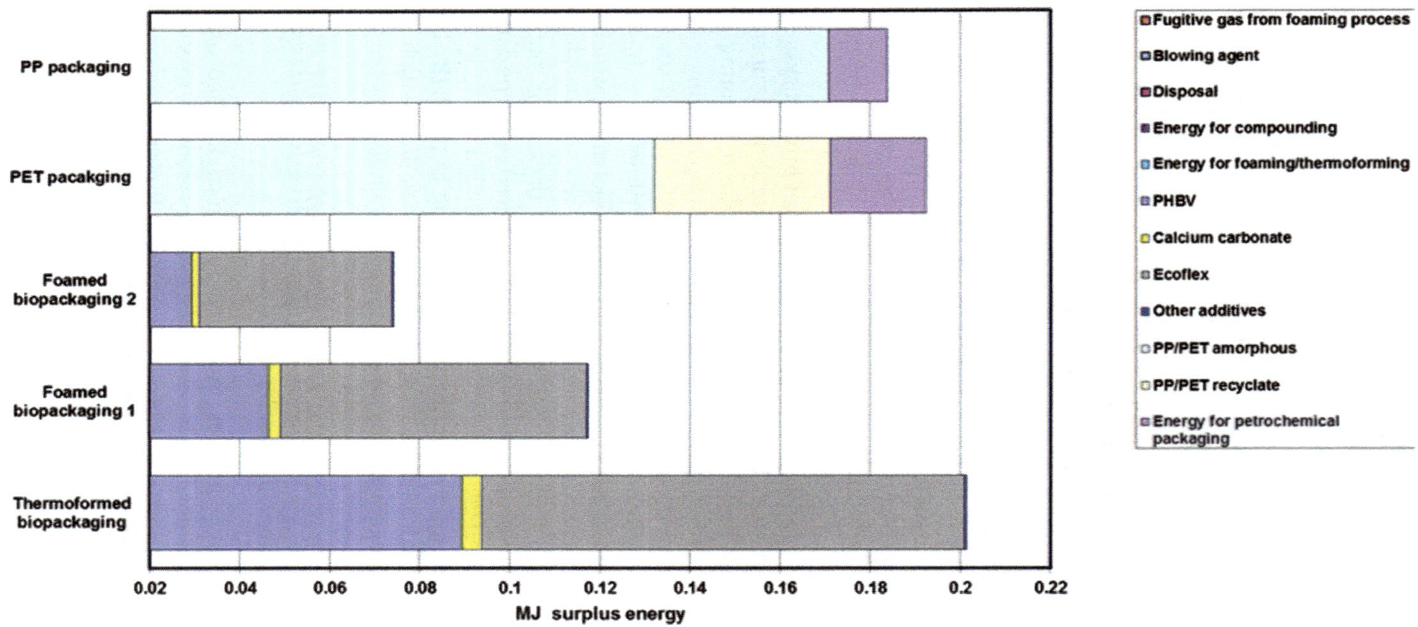

Figure 7.7 'Cradle-to-factory-gate' characterized fossil fuel profiles of bio-packaging *vs.* fossil packaging (unit functional unit, method Eco-indicator).

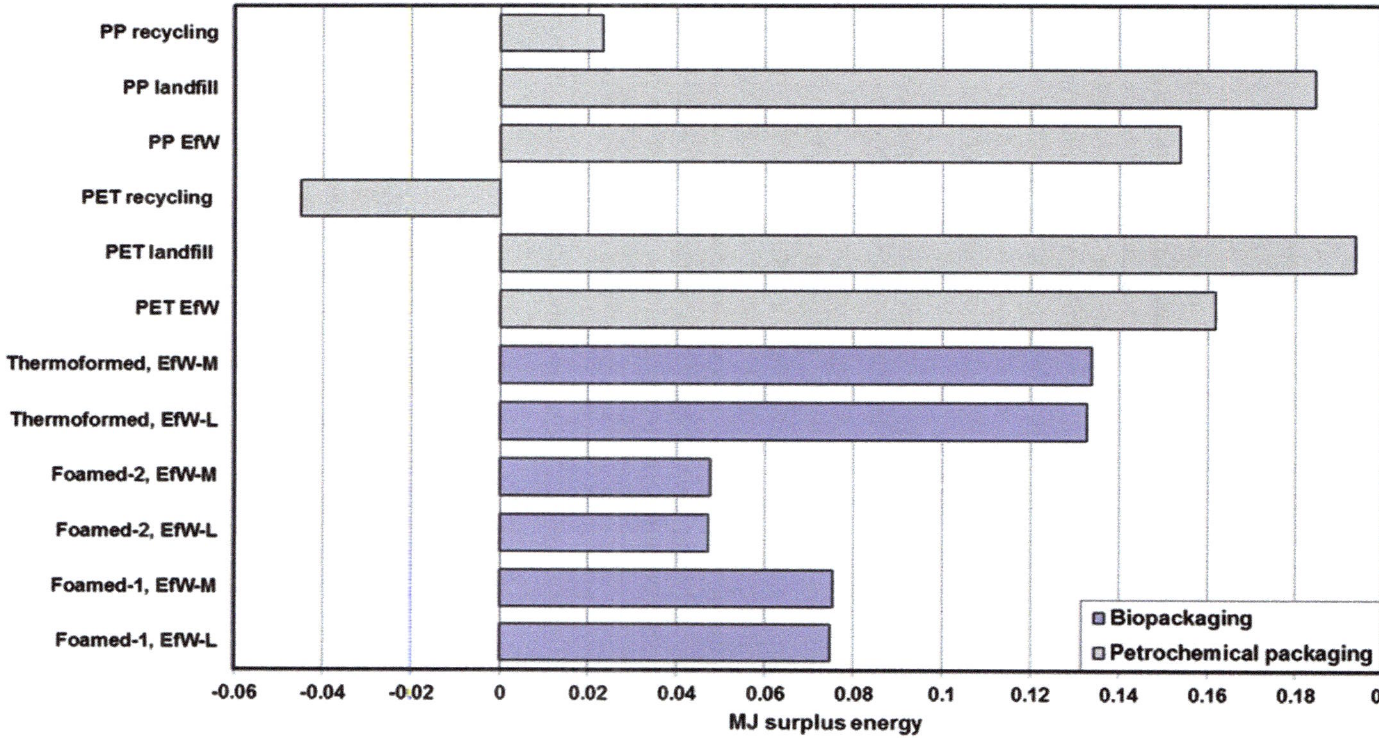

Figure 7.8 'Cradle-to-grave' characterized fossil fuel profiles for bio-packaging *vs.* petrochemical counterparts (unit functional unit, method Eco-indicator, EfW-M = medium scale incineration, EfW-L = large scale incineration).

substitution. These effects were sufficient to offset the burdens caused by the incineration process, leading to bio-packaging end-of-life phases with resource savings on fossil fuel surplus energy demand.

In general, the environmental advantages of foamed bio-packaging over PET and PP in fossil fuel demand were carried through from the packaging production stage to the end-of-life stage (except PP/PET recycling scenarios). The inclusion of end-of-life led to the reversal of the LCA comparisons between the thermoformed bio-packaging and the petrochemical packaging disposed of in landfill or incineration. However, the PET and PP life cycles with recycling end-of-life scenarios delivered superior environmental scores to bio-packaging product systems due to the avoidance of virgin PET/PP manufacture. The LCA results demonstrate that the foaming technology provides a promising option to further improve the environmental profiles of bio-packaging, *e.g.* scenario 2 (foamed bio-packaging with 60% density reduction) incurred up to 75% fossil resource savings compared to a conventional product over the life cycle.

A range of processing and technology optimizations projected for future production of the PHBV biopolymer have been explored by Guo *et al.*[40] Fossil energy is currently utilized for the corn-glucose production and for PHBV plant operation. To investigate the potential beneficial effects on bio-packaging from introducing bioenergy, a future scenario with 100% bioenergy generated from wood biomass combustion in a CHP system was assumed for glucose and PHBV production processes. As shown in Figure 7.9, by introducing bioenergy technologies for PHBV and glucose production, 12–15% resource reduction could be achieved on fossil fuel demands, which led to bio-packaging with significant (40–79%) environmental savings over petrochemical counterparts.

Overall, the LCA findings highlight the importance of waste resource recovery at bio-packaging disposal stage in shifting system from energy sinks to sources, and the available opportunity to develop promising PHBV-based bio-packaging that can have very low requirements for fossil resources.

7.4 Remarks on Sustainable Biopolymers

Regardless of processing technologies, both starch-based bioplastic case studies suggest the environmentally damaging steps include feedstock resources, applications and end-of-life. Despite the environmental advantages of bioplastic at the refinery gate, starch-based bio-packaging in general cannot compete with petrochemicals with closed-loop recycling at end-of-life.

Other sustainable biopolymers offering promising solutions as carbon sinks or energy sources include upcycling CO_2 into polymers and recyclable bioplastics.[5] Taking bio-based HDPE as an example, based on the published studies,[26,58] bio-HDPE can be produced as a pure carbon sink with the cradle-to-factory-gate GHGs ranging from -0.75 kg to -2 kg kg^{-1} PE.[58] Our recent study also shows the GHG saving potential from CO_2 co-polymerization with bio-based limonene oxide to produce poly(limonene carbonate).[59]

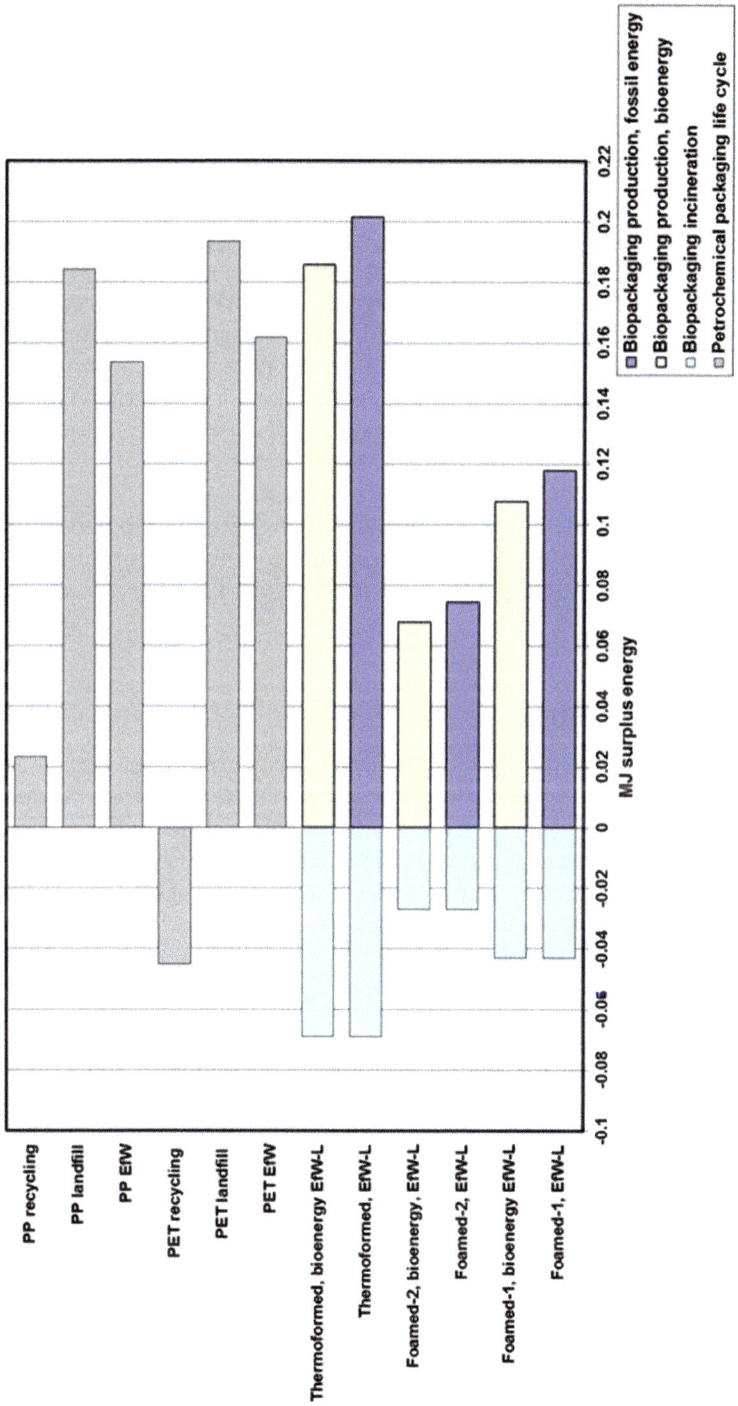

Figure 7.9 Prospective fossil fuel profiles for bio-packaging (unit functional unit, method Eco-indicator, EfW-L = large scale incineration).

Over the life cycle of sustainable biopolymers, a range of carbon sinks and emitters, energy sinks and sources emerge, including:

- Carbon capture and storage or utilization processes integrated with biopolymer manufacturing could potentially lead to a biogenic carbon sink system.
- Sustainable end-of-life management is essential. Waste recovery options including anaerobic digestion or biopolymer recycling can not only generate carbon sink effects by recovering biogenic carbon as biogas or petrochemical material substitution, but also have the potential to shift low-grade waste from energy sinks to energy sources.
- Other carbon emitters evolved during bioplastic life cycles include agricultural (fuel combustion over field operations, carbon released agricultural lands due to microbial soil respiration), platform chemical and biopolymer manufacturing stages, due to chemical/biochemical/thermochemical reactions or operational inputs.

By process integration and optimization, in particular by minimizing energy sinks and carbon emitters, sustainable polymers produced from renewable resources have great potential to contribute to a sustainable circular economy.

Acknowledgements

We would like to acknowledge the research consortium under project 'Farm Produced Ultrathin Lightweight Packaging' (TS/G000859/1) funded by EPSRC. Authors would like to acknowledge EPSRC supporting our research through projects 'Bioenergy value chains: Whole systems analysis and optimisation' (EP/K036734/1) and Resilient and Sustainable Biorenewable Systems Engineering Model (EP/N034740/1).

References

1. S. M. Al-Salem, P. Lettieri and J. Baeyens, *Waste Manage.*, 2009, **29**, 2625–2643.
2. *Plastics - the Facts 2017, An analysis of European plastics production, demand and waste data*, PlasticsEurope, 2008.
3. *Plastics Resin Production and Consumption in 63 Countries Worldwide 2009–2020*, EUROMAP, 2018.
4. J. B. Lamb, B. L. Willis, E. A. Fiorenza, C. S. Couch, R. Howard, D. N. Rader, J. D. True, L. A. Kelly, A. Ahmad, J. Jompa and C. D. Harvell, *Science*, 2018, **359**, 460–462.
5. Y. Zhu, C. Romain and C. K. Williams, *Nature*, 2016, **540**, 354.
6. X. L. Wang, K. K. Yang and Y. Z. Wang, *J. Macromol. Sci., Polym. Rev.*, 2003, **C43**, 385–409.

7. A. A. Shah, F. Hasan, A. Hameed and S. Ahmed, *Biotechnol. Adv.*, 2008, **26**, 246–265.
8. J. H. Song, R. J. Murphy, R. Narayan and G. B. H. Davies, *Philos. Trans. R. Soc., B*, 2009, **364**, 2127–2139.
9. E. Chiellini, A. Corti, S. D'Antone and R. Solaro, *Prog. Polym. Sci.*, 2003, **28**, 963–1014.
10. G. Davis and J. H. Song, *Ind. Crops Prod.*, 2006, **23**, 147–161.
11. J. Guan and M. A. Hanna, *Bioresour. Technol.*, 2006, **97**, 1716–1726.
12. R. L. Shogren, J. W. Lawton and K. F. Tiefenbacher, *Ind. Crops Prod.*, 2002, **16**, 69–79.
13. L. G. Carr, D. F. Parra, P. Ponce, A. B. Lugao and P. M. Buchler, *J. Polym. Environ.*, 2006, **14**, 179–183.
14. M. Avella, J. J. De Vlieger, M. E. Errico, S. Fischer, P. Vacca and M. G. Volpe, *Food Chem.*, 2005, **93**, 467–474.
15. J. Zhou, J. Song and R. Parker, *Carbohydr. Polym.*, 2006, **63**, 466–475.
16. M. Kim and S. J. Lee, *Carbohydr. Polym.*, 2002, **50**, 331–337.
17. S. Kiatkamjornwong, M. Sonsuk, S. Wittayapichet, P. Prasassarakich and P. C. Vejjanukroh, *Polym. Degrad. Stab.*, 1999, **66**, 323–335.
18. M. Flieger, M. Kantorova, A. Prell, T. Rezanka and J. Votruba, *Folia Microbiol.*, 2003, **48**, 27–44.
19. P. Aggarwal and D. Dollimore, *Thermochim. Acta*, 1998, **324**, 1–8.
20. A. Hebeish, M. H. Elrafie, A. Higazy and M. A. Ramadan, *Starch-Starke*, 1992, **44**, 101–107.
21. A. Hebeish, M. H. ElRafie, A. Higazy and M. Ramadan, *Starch-Starke*, 1996, **48**, 175–179.
22. R. Jayasekara, I. Harding, I. Bowater and G. Lonergan, *J. Polym. Environ.*, 2005, **13**, 231–251.
23. L. J. Mao, S. Imam, S. Gordon, P. Cinelli and E. Chiellini, *J. Polym. Environ.*, 2000, **8**, 205–211.
24. C. Bastioli, *Properties and applications of Mater-Bi starch-based materials*, presented at Symposium H on Biodegradable Polymers and Macromolecules at the E-MRS 97/ICAM 97 Conference, 1997, Strasbourg, France, Elsevier Sci. Ltd.
25. H. Y. Ren, F. Q. Qiao, Y. Shi, M. W. Knutzen, Z. Wang, H. Z. Du and H. Y. Zhang, *J. Renewable Sustainable Energy*, 2015, **7**, 5.
26. E4tech and LCAworks, *Summary of the life cycle assessment, land-use change and water footprint reports*, 2013.
27. ISO, *Environmental management—Life cycle assessment—Principles and framework*, 2006, British Standards Institution.
28. UNEP/SETAC Life Cycle Initiative, *Towards a Life Cycle Sustainability Assessment*. 2011, UNEP, https://www.lifecycleinitiative.org/wp-content/uploads/2012/12/2011%20-%20Towards%20LCSA.pdf (accessed Jan 2021).
29. ISO, *ISO 14042 Environmental management—Life cycle assessment—Life cycle impact assessment*, 2000, p. 28.

30. G. Finnveden, M. Z. Hauschild, T. Ekvall, J. Guinee, R. Heijungs, S. Hellweg, A. Koehler, D. Pennington and S. Suh, *J. Environ. Manage.*, 2009, **91**, 1–21.
31. Y. Lin, M. Guo, N. Shah and D. C. Stuckey, *Bioresour. Technol.*, 2016, **215**, 227–238.
32. L. Luo, E. van der Voet, G. Huppes and H. A. U. de Haes, *Int. J. Life Cycle Assess.*, 2009, **14**, 529–539.
33. International Organization for Standardization, *ISO 14044 Environmental management—Life cycle assessment—Requirements and guidelines*, 2006.
34. B. Weidema, *J. Ind. Ecol.*, 2000, **4**, 11–33.
35. International Organisation for Standardization, *ISO 14041 Environmental management—Life cycle assessment—Goal and scope definition and inventory analysis*, 1998, p. 31.
36. British Standards Institution, *PAS 2050: 2011 Specification for the assessment of the life cycle greenhouse gas emissions of goods and services*, B. Standards, 2011.
37. R. J. K. Helmes, A. M. Lopez-Contreras, M. Benoit, H. Abreu, J. Maguire, F. Moejes and S. W. K. van den Burg, *Sustainability*, 2018, **10**, 15.
38. R. Haylock and K. A. Rosentrater, *J. Polym. Environ.*, 2018, **26**, 1484–1503.
39. K. Changwichan, T. Silalertruksa and S. H. Gheewala, *Sustainability*, 2018, **10**, 15.
40. M. Guo, D. C. Stuckey and R. J. Murphy, *Green Chem.*, 2013, **15**, 706–717.
41. K. Dietrich, M. J. Dumont, L. F. Del Rio and V. Orsat, *Sustain. Prod. Consump.*, 2017, **9**, 58–70.
42. F. Razza, F. Degli Innocenti, A. Dobon, C. Aliaga, C. Sanchez and M. Hortal, *J. Cleaner Prod.*, 2015, **102**, 493–500.
43. T. Semba, Y. Sakai, T. Sakanishi and A. Inaba, *J. Cleaner Prod.*, 2018, **195**, 932–938.
44. L. Y. Chen, R. E. O. Pelton and T. M. Smith, *J. Cleaner Prod.*, 2016, **137**, 667–676.
45. Y. Akanuma, S. E. M. Selke and R. Auras, *Int. J. Life Cycle Assess.*, 2014, **19**, 1238–1246.
46. H. L. Bos, K. P. H. Meesters, S. G. Conijn, W. J. Corre and M. K. Patel, *Biofuels, Bioprod. Biorefin.*, 2012, **6**, 146–158.
47. M. Guo, *Life Cycle Assessment (LCA) of Light-weight Eco-composites*, Springer, 2012.
48. M. Guo, C. Li, J. N. B. Bell and R. J. Murphy, *Environ. Sci. Technol.*, 2012, **46**, 320–330.
49. C. A. Finch, *Polyvinyl Alcohol-developments*, John Wiley & Sons Ltd., 1972.
50. M. Lewin and E. M. Pearce, *Handbook of Fiber Chemistry*, Marcel Dekker, Inc., New York, 2nd edn, 1998.
51. F. L. Marten, in *Kirk-Othmer Encyclopedia of Chemical Technology*, John Wiley & Sons, Inc., 2002.

52. H. ter Jung, L. Reihs and G. Roh, Process for the manufacture of polyvinyl alcohol by alcoholysis of a polyvinyl ester, *U. S. Pat.*, US4401790A, Hoechst Aktiengesellschaft, 1983.
53. R. Estermann, B. Schwarzwälder and B. Gysin, *Life cycle assessment of Mater-Bi and EPS loose fills* Study prepared by COMPOSTO for Novamont, Novara, Italy 2000.
54. M. Guo and R. J. Murphy, *Sci. Total Environ.*, 2012, **435–436**, 230–243.
55. M. Guo, D. C. Stuckey and R. J. Murphy, *Bioresour. Technol.*, 2013, **127**, 256–266.
56. M. Goedkoop and R. Spriensma, *The Eco-indicator 99 A damage oriented method for Life Cycle Impact Assessment Methodology Report*, 2001.
57. L. Ingram, D. Mohan, M. Bricka, P. Steele, D. Strobel, D. Crocker, B. Mitchell, J. Mohammad, K. Cantrell and C. U. Pittman, *Energy Fuels*, 2008, **22**, 614–625.
58. I. Tsiropoulos, A. P. C. Faaij, L. Lundquist, U. Schenker, J. F. Briois and M. K. Patel, *J. Cleaner Prod.*, 2015, **90**, 114–127.
59. A. Durkin, I. Taptygin, M. F. M. Gunam Resul, A. Rehman, Q. Y. Kong, A. M. L. Fernández, A. P. Harvey, M. Guo and N. Shah, *ChemistryOpen*, 2019, **8**(6), 668–688.

CHAPTER 8

Case Study: Textiles

Ş. ALTUN KURTOĞLU

Artisan Technical Consultancy, Formerly at Bursa Technical University, Department of Fibre and Polymer Engineering, Bursa, Turkey
Email: slaltun54@gmail.com

8.1 Introduction

Global fibre production reached 103 million tonnes in 2017 (surpassing 100 million tonnes for the first time)[1] and textile production is expected to continue to increase annually by 6.5% in the near future.[2] This rapid growth in production and consumption brings forth various problems. In recent years, the significant environmental impacts of textile production and use, consumer attention, globalization, media accessibility, powerful NGOs and market pressure have made environmental issues vital for the textile industry.[3,4] Textiles are produced from natural raw materials, such as wool and cotton, or can be semi-synthetic, produced from regenerated cellulose, *e.g.* viscose, or entirely synthetic, produced from fossil-oil-derived chemicals, *e.g.* polyesters. The textiles available in the market today display a wide range of properties. Many different processes are used in textile production, such as knitting, weaving, dyeing, finishing *etc.*, and production takes place in many different countries, such as Bangladesh, China and Turkey, to name a few. Textile production is complex and different process steps are practiced, even for the same products, globally and locally.[5] All of these factors complicate the assessment of environmental problems associated with textile production, and the solutions that may be offered.

Textile production, aside from being an energy-intensive process, also requires copious amount of water, from raw material production to the end product. A large quantity and variety of chemicals are also used, from

Life Cycle Assessment: A Metric for the Circular Economy
Edited by Aiduan Borrion, Mairi J. Black and Onesmus Mwabonje

Published by the Royal Society of Chemistry, www.rsc.org

agricultural processes to textile production. Recently, the enormous post-consumer textile wastes that are generated by the industry have been highlighted, and the health and safety of labour forces and working conditions are other significant problems facing the industry. Europe is the largest clothing consumer, together with the USA, and the environmental impacts of textile consumption in Europe amounts to 1–16%.[6]

Two fibres make up the backbone of textile production and trade, *i.e.* cotton and polyester (PET). They account for more than 70% of the consumption of textile raw materials.[1] In terms of the environmental effects of textiles, cotton is the most studied fibre. Large amounts of water, chemical and energy consumption are associated with the field production of cotton, as well as in the manufacturing of cotton textiles and in the use phase (*i.e.* from consumer laundering of clothes). There is also intense interest from consumers regarding labour rights, placing cotton at the focal point of studies, among all the fibres. World cotton production is expected to grow 2.0% annually in the near future[7] and water-related concerns are increasing globally, with cotton consumption accounting for 2.6% of global water use.[8]

The textile industry is a multi-stage sector. Production is carried out in different regions, and consumption takes place all over the world. It has considerable environmental impacts and thereby has a social dimension. All of these features have made all aspects of the textile industry an issue for LCA studies. In particular, cradle-to-grave work has yielded important results, including the neglected use phase effects, and provided important clues for environmentally friendly production. Although not yet widely applied, the estimation of potential environmental impacts in the design phase, with different scenario options, provides significant advantages in this sector where speed and fast-fashion are *de rigueur.*

Many studies have been carried out to assess the environmental effects of textiles and, in addition to academic research, global brands, textile associates, governmental authorities, NGOs and other stakeholders are involved. In this context, for example, chemical producers and machinery manufacturers are also working to reach an environmentally friendly textile industry. However, unless improved resource efficiency during production is reached and the value of products, materials, and resources is maintained in the economy as long as possible – in other words, if the current economic system is still pursued – it has become apparent that efforts to improve the environmental impact of textiles will fail. If fashion brands continue with "business-as-usual" models, apart from continuing to create the environmental impacts of today, a 3% decline in EBIT[†] margins may occur by 2030. Labour, water, and energy costs are projected to be the dominant factors in this decline.[9] The circular economy could be an effective remedy to today's unsustainable situation.[10]

In following section, the environmental effects of textiles are summarized and examples of LCA studies on this subject matter are given.

[†]EBIT – Earnings Before Interest and Taxes.

8.2 Fast Fashion, Textile Wastes and Recycling

Quota restrictions for textiles and clothing were eliminated on 1st January 2005, after the termination of the World Trade Organization (WTO) agreement, which was signed in 1995.[11] Lower-cost producers were then able to access the euro area market, the United States and other member countries of the WTO. As a result, consumer prices have fallen in those countries. Production in different countries, and global distribution have become imperative and the importance of transnational corporations in world textile and apparel trade has increased. Large retailers based predominantly in developed countries have controlled prices, sourcing locations, *etc.*[12] In addition, new, demanding, avid, young consumer profiles exert pressure on clothing companies and fashion retailers. Short production and distribution lead times, compressed turnaround times, frequently changing collections, lower prices and high volume, are all determinates of "Fast Fashion".[13,14] Furthermore, the growing middle-class population across the globe have driven sales of textiles in recent years.[10] These are all factors which have encouraged and increased textile consumption in recent years; *e.g.* in Sweden, the quantity of products put on the market increased by 40% from 2000 to 2009.[15] Global fibre consumption, which was 56.6 million tonnes in 2004, has increased to 103 million tonnes in 2017.[1,16]

Low prices, low quality, short average life of products and new customer profiles[17] have also caused vast amounts of pre- and post-consumer waste, together with the associated ethical problems. The take-make-use-dispose model and planned obsolescence, which are parts of a linear economy, are also encouraging disposing instead of mending and re-using.[18] According to the survey of the Waste and Resources Action Programme (WRAP), the average active use of clothes is 3.3 years.[19] Thus, textile waste comprises 6.1% of total municipal solid waste (MSW) in the USA, an increase from 5.8 million tonnes in 1990 to approximately 16 million tonnes in 2015.[20] The amount of discarded textiles in the European Union (EU) is estimated to be 5.8 million tonnes per year,[21] of which about 15–20% is collected, the remaining part being landfilled or incinerated.[22] Different methods for waste disposal come to the fore in different countries. The most common method in the UK is landfilling, while in Sweden it is incineration.[21] The recycling rate is only 15% for all textiles in the USA.[20] Collected (used) textiles are sold in charity/second hand shops, exported or recycled. Trade of used textiles globally amounted to 4.2 million tonnes in 2014. The USA, Germany and England are the top 3 exporters. Used textiles are exported mainly to Africa, then to EU and to Asia.[23] Some African countries want to ban used textile import since, among other reasons, they want to develop their own textile industries.[24]

In the EU, approximately 6.4 million tonnes of clothing was consumed in 2015 and the carbon footprint of this consumption was roughly 195 million tonnes CO_2e.[25] When primary resources are replaced with waste material in reuse process, almost 8 tonnes of CO_2e can be saved for each tonne of textile waste.[21] The transition from the take-make-use-dispose model (linear economy) to the circular economy can be a radical solution to fast fashion related environmental problems. It is well known that in a circular economy, "the value of products, materials and resources is maintained in the economy for as long as possible, and the generation of waste minimised".[26]

Pre-consumer and post-consumer wastes are generated during the life cycle of a textile. During textile production (pre-consumer), especially in spinning and in the making-up phases, a significant amount of textile wastes are produced. When natural fibres are used, wastes produced in the spinning phase can amount to up to 40–50%, when all phases are taken into consideration. Wastes produced in the making-up phase can be around 20%.[27] Some of the wastes are recycled at factories internally, whilst the rest may be sent to recycling firms. Scraps produced in garment manufacturing, spinning and trimming wastes are valuable sources for textile recycling facilities. Common paths of textile wastes are given in Figure 8.1. Different paths are also available, depending on the applied methods.

The issue of waste does not only cover the disposal of waste generated, but also considers the need to reduce waste generation during the production phase, *i.e.* improving the efficiency of processes in textile production. Measures such as the selection of the right technology and proper raw materials, correct machine settings, timely maintenance of the machines and precautions against power outages can all contribute to the reduction of industrial waste.

Two fundamental methods of regaining textile wastes are reuse and recycling. Reuse is a non-destructive process, where textiles undergo some minor mending and are then reused by other users. This often happens in the context of charity, vintage, second hand shops, export to other countries and renting with new owners. In recycling, textile wastes are transformed to other products that are used in textile or used for other purposes.[22]

Textile recycling methods are generally classified as mechanical, chemical and thermal.[22] The most common recycling method, which is applicable to all raw materials, is the mechanical one. With this method, wastes are transformed to yarn or non-woven material. Building materials such as insulation felt, automotive textiles, carpet underlays, blankets, upholstery *etc.* are application examples of mechanically recycled products.[28] Mechanical recycling is also the most important recycling method for post-consumer wastes. Mechanically recycled products are also used by some global brands.[29–31] Recycling through melting is another commonly used recycling method of textile garments. Fibres such as polyester, polyamide and polypropylene have thermoplastic properties that allow melting as a means of recycling. As in PET bottles, production wastes can be melted in extruders and transformed to yarns to be used in re-textile processes.[28] In chemical recycling methods, the waste polymer can be recovered up to its monomer or raw materials status. However, the commercial application of chemical recycling for textiles is lower than other methods.

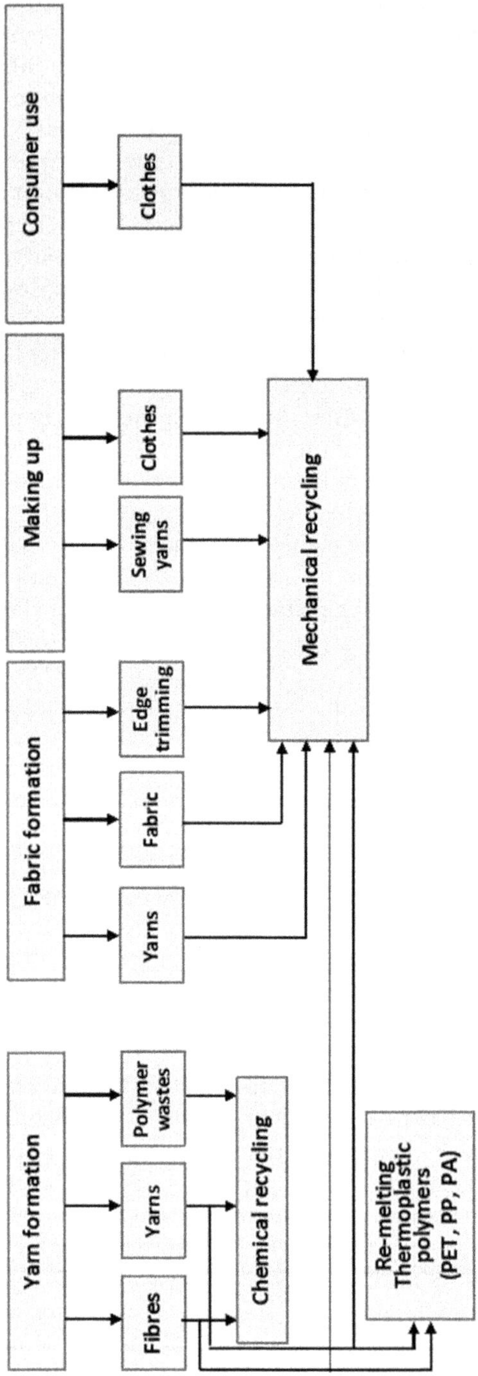

Figure 8.1 Paths of textile waste.

Increasing use of technical textiles in the present and future will cause additional problems related to waste. Electronic textiles (e-textiles) comprise lighting elements, dial pads, mp3 players, solar cells, rechargeable batteries, speaker, embedded computers *etc.* Electronic textiles have relatively short service lives.[32] Elements of electrical and electronic equipment that are used in e-textiles can cause some ecological risks, for example, heavy metals. Improper disposal or recycling of these wastes can lead to the emission of toxic substances that can pollute water, soil and air.[33] Since both fibres and valuable electronic elements are present in e-textiles, it will be difficult to recycle used e-textiles with current recycling methods.[32]

The most important benefits of textile recycling are the possibilities for eliminating all environmental effects of raw material production, reducing manufacturing phase effects, less landfill need, minimizing the damage caused by non-biodegradable wastes and reducing the use of raw materials obtained from non-renewable sources.

8.3 Environmental Impacts of Fibres

8.3.1 Environmental Impacts of Natural Fibres

Natural fibres account for 31% of the total global fibre production, of which cotton makes up 80% of natural fibres produced in 2017.[1] Other plant fibres such as jute, flax, kenaf and sisal are also used in textiles, however, not as significantly as cotton, and some are only used locally. Compared with cotton, the environmental effects of these fibres are less, due to the fact that they need less energy, water, fertilizer and pesticide for growing.[34,35] However, unlike cotton, some plant fibres require a retting process, which results in an additional burden on water resources and higher BOD and COD values.[36,37]

8.3.2 Cotton

Cotton is the most important natural fibre and the second largest fibre used in the textile industry. World cotton production in 2017/18 was about 26.42 million tonnes and the global harvested area was about 33.4 million hectares, almost 2.3% of the world's arable land.[38,39] Cotton is the most studied fibre among all fibres in LCA studies.[6] The most important reason cotton is so studied is the importance of cotton cultivation to agricultural communities (over 100 million farms are involved in cotton production globally). This means that cotton plays a significant role in shaping the economy of many countries, and highlights the importance of the social aspects and environmental impacts of its production.

Cotton lint production has significant environmental impacts due to its heavy use of pesticides, fertilizers, energy, water and land. According to different studies, the production of one tonne of cotton lint emit 1.8–12 tonnes of

CO_2e.[40,41] The location of production and applied methods affect all aspects of environmental impacts. India is the largest cotton producer, together with China. According to a study in India conducted by WWF, crop management practices, field energy and transport of fertilizers are the main sources of GHG emissions in cotton production (Cool Farm Tool version 373 was used to calculate GHG).[42] Another comprehensive LCA study for cotton products was conducted by Cotton Incorporated and PE International, covering 5 regions; namely, China, India, USA, Turkey and Latin America. According to this study, the major contributors to GWP in the fibre production stage are fertilizer production and field emissions. Field emissions are the main contributor to Eutrophication Potential (EP) and Acidification Potential (AP). Nitrogen-based fertilizers are the most important source of these environmental impacts.[41]

The main environmental impacts of cotton and LCA studies based on the literature will be briefly discussed in the following section.

8.3.2.1 Labour Rights

Cotton production can be highly labour-intensive. Hazardous work, use of forced labour and child labour are among the major issues in cotton production.[44,45] Specifically, with the work of NGOs and the influence of consumers on brands and retailers, effective campaigns could be created to stop outsourcing to the countries where forced child and adult labour are used in the cotton fields.[46–48]

8.3.2.2 Water

Water consumption and other detrimental effects related to water consumption constitute the most important environmental problems in cotton production, aside from the use of chemicals. Cotton consumes a significant amount of water in both fibre production and in textile processes. The cotton plant originated from sub-tropical regions, but has now been modified and adapted to grow in many places. Water consumption figures for cotton fibre production vary from country to country. For example, the average virtual water content for cotton lint is 5733 $m^3\,t^{-1}$ in the USA, 18 694 $m^3\,t^{-1}$ in India and 3210 $m^3\,t^{-1}$ in China. The world average is 8242 $m^3\,t^{-1}$.[50] Large amounts of water are also consumed in the process stages of cotton textile production. Production of a dyed cotton T-shirt or a pair of cotton jeans consumes approximately 11 000 $m^3\,t^{-1}$ of virtual water.[8]

Cotton production also affects water quality. Nutrients and pesticides can contaminate groundwater and surface water. Accumulation of chemicals in the soil, extensive irrigation problems such as depletion of surface or groundwater, run-off from fields, falling water tables, salinization and land reclamation have serious effects on the ecosystem. Freshwater eutrophication, wildlife contamination and habitat destruction are examples of these impacts.[49]

8.3.2.3 Chemicals

Pesticides are commonly used in cotton as seed treatments, soil treatments and foliar applications. Although pesticide use in cotton cultivation has been reduced in recent years, its impacts on ecosystems and human health are still very high.

Most pesticides used in cotton are highly hazardous.[49] Chemical pesticides can cause dermatological, gastrointestinal, neurological, carcinogenic, respiratory, reproductive and endocrine effects on humans. High exposure to pesticides can result in hospitalization and death.[49,52] Acute pesticide poisoning is a major problem for field workers and pickers, especially in developing countries.[53,54] Contamination of groundwater, surface and drinking water, poisoning of wildlife, air pollution and biodiversity losses are some of the major environmental impacts of pesticides.[39,49] Some efforts have been made to reduce the effects of heavy chemical consumption in cotton farming, such as use of the Better Cotton Standard System,[55] growing organic cotton and genetically modified (GM) cotton. However, organic cotton shares less than 1% of all cotton production and the use of GM cotton has proven to be highly contentious for various reasons.[57–60]

Fertilizers used in cotton agriculture also contribute to the environmental effects. Fertilizer loss through run-off with excessive irrigation leads to eutrophication and damage to the stability of the ecosystem. One of the major problems is the risk of mixing nitrate into drinking water sources. In addition, nitrite can transform to gaseous nitrogen compounds such as NO and NO_2.[51,61] The denitrification of N-fertilizer causes N_2O emissions, and is one of the main greenhouse gases associated with agricultural production systems. Cleaner production efforts, such as organic cotton production contribute to reduce burden. If organic cotton replaced conventional cotton, AP could be reduced by 70%, EP by 26% and GWP by 46%.[56]

8.3.3 Other Natural Fibres: Wool

When dealing with natural fibres, it is necessary to open a separate parenthesis for wool. Wool, being an animal by-product, has a different production regime to vegetable fibres and has the highest global warming potential (GWP) due to the high methane formation, which results from the livestock production phase.[62] Fertilizer and pesticide use, water consumption (for sheep and pastures), land use, mulesing and animal welfare practices are other factors that cause environmental impacts. The surface of wool is covered with overlapping cuticle cells, which are the source of "felting shrinkage". Several different surface modification methods have been used to prevent undesirable felting shrinkage and the chlorination process, which has been used widely in anti-felting, cause environmental concerns because of adsorbable organic halogens (AOX) content. New environmentally friendly methods are being developed, such as plasma treatment and enzyme treatment.[63,64]

Raw wool also contains impurities such as suint, wool grease, cellulosic materials and dirt. These impurities can be up to 40% by weight and must be washed off before subsequent processes. Cellulosic materials are removed *via* a carbonization process in which the wool is treated with acids, mainly sulphuric acid. High acid (5–7%) and temperature (120–130 °C) requirements result in a non-environmentally friendly process. New efforts, such as enzyme applications, have been practised.[65] Other impurities can be eliminated by the process of scouring, whereby wool is washed in hot water with detergents and then dried. This process contributes high organic content (2–15 L kg^{-1} of greasy wool at about 150–500 g COD kg^{-1} of wool),[37] pesticide residues, soap and detergents to wastewater. Wastewater from wool treatment may also contain chromium (Cr) complexes released by dyes.[66] Woollen products are also very vulnerable to moth attacks and wool moth damage is estimated to cost millions of dollars every year globally. Moth-proofing can be applied during the dyeing or finishing processes. Anti-moth agents are generally based on two main chemical structures: polychlorinated aromatic compounds and synthetic pyrethroid insecticides and the majority of them are toxic.[67] Anti-moth agents based on permethrin and cyfluthrin are highly toxic to aquatic life.[37] These chemicals are included in restricted substance lists (RSL) of some companies and in some eco-labels, such as M&S RSL and EU Ecolabel RSL.[68,69]

8.3.4 Environmental Impacts of Man-made Fibres

Man-made fibre production in 2017 was 72 million tonnes, which corresponded to 69% of global fibre production. Petroleum-based fibres, such as polyesters, polyamides, olefins and acrylics are the largest man-made fibres by volume. Others are mainly cellulose-based *e.g.* viscose, and are sometimes referred to as "regenerated" or "semi-synthetic".

Environmental issues associated with fossil oil extraction and refining (such as gas flares, spills, destruction of wildlife and biodiversity, pollution of air, soil and drinking water, emission of NO_x, CO_2, SO_x and volatile organic compounds) contribute to the environmental burden of synthetic fibres. Other issues associated with the production of man-made fibres are critical to human health and the environment and include: antimony and volatile organic compounds (VOCs) levels from polyester production; N_2O emissions from polyamide production; organotin compounds for elastane production; emission of acrylonitrile to the air and acrylonitrile residue in fibre for acrylonitrile fibre production; and lead-based pigments for polypropylene production.[22]

Man-made fibres undergo a wide variety of processes, for example, spinning, texturizing, twisting, heat setting, interlacing *etc.* Energy consumption is a significant contributor in all of these steps. For example, 104–126 GJ energy is required to spin 1 tonne of PET fibre.[14] Preparation agents may also be applied to man-made fibres during some of these processes, *e.g.* in spinning, interlacing and texturing. Preparation agents are applied to synthetic fibres and also to natural fibres for several reasons, but mainly to reduce static electricity. They generally comprise lubricants, emulsifiers, antistatic agents, wetting agents and some additives such as biocides.[37] These agents are removed later during wet processes, and then contribute to wastewater pollution and air emissions. Lubricants can contain mineral oils, which are highly non-biodegradable and since mineral oils are cheap, they are used widely in the textile industry. The presence of polyaromatic hydrocarbons in mineral oils causes water pollution during wet processes. Ethylene oxide (EO)/propylene oxide (PO) copolymers are also used as lubricants for texturizing processes and EO/PO products are highly non-biodegradable products, which contribute to water pollution. Anionic and non-ionic surfactants are used as emulsifiers and can contain alkyl phenol ethoxylates (APEO).[14] APEOs appear in most of the restricted lists in ecological certificates. Adopting preparation agents that are readily biodegradable or eliminable in wastewater plants, are required for some ecolabels.[69]

Microfibres, which are a subcategory of microplastics, raise pollution concerns particularly in marine habitats, although they are also found in terrestrial environments and in the atmosphere. Microfibres are shed from clothes during regular washing in the use phase and are released into the environment, where they accumulate. Discharged water from washing machines is treated by wastewater treatment systems in many countries and then released into rivers, oceans *etc.* However, wastewater treatment systems cannot remove microfibres completely. Microfibre concentrations can be in the range of 4–160 000 fibres m^{-3} in effluents (after wastewater treatment plant).[70] Marine species, which are in the human food chain, can consume microfibres directly and indirectly.[71,72] Microfibres can also transport hazardous process chemicals, nanomaterials, and other process substances to the aquatic environment.[70,72] Sewage sludge, which also contains microfibres, may be applied to land as fertilizer and it has been found that microfibres can still be detected in soils some 15 years after sludge application.[73]

8.4 Environmental Impacts of Textile Processes

8.4.1 Environmental Impacts of Yarn-spinning Processes

Ginning is one of the main pre-processes of cotton yarn production. It is the process step after harvest of cotton, where seeds and fibres are separated mechanically from the cotton boll. Besides electricity consumption, the emission of large amounts of particulate matter (PM) is also significant in this process. About 37% of the total PM emitted during the ginning process is less than or equal to 10 microns.[75] Overexposure to dust from cotton can cause byssinosis, which is an obstructive lung disease.[76]

Yarn is a semi-finished product between raw material and fabric and its production is energy intensive. In a study, real specific energy consumption

(SEC) values for yarn spinning (cotton) were found to be 3.24–3.47 kWh kg^{-1} for a yarn spinning plant.[74] When all production steps (spinning, knitting, wet processes, make-up) are considered, yarn spinning shares almost 55% of the production phase's consumption of primary energy which originates from electricity.[43] Preparation agents are also used in yarn spinning and contribute to water pollution, as mentioned in Section 8.3.4.

8.4.2 Environmental Impacts of Weaving

Weaving is a method of forming a fabric and it is an energy-intensive process that mainly uses electricity. Electricity consumption depends on weaving machine type, fabric and yarn properties, working conditions, *etc.* The weaving process simply comprises two basic steps; weaving preparation, which includes warping and sizing, and weaving. Electricity consumption of the preparation phase is quite low, with an energy consumption of 0.03[77]–0.0073[74] kWh kg^{-1} compared with the weaving phase, which is 1.58–5.8 kWh kg^{-1}.[74] Apart from electricity consumption, the sizing process is also a main contributor to the environmental burden. Sizing is a process to improve the weavability of warp yarn. The yarns are treated with sizing agents in aqueous medium at elevated temperatures and then dried. Common sizing agents are synthetic or natural. Starch and starch derivatives account for almost 70% of total consumption.[78] Some auxiliaries such as wetting agents, viscosity regulators, softeners, lubricants and anti-static agents are also used in the sizing process. After the weaving process, sizing agents are removed, a process which is called desizing. Desizing methods mainly depend on the sizing agent used. Oxidative desizing, enzymatic desizing and hot water washing with suitable auxiliaries are widely used desizing methods.[37] Starch and starch derivatives are generally biodegradable, and the desizing process leads to high COD values in effluents.[66] The COD of starch-based sizing agents is *ca.* 900–1000 mg O_2 g^{-1}. PVA is also biodegradable and its COD value is *ca.* 1700 mg O_2 g^{-1}. Poly(methyl acrylates) are not biodegradable and their COD value is *ca.* 1300–1700 mg O_2 g^{-1}.[78] The desizing process also leads to the highest BOD content, which is about 45% of the total BOD load for a textile mill effluent. The main contribution to textile wastewaters in terms of COD, BOD and Total Organic Carbon (TOC) values come from the desizing process.[79–81] Many efforts have been made to reduce the environmental impacts of the desizing process, such as recovery of the sizing agents, using different sizing agents such as chitosan and making the process in one-step.[79,82]

8.4.3 Environmental Impacts of Knitting

Knitting, together with weaving, is one of the main methods of making a fabric. Significant amounts of oil are used in knitting machines, specifically for the needles, to reduce friction and prevent wearing, to avoid corrosion and to facilitate working at high speeds. The amount of oil used ranges between 4% and 10% of the weight of the fabric.[37] The oil in the fabric is washed out during the following wet processes and, thereby, the wastewater load of BOD increases. Mineral oils are widely used in knitting oils and are non-biodegradable, and can cause emissions at elevated temperatures.[37] The knitting process is also an energy-intensive process.

8.4.4 Environmental Impacts of Nonwoven Production

Nonwovens are special textile products that have a wide usage area, especially in the field of technical textiles. In nonwoven production, the fibres do not need to be converted to yarn. Technical textiles, such as absorbent hygiene products, geotextiles and medical textiles are the main application areas of nonwovens. Besides electricity consumption, bonding agents and off-gases emitted during the thermal and chemical bonding steps are the main environmental concerns.[37]

8.4.5 Environmental Impacts of Wet Processes

Wet processes are among the basic processes in order to dye textile materials and provide different properties. The main wet processes are:

- Pre-treatments (scouring, bleaching, desizing, merserization);
- Dyeing/printing;
- Drying/heat setting; and
- Finishing.

Different methods are used in wet processes. The method chosen mostly depends on fibre type, required properties of the end product and site-specific manufacturing practices.[83] Wet processes can be applied to fibres, yarns and fabrics and can be implemented as continuous or as batch processes.

An enormous variety of chemical agents are used in textile wet processes, for different aims. Examples of chemical agents are wetting agents, bleaching agents, detergents, flame retardants, pigments, carriers, resins, lubricants, emulsifiers, salts *etc.* In addition, there may be impurities on fibres such as pesticides, waxes and organic matter. There may also be process chemicals on the yarn and fabric such as preparation agents, texturizing oil, sizing agents and mineral oils. In pre-treatment processes, those natural impurities and chemicals that can interfere with dyes and other chemicals have to be removed. In a wet process plant, together with process chemicals such as dyes, the pre-treatment loads that come from fibre, yarn and fabric also contribute to the effluent. If sizing processes are applied, most of the total organic load value of the effluent can come from the desizing process.[84] If sodium hypochlorite bleaching is applied, halogenated

Table 8.1 Common restricted chemicals used in textile processes and some potential effects of them.

Chemicals	A few examples of the application areas	Some effects
Perfluorinated Compounds (PFOA, PFOS, and others)	Soil and water repellent agent	PFOS are persistent, bioaccumulative and toxic to mammalian species. PFOS exposure can cause bladder cancer.[127]
Phthalates	Plasticizers, softeners, solvents, dispersing agent	Some phthalates can be very toxic to aquatic life, may damage human fertility and can cause cancer
Isocyanates	Polyurethane production	Isocyanates are classified as potential human carcinogens.[128]
Organotin compounds	Stabilisers, catalyst, pesticides	Organotin compounds are among the most hazardous organic toxicants for aquatic systems and also accumulated in aquatic organisms.[129]
Azo dyes	Textile coloration	Majority of all dyes used in industry are azo derivatives.[130] Certain azo dyes in reductive conditions can release aromatic amines which can be carcinogenic or potentially carcinogenic
Metal Complex Dyes	Mainly coloration of wool and nylon	Chromium and cobalt content
Reactive dyes	Coloration textile products	Dye fixation is in general below 70%. Unfixed dyes released into dye baths. Reactive dyes have poor degradability and it is difficult to eliminate them from wastewaters.[131] High load of soluble salt (NaCl, Na_2SO_2) and coloured wastewaters are the main concerns.[66]
Disperse Dyes (Some of them)	Synthetic fibre coloration	Restricted disperse dyes are suspected of causing allergic reactions.[91]
Formaldehyde	Anti-creasing and anti-shrinking process, polymeric resins	Formaldehyde is a human carcinogen.[132]
Alkylphenols (APs), Alkylphenol Ethoxylates (APEOs) and all isomers	Industrial laundry detergents, scouring agents, softeners, spinning oils, wetting agents, emulsifying/ dispersing agents for dyes and prints	Nonylphenol ethoxylate (NPE) is a member of APEOs degrades into Nonylphenol (NP) in the environment or wastewater plants. NP is bioaccumulative, extremely toxic to aquatic organisms and persistent in the aquatic environment. NPEs are used in a wide variety of textile applications as wetting agents, scouring agents, detergents, spinning oils.[90,133]

Table 8.1 (*Continued*)

Chemicals	A few examples of the application areas	Some effects
Heavy Metals; cadmium (Cd), chromium (Cr), cobalt (Co), nickel (Ni), arsenic (As), lead (Pb) and silver (Ag)	Dyes, catalyst, preservatives, pesticides, pigments	Heavy metals have toxic effects on plants, animals and other living organisms. Some heavy metals are bioaccumulative.[134]
Organohalogens	Flame retardants, pesticides	Some organohalogens have neurotoxic potentials, extremely persistent and bioaccumulative.[135]
Chlorophenols (CPs)	Anti-mold agents, pesticides, dyes, print pastes	The chlorophenols are toxic to aquatic organisms, known or suspected as endocrine disrupting compounds, possibly carcinogenic.[136]
Polycyclic Aromatic Hydrocarbons (PAHs)	Printing pastes, softener, thermal decomposition product of recycling process	Some PAHs are highly toxic, carcinogenic, mutagenic and teratogenic.[137]
Volatile Organic Compounds (VOCs)	Solvents, polyurethane (PU) coatings, adhesives	Some VOCs are suspected or known to cause cancer in humans.[138]

organic compounds (AOX) can be formed[37,85] and usage of metal complex dyes can cause transferring of metals and AOX to wastewater.[86] Cotton textiles are treated with high alkaline solutions to increase lustre, dye uptake, strength and dimensional stability (typically, 150–300 g L^{-1} NaOH is used in the process).[66] This process is called mercerization and it significantly increases the alkali load in wet processing effluents. Most synthetic dyes are toxic and non-degradable. Effluents containing synthetic dyes reduce light penetration in water, increase BOD value of the contaminated water, and thus adversely affect aquatic life.[87] Typical dye concentrations values of 0.01–0.25 g L^{-1} in textile effluents[88] constitute a non-negligible sum. Taking the total production into consideration, however, colour is also a problem for esthetical reasons in nature and aquatic life sustenance.

In summary, wet processes consume copious amount of water and have the potential to discharge heavily polluted wastewater. To make 1 kg of garment, between 1.5–6.9 kg of chemicals are used.[89] Together with consumer pressure created by the negative effects of textile chemicals on the environment and on human health, the efforts of NGOs and state mechanisms, many companies have started to publish restricted/prohibited substance lists, which are constantly updated. Table 8.1 lists the chemicals that are commonly found in restricted substance lists and their possible effects.[68,69,90–92] Registration, Evaluation, Authorisation and

Restriction of Chemicals (REACH) Regulation and Substances of Very High Concern (SVHC) candidate lists are also taken into consideration in textile processes.[93] The Swedish Chemical Agency has prepared a report on the hazardous chemicals that are possibly present in textiles on the Swedish market. Many chemical substances (165 of 1900), in their non-exhaustive list are, to a certain degree, carcinogenic, mutagenic and toxic to reproduction (CMR), aquatic chronic, respiratory and/or skin sensitising and presented in REACH candidate list.[94]

Textile wastewaters can also be very warm and have high pH values. Wet processes require thermal energy to heat baths, for drying and curing processes and to produce steam for these steps. Since fossil energy sources emit SO_x, CO_2 and NO_x, the use of them also contributes to environmental impacts.

8.5 Eco-design

The design stage in textile products is a very important opportunity to significantly reduce the environmental impact of all stages of the product, including the use and disposal phases. Selection of materials, the choice of production processes, use phase requirements, and end-of-life treatments can be determined in the design phase. Thus, this phase is decisive in sustainable product development.[95]

One of the most important issues in efficient recovery of used products/wastes is also the design phase. Recycling complications of composite textiles and the fact that industrial textiles can only be reused in very limited areas make the recovery of such wastes difficult. An LCA study, which was focused on electronic textiles, showed that the implementation of eco-design can improve environmental impacts by at least 25% (expressed in eco-cost).[96]

8.6 Environmental Certification

In recent years, many efforts have been made to decrease the environmental impacts of the textile and garment industries.[97] Environmental performance systems have come to the forefront as these issues have become more important. Consumers, companies and state administrations increasingly want to see certificates of environmental assessments of products and production systems. In addition to labels that are specific to countries and regions, such as Nordic Swan, Blue Angel, EU Ecolabel, textile-specific new ecolabels have also been created, such as Global Organic Textile Standard (GOTS), Blue Sign and STeP. Furthermore, among others, the Responsible Wool Standard and Responsible Down Standard have been developed to endorse animal welfare; the Environmental Product Declaration (EPD) to disclose life cycle environmental impacts of products; Recycled Claim Standard to document the content of recycled material in a product; Organic Content Standards to declare organic material used in products; Better Cotton Initiative (BCI) to promote better standards in cotton farming and practices and the ISO 14000 family, especially ISO 14001 (2015), have been developed for guidance on the environmental management of companies and organizations.

In addition, brands and coalitions publish environmental criteria, expecting their suppliers to comply with them. Three examples are the Higg Index, published by the Sustainable Apparel Coalition (SAC), Nike Material Sustainability Index (MSI), and the Zero Discharge of Hazardous Chemicals (ZDHC) programme.[98–100]

8.7 Life Cycle Assessment in Textile Production

Life Cycle Assessment is a methodology that identifies potential environmental impacts of products and services, and it is well documented in ISO 14040 series, in many other sources and in Chapter 3 of this book. Once the goal and scope of an LCA are well defined, it is a very useful multi-purpose tool. As LCA is a practical toolset, it can also be considered as a conceptual framework.[101]

When applied correctly, LCA has the capability to identify environmental impacts in production systems, in global supply chains and in global brands, as well as allowing comparisons to be made between products in the supply chain, between different raw materials and sources, and company practices, with the potential for creating a systematic advantage within a company.

LCA can be applied in different life phases of products, such as cradle-to-grave, *e.g.* from cotton farming to cotton product disposal; cradle-to-gate, such as from cotton farming to textile factory gate; gate-to-gate, such as from yarn input to finished fabric delivery. Cradle-to-grave is the most frequent implementation. However, gate-to-gate is very useful to disclose hot spots in production systems.

Results obtained from LCA studies are used as indicators of environmental sensitivity in textiles. Several companies have started to publish LCA results of all models in their product ranges. LCA in textiles is also the base of many labels used, such as GOTS, EU Ecolabel and, as it is an holistic approach, its importance in labelling is increasing.

In summary, LCA has four phases (ISO 14044):[102]

1. The goal and scope definition phase
2. The inventory analysis phase
3. The impact assessment phase
4. The interpretation phase.

According to ISO 14044, "the goal and scope of an LCA shall be clearly defined and shall be consistent with the intended application".[102] A goal that cannot be clearly defined causes system boundaries not to be determined well and may lead to lower-quality data or repetition of work already done. Before determining the scope, it is essential to carry out a pre-study on the processes used and to take the audience of the intended study into consideration. The functional unit is determined according to the characteristics of the process to be studied. For the textile example, if the fibre production process is the focus, then kg or tonnes are appropriate units, whereas, for fabric production, meter (length) may be a more suitable unit, but should also consider the fabric weight if comparisons are to be made with similar products. In addition, the functional unit, depending on the intention, can be used for physical products but also for functionality, *e.g.* life time of a garment.[103]

The LCA inventory phase is a data- and labour-intensive, and error-prone phase. LCA studies of production phases of natural-based textile raw materials (such as cotton, linen and wool) and synthetic-based raw materials (such as polyester and polypropylene) have been studied by chemists and agricultural scientists, as well as textile researchers, and there are a lot of data to consider at the inventory phase for input into LCA studies of textile products. However, the manufacturing stages of textiles are almost unique to every product and company and consequently, each company has its own "style". Depending on the production practices companies use, the inputs, outputs and results may vary even for the same product.

In the data collection process, it is an absolute necessity for data collector(s) to collaborate with the related personnel of all departments and with the technical staff closely. In the textile manufacturing phase, there may be hundreds of inputs (*e.g.* chemicals used in developing the functionality of a textile, colour, durability *etc.*), a large number of processes and numerous machines to be taken into consideration. A large number of emissions may occur at different stages, which constitutes one portion of outputs. It is very simple for inexperienced practitioners to overlook emissions produced during particular parts of textile processing, *e.g.* drying stages between textile treatments (stenters), as well as solvents/chemicals used in mg levels at the quality control stage of the final product. Engineering calculations, such as the modelling of steam boilers used in wet processes, calculation of electricity consumption are also important in the textile production phase.

Use of annual average data for a company/factory might give better results than data collected during the process. In this case, using the last three years of data obtained in factories carrying out routine production would reduce the probability of error.

In the life cycle impact assessment (LCIA) phase, the selection of impact categories, category indicators, characterization models, classification and characterization are mandatory elements (ISO 14044). Impact categories should cover all work-related environmental impacts, but should not contain too many categories for practicality. The purpose and scope of the study are also important. A research study carried out for cotton fibre and a company's self-declaration of that may not contain the same categories. When needed, specific approaches (gravity analysis, sensitivity analysis and uncertainty analysis) can be applied to fulfil the goal and scope of the study.[104]

The results of LCA are evaluated in the interpretation phase. According to the aim of the study, recommendations can be made according to the outcomes. Since LCA is an iterative process, the system can be revised if the conclusions are not consistent with the requirements. When this occurs, the scope of the study, system boundaries and collected data should be revised.

In LCA, different scenarios can also be studied: *e.g.* the environmental impacts of producing the same trousers with recycled, organic or virgin materials can be evaluated; or the effects of different raw material ratios can be compare; and the effects of different dyes or chemicals can be calculated, to highlight a few examples. In the following sections, examples of various LCA studies are given. These examples should be read considering that LCA studies may vary depending on system boundaries and application conditions and different results can be obtained for the same study under different conditions.

8.7.1 LCA Studies in Fibre Production

The environmental impacts of a textile product can only be determined by taking the whole life cycle into account. Focusing only on the production phase can be misleading and thus the real environmental impact can be overlooked. Figure 8.2 depicts the life cycle of a typical textile production.

When the history of LCA studies on fibres are scrutinized, a study from 1978 draws particular attention. With the effect of the oil crisis in the 1970s, cotton and polyester were compared, to promote the use of natural raw materials instead of petroleum-based raw materials. In this study, the environmental impacts of the textile in whole life cycle were evaluated in a similar way as present-day practices.[105,107] Franklin Associates conducted also a life cycle inventory (LCI) study for a polyester blouse; emissions, solid wastes, energy consumption and use phase were evaluated in this pioneer study.[108] Taking a different approach, GWP, acidification potential, eutrophication and human health impacts were calculated in Kalliala and Nousiainen's study, in which the materials were also cotton/polyester and cotton.[109]

The selection of raw materials, which is one of the most important parts of the design phase, plays an important role in determining the environmental impacts of the product. According to an LCA study, for a 100% cotton T-shirt, 98% of the total ecotoxicity and 90% of the total persistence toxicity come from cotton cultivation. Furthermore, if conventional cotton is replaced with organic cotton, persistent toxicity is reduced by 85%, while ecotoxicity is reduced by 95%.[43] When cotton and viscose blouses are compared according to their energy profile, it was shown that the use stage is most significant for cotton and the raw material production is most significant for viscose.[110] Energy use and CO_2 emissions of a cotton T-shirt occur mostly in the consuming country, whereas that of a polyester jacket occurs mostly in the producing country since it is washed less regularly in use.[111]

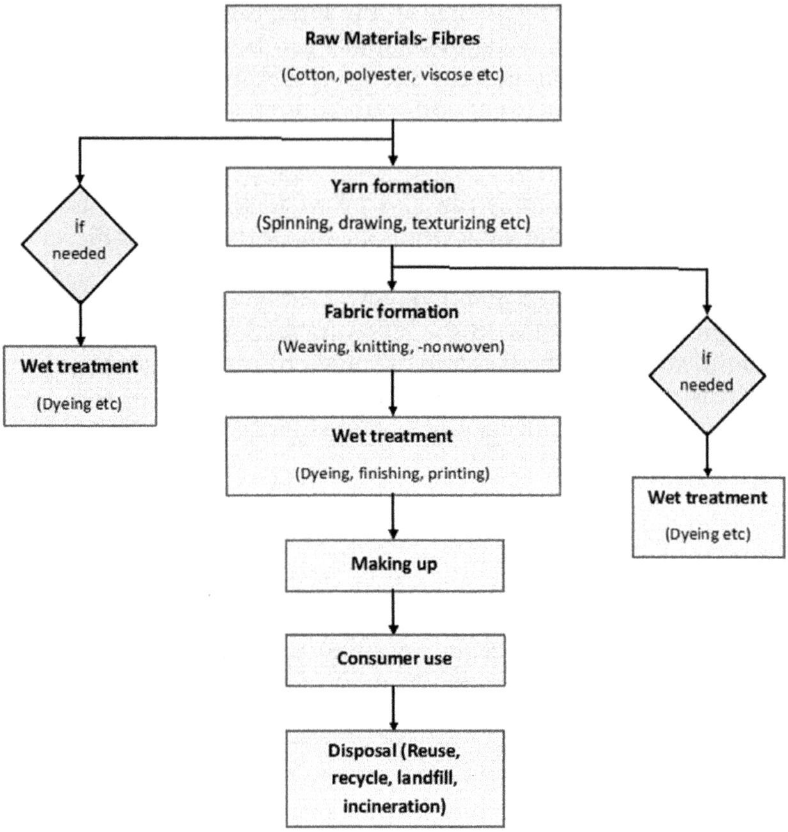

Figure 8.2 A typical textile product life cycle.

GWP of wool is significantly higher than that of cotton and polyester.[112] Wool also has the highest AP because of the ammonia that is released during sheep farming.[113]

Agricultural practices and technologies, such as continuous or discontinuous production, energy efficient technology use, *etc.* also affect environmental impacts of fibres. According to a study, viscose fibres derived from crop-based cellulose and manufactured in Asia consume 3.2 times more non-renewable energy than the fibres produced in Austria. Process energy was supplied from biomass and municipal solid waste incineration in Austria, whereas coal, gas and oil were used in Asia.[114] The total energy required for the cultivation of 1 tonne of conventional cotton is almost double in the USA than in India (Punjab). However, CO_2 emissions from energy consumption are lower in the USA than India because of the proportion and type of fuel used.[35]

Petroleum-based fibres are non-biodegradable and consume huge amounts of energy during fibre production.[97] Biodegradable and plant-based fibres, such as polylactic acid (PLA), have been suggested as an alternative to synthetic fibres, however, PLA has higher values of eutrophication and ozone depletion than PET because of fertilizer, pesticide and land use in plant production.[115]

8.7.2 LCA Studies in Textile Processes

Electricity is the major energy source in yarn spinning.[74] In a cotton T-shirt's life cycle study, it was calculated that 60% of energy-related greenhouse gas effect came from electricity consumption.[43] Spinning was found to be the main contributor for GWP, AP, Photochemical Ozone Creation Potential, and Primary Energy Demand due to excessive electricity consumption for cotton garment life cycle.[41] However, yarn count should be taken into consideration: the energy consumption per kilogram of yarn is inversely proportional to the yarn count in decitex.[‡] As the amount of energy required in weaving and knitting processes with coarse yarns is lower than that of fine yarns, environmental impact of coarse yarn will therefore be lower.[116] Two different spinning processes – carded and combed ring spinning – were compared in a LCA study. While there was no difference in other categories, water consumption, water eutrophication and air acidification were slightly higher in combed spinning, since more cotton is required to produce the same amount of yarn in this process, as the result of high waste formation during manufacture.[117]

China is the largest manufacturer of textiles. An LCA study conducted in China showed that, for a polyester/cotton production system, the printing and dyeing processes contribute to over 80% of all impact categories. Desizing was the major contributor to dyeing and printing. It should be taken into consideration that the amount of coal used by the textile industry of China is 2–3 times that of developed countries, and the total wastewater emissions are 1.2–1.8 times that of developed countries.[118]

Absorbent hygiene products have considerable market share in nonwoven products. The environmental effects of these products were studied to compare disposable and reusable diapers,[119] cradle-to-grave analysis of a baby's diaper and environmental improvement analysis of hygiene products.[120,121]

According to a study on the Swedish apparel sector, using renewable energy to replace fossil energy in textile manufacturing processes is the most effective effort to reduce climate impacts.[122]

8.7.3 LCA Studies in Use Phase, Wastes and Recycling – Some Examples

According to some studies, in a textile product life cycle, the use phase is the largest contributor to environmental impacts.[43,111] For example, in an LCA

[‡]Decitex (dtex): The weight in grams of 10 000 m of yarn.

study for an organic cotton T-shirt, the use phase constitutes 48% of the total carbon footprint (CFP). Tumble drying (45%) and ironing (31%) share the largest ratios. For applications that do not require frequent washing, *e.g.* a polyester jacket, the environmental burden of the production phase can be higher than the consumer phase.[111] In a study conducted in China, the use phase contribution was found to contribute 4.21% to the CFP, through the life cycle of a pure cotton T-shirt.[123] In another LCA study, water consumption was found to be 35% and 26% of the total water used during the life cycle of a woven and knitted garment, respectively. It was assumed that woven fabric was washed more often than knit fabric.[41] Fibre type, consumer behaviour (frequently/rarely washing, cold/warm washing, drying type, *etc.*), and equipment efficiency are primary causes of use phase impacts. In a cotton T-shirt's use phase, if tumble drying is not used, primary energy (and related resource consumption) reduce almost 70%.[43] According to an LCA study on cotton woven trousers, 37% of EP, 63% of GWP and 37% of AP of the product came from use phase.[41]

Reuse is the method that alleviates the environmental impacts of textile products the most. Research has demonstrated that the amount of energy required for the reuse of 1 tonne of polyester garments was 1.8% that of the use of virgin polyester, and 2.6% that of the virgin cotton.[124] In another study, it was found that cloth reuse can reduce 14% of GWP for a cotton T-shirt and 45% of human toxicity for a pair of polyester/cotton trousers.[125]

The raw material of a cotton garment could be harvested in Benin, manufactured in China and sold in Europe. Transportation processes can occur several times throughout the production process. Of Europe's woven fabric imports, 44% are made in China and 15% in South Asia. The average distance from Europe to China is 19 601 km (by sea) and 9262 km (by air) and to South Asia, 12 354 km (by sea) and 7482 km (by air). Transportation constitutes a significant contribution to the overall environmental burden of the industry.

According to the ReCiPe[§] midpoint indicators, the production phase contributes 52%, the use phase 45%, and transport 5% of the impacts of textile consumption in the EU 27, in terms of climate change [the end-of-life phase can contribute a negative effect (−2%) if energy is recovered through incineration of the textile as post-consumer waste management].[6] Increasing the collection of used clothing for reuse and recycling can decrease climate change up to 8%, whilst replacement of traditional cotton by GM cotton can decrease terrestrial ecotoxicity up to 45%.

Several LCA studies have been conducted to analyse the disposal phase of textiles. Since waste characteristics, energy sources and disposal options vary depend on region/country, waste management systems should be determined locally.[126]

[§]ReCiPe is a method for the life cycle impact assessment (LCIA). Detailed information can be obtained from ReCiPe 2016 v1.1, RIVM Report 2016-0104a and from relevant literature.

8.7.4 Case Study – Polyethyelene Terephthalate (PET)

Polyethylene terephthalate (PET) textured yarn has also been selected as a case study for this chapter. The case study is based on original LCA studies by Altun *et al.*, 2018,[106] for direct and indirect extrusion spinning of PET and derived from the project cited in Table 8.2 in Bursa, Turkey. PET is the most widely produced and consumed textile fibre in the world and textured yarn is a preferred option in textile products, because of the unique properties it imparts. Texturing is a process that gives yarn crimp, loops, coils or crinkles. Synthetic yarns add properties such as softness, bulkiness and insulation.

The goal and scope of this LCA study was to evaluate the environmental impacts of textured PET filament yarn production. The functional unit is defined as one kg of textured yarn. The system boundary of the LCA is gate-to-gate (Figure 8.3) and the study includes all of the steps from the input of raw materials to the manufacturing plant, to the delivery of product (textured yarn). The raw materials of PET in this study are purified terephthalic acid (PTA) and ethylene glycol (EG), and PET was obtained from polycondensation polymerization of PTA and EG. Although PET polymer production is most often taken as a single process, production comprises some auxiliary processes that should also be taken into account. These auxiliary processes are: the preparation of catalyst; preparation of titanium dioxide (TiO_2) and preparation of the polymer mixture. The polymer mixture is fed into reactors for the polymerization stage and all of these steps were included in the study.

After polymerization, the polymer can be fed to the yarn production step as a continuous process, or in a discontinuous process, where the polymer is first directed to granule production, following which the granules are re-melted and converted into yarn. In this study, the production was continuous and polymer was directly pumped to nozzles to obtain partially oriented yarn (POY), then a "false twist texturing" process was applied to the POY. In false twist texturing, the filaments are heated and twisted simultaneously, then cooled and untwisted. As a scenario, discontinuous yarn production was also examined and a comparison of environmental impacts of continuous and discontinuous process was given.[106]

8.7.4.1 *Life Cycle Inventory Analysis*

Site-specific data were collected from the production site in Bursa, Turkey. Secondary data were obtained from the Ecoinvent database. The inputs and outputs of all packaging materials used throughout the process were evaluated separately, but included in the study. Electrical data were

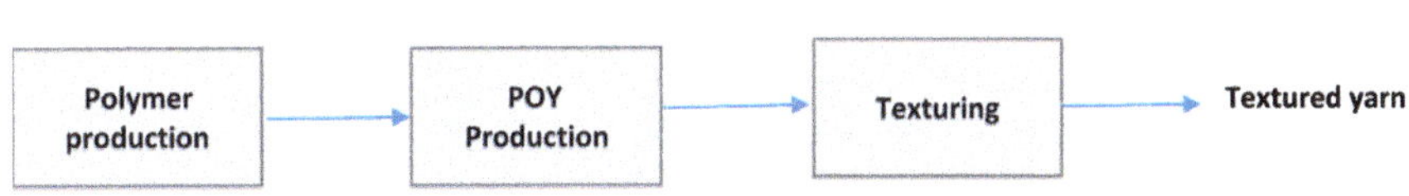

Figure 8.3 The main production processes in PET textured yarn (gate-to-gate).

calculated from the company's own energy data for the production line. Data input to the inventory was made according to the assumption that: contaminated wastes are taken by a licensed company for treatment; packaging wastes are sent to a recycling facility; and 95% of the "general wastes" are sent to the urban waste site. Maintenance was also included in inventory and annual average data were used. Due to confidentiality issues, specific details/data are not given for monoethylene glycol (MEG) management, however, waste steams and waste water that were generated on this production line were recycled and reused in the same process. The recycling processes were modelled and the related data were used in the inventory analysis phase.

The inputs and outputs of the main processes are given in Figures 8.4 and 8.5.

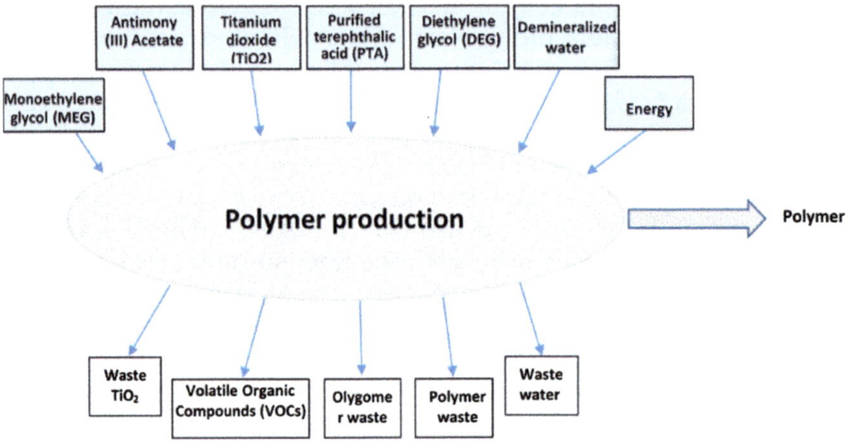

Figure 8.4 Input and output of polymer production.

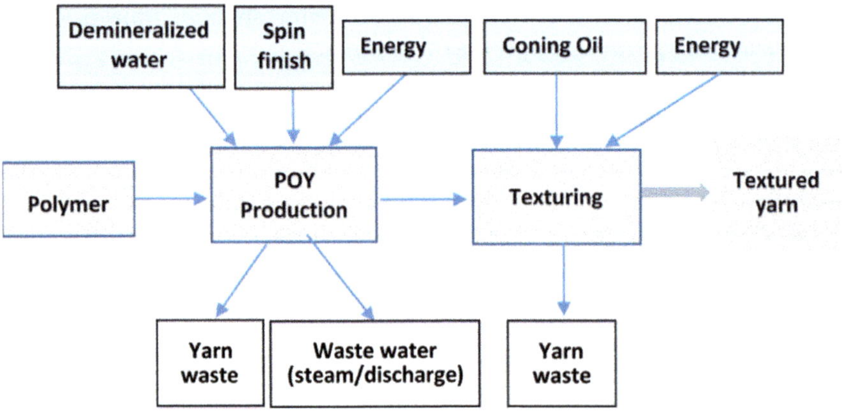

Figure 8.5 Inputs and outputs of textured yarn processing.

8.7.4.2 *Environmental Impact Assessment*

The LCA study was carried out according to the ISO 14044 standard. The LCA software SimaPro 7.3.3 was used to perform the impact assessment stage. The impact categories, according to the CML 2 baseline 2000 V2.05 method were used. These were: global warming potential (GWP); ozone layer depletion (OLD); photochemical oxidants formation (POF); acidification (A); eutrophication (E); human toxicity (HT); fresh water aquatic ecotoxicity (FE); marine aquatic ecotoxicity (ME); and terrestrial ecotoxicity (TE). The impact assessment results are given in Table 8.2.

The use of a continuous process for textured yarn production is one of the factors that reduce environmental impacts. The reason is that in the discontinuous process, the polymer obtained is additionally melted and fed back to the yarn production line again. However, the discontinuous process *i.e.* production of granules provides product versatility and flexibility, and may be a preferred method. An LCA study comparing the environmental impacts of the continuous process *vs.* the discontinuous process is also shown as part of this case study. In this case, the functional unit is 1 kg POY. The systems boundary is again a gate-to-gate assessment. The continuous part of the process is the same as that of the first study. The process flow of the discontinuous part is given in Figure 8.6. In the discontinuous process, polymer feeds to the granule production step where polymer ribbons are cooled and cut to obtain granules. Then granules are dried and fed to the extruder. In the extruder, the granules are melted and pumped through spinning lines where the molten threadlines are solidified, drawn and sent to take-up unit to produce POY.

Results of the life cycle impact assessment of the study are given in Table 8.3.[106]

The spent energy of the discontinuous process is higher than continuous process and an additional 13.1 MJ of energy is required in granule production, drying and extruder stages. As can be seen, the discontinuous process has higher environmental impacts in all impact categories than the continuous process, mainly because of higher energy consumption.

Table 8.2 Impacts assessment of textured yarn.[a]

Impact categories	Units	Textured yarn
Acidification (A)	kg SO_2 eq	0.0168
Eutrophication (E)	kg PO_4 eq	0.037
Global warming (GWP)	kg CO_2 eq	3.15
Human toxicity (HT)	kg 1.4-DB eq	3.62
Fresh water aquatic ecotox.(FE)	kg 1.4-DB eq	0.10
Marine aquatic ecotoxicity (ME)	kg 1.4-DB eq	333.22
Terrestrial ecotoxicity (TE)	kg 1.4-DB eq	0.0066
Photochemical oxidations formation (POF)	kg C_2H_4 eq	0.0005

[a]These figures are derived from the Project "Calculating Environmental Impacts of Textile and Clothing Industry in Bursa *via* LCA, and Preparing the Companies for Ecolabels", funded by Bursa Eskisehir Bilecik Development Agency (BEBKA) and managed by Uludağ Textile Exporters' Association(UTIB).

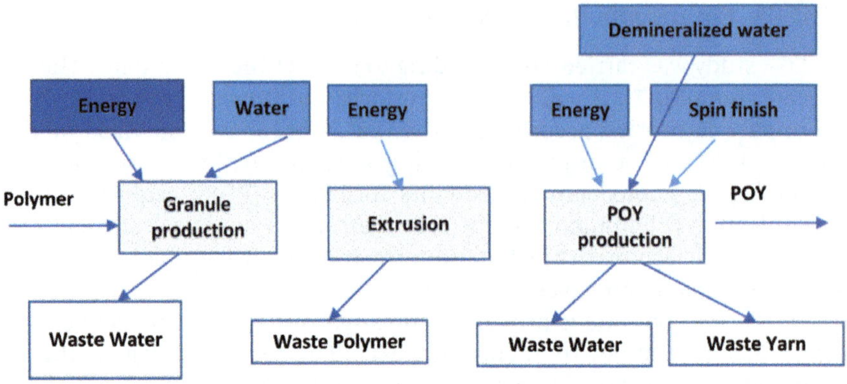

Figure 8.6 Inputs and outputs of the discontinuous production.

Table 8.3 Comparison of continuous and discontinuous processes.[106]

Impact categories	Units	Discontinuous	Continuous
Acidification (A)	kg SO_2 eq	0.0102	0.0097
Eutrophication (E)	kg PO_4 eq	0.038	0.037
Global warming (GWP)	kg CO_2 eq	3.018	2.334
Human toxicity (HT)	kg 1.4-DB eq	3.579	3.507
Fresh water aquatic ecotoxicity (FE)	kg 1.4-DB eq	0.089	0.085
Marine aquatic ecotoxicity (ME)	kg 1.4-DB eq	280.101	248.726
Terrestrial ecotoxicity (TE)	kg 1.4-DB eq	0.0066	0.0065
Photochemical oxidation (POF)	kg C_2H_4 eq	0.00048	0.00043

8.7.4.3 LCA Results

Gate-to-gate LCA studies offer great opportunities for companies to identify "hot spots" in their processes. This study was conducted to evaluate the environmental impact of a company's own production line and to calculate the total carbon footprint of 1 kg of textured yarn. Data from published literature were avoided as much as possible and actual data obtained from the company were used.

Synthetic fibre production is an energy-intensive process and electricity consumption and generation of electricity are the main contributors for all categories. In this LCA study, natural gas was used to generate electricity and a continuous process was used to produce over 200 tonnes/day of PET polymer. The environmental impacts are relatively low, however, and since the main contributor to the environmental impacts categories is the electricity produced from natural gas, renewable sources of electricity are an option to reduce environmental impacts.

Although gate-to-gate studies are suitable for capturing hot spots in the production of textiles, they are only an intermediate phase in determining the environmental impact of a product. In order to evaluate the environmental impacts of a fibre, the environmental impacts in the production stage are just one component of the overall impact of PET fibres. Products made from polyester yarn are washed and dried with less energy in the use phase of the textile, and do not require ironing. Therefore, the energy consumption of the use phase is lower compared with natural fibres.[111]

8.8 The Textile Industry and the Circular Economy – Discussion

The textile industry, which produces a wide range of products from everyday clothing to space suits; household goods to car and airplane interiors; as well as technical innovations in medical devices, generates significant environmental impacts during the production, use and disposal phases. The industry uses a wide range of fibres, accessories and chemicals. Production takes place in many different countries with different approaches and technologies. The impacts of the production phase, such as the chemicals used, may also be effective in the use phase (washing machines effluents), and this is a sector where local practices also affect the global impact. Fast fashion and accelerated textile production cause high water and energy consumption; millions of tonnes of effluents with high BOD, COD, temperature and PH values are discharged every year. Discarded textiles share almost 5% in landfills and recycling rates are very low.

In recent years, several works have been carried out to reduce the environmental impacts of textiles, especially with the demands of conscious consumers, NGOs and brands. There are promising developments in many areas; from alternative fibre use, use of effective wastewater treatment technologies, use of machines with low energy and water consumption, use of green chemicals, to improved labour rights. The active role of ecological labels and certificates cannot be ignored (although labels should be based on similar criteria whenever possible). LCA methodology offers an important possibility in this field, for instance, to determine appropriate criteria for labelling and for product design. Nevertheless, these individual efforts do not constitute a holistic approach and are inadequate due to the complex structure of the textile industry. There is a need for a remarkably compelling systemic change in the industry. The principles of the Circular Economy (CE) can give significant contributions to solving the problems. Among others, better resource efficiency, less waste, extending lifespan, redesigning products are better opportunities for the industry.

A recent report by the Ellen MacArthur Foundation (2017), titled "A New Textiles Economy: Redesigning Fashion's Future", explains how the textile industry should concentrate on CE principles.[10] According to this report, the new textile economy relies on four ambitions: phase out substances of concern and microfibre release; increase clothing utilization; radically improve recycling; and make effective use of resources, moving to renewable inputs. A new level of alignment and collaboration is required to succeed in this transition. Businesses of all sizes throughout the textiles value chain, education and research institutions, policymakers, industry associations, NGOs and others, all have an important role in the transition towards a new textiles economy.[10]

Another opportunity to constrain environmental impacts can be derived from a future possible transition to Industry 4.0. Industry 4.0, put simply, will connect several actors in the value chain together.[10] Thus, effective international cooperation can be very well established. This approach is also supported by CE thinking.

LCA is an important tool in transitioning to the CE. LCA studies carried out to date have made important contributions in determining the urgent hot spots to be addressed in the textile sector, in order to better suit CE thinking. However, as it is known well, factors such as choosing different system boundaries, impact categories, the origin of energy sources, choosing or avoiding allocation, can give different results even for the same fibre or process. Additionally, it should also be observed that LCA studies are not always in concurrence with each other in terms of, among others, obtained results and focusing on different phases objectively in the value chain.[9] Although it is very useful for companies to perform LCA studies, these studies may be presented with information that is not always representative of competitors, thus an unfair advantage could be perceived. However, peer review of LCAs are a critical component if market claims are to be made on the basis of LCA studies.CE thinking is also a comprehensive approach and can also be helpful to lessen such occurrences.

Pre-consumer and post-consumer waste management, particularly in recent years, is one of the biggest challenges of the textile industry. The most crucial contributions of the CE thinking approach, aside from resource efficiency, can be in the areas of waste management and recycling. Recycling and reuse are vital opportunities to reduce the environmental impacts of textiles. However, studies on recycling need to be improved to assure the quality of recycled products and to make recycling methods cheaper. The rapid increase in textile consumption and post-consumer waste have made studies in this area very valuable.

However, due to the complex structure of the textile industry, the transition to CE will be more difficult compared with other industries. The adoption and application of CE thinking by global brands, which have both manufacturing and stores in different parts of the world, is the most critical step in this transformation.

References

1. A. W. Engelhardt, *Int. Fiber J.*, 2018, **38**(3), 16.
2. B. Xu and Y. L. Jiang, *China Econ. Rev.*, 2009, **20**, 425.
3. B. Resta, P. Gaiardelli, R. Pinto and S. Dotti, *J. Cleaner Prod.*, 2016, **135**, 620.

4. N. C. Smith, *Calif. Manag. Rev.*, 2003, **45**(4), 52.
5. S. S. Muthu, Introduction, in *Sustainability in the Textile Industry*, Springer. 2017, pp. 1–8.
6. European Union and JRC, *Environmental Improvement Potential of Textiles*, cited 2018 November, Available from: https://publications.europa.eu/en/publication-detail/-/publication/f8d0def8-4fd5-4d84-a308-1dfa5cf2e823/language-en.
7. OECD, Cotton, in *OECD-FAO Agricultural Outlook 2017–2026*, Paris, 2017.
8. S. M. Burkinshaw, *Physico Chemical Aspects of Textile Coloration West Yorkshire: Society of Dyers and Colorists*, Wiley & Sons, 2016.
9. J. Eder-Hansen, C. Chalmer, S. Tarneberg, T. Tochtermann, J. Seara and S. Boger, *et al.*, *Pulse of the Fashion Industry*, Global Fashion Agenda & The Boston Consulting Group, 2017.
10. Ellen MacArthur Foundation, *A new textiles economy: Redesigning fashion's future*, 2017, cited 2018 December, Available from: http://www.ellenmacarthurfoundation.org/publications.
11. WTO. Textiles, cited 2018 December, Available from: https://www.wto.org/english/thewto_e/minist_e/min96_e/textiles.htm.
12. United Nations, cited 2018 November, Available from: https://unctad.org/en/Docs/iteiia20051_en.pdf.
13. J. Annamma, J. F. Sherry Jr, A. Venkatesh, J. Wang and R. Chan, *Fashion Theor.*, 2012, **16**(3), 273.
14. N. J. Tokatlı, *Econ. Geogr.*, 2008, **8**(1), 21.
15. N. Tojo, B. Kogg, N. Kiorboe, B. Kjaer and K. Aalto, *Prevention of Textile Waste: Material Flows of Textiles in Three Nordic Countries and Suggestions on Policy Instruments*, Copenhagen, Denmark, 2012.
16. ICAC, *International Cotton Advisory Committee*, cited 2018 November, Available from: https://www.depts.ttu.edu/aaec/icac/archive/pdf/Demand_optimized/Textile_Demand/World_Textile_Demand-September_2005.pdf.
17. K. Niinimäki and C. Armstrong, *Int. J. Fashion Des., Technol. Educ.*, 2013, **6**(3), 190.
18. D. Andrews, *Local Econ.*, 2015, **30**(3), 305.
19. E. Langley, S. Durkacz and S. Tanase, *Clothing longevity and measuring active use*, Oxon: WRAP, cited 2018 November, Available from: http://www.wrap.org.uk/sites/files/wrap/priv_download/Clothing%20longevity%20SUMMARY%20REPORT.pdf.
20. US EPA, *Textiles: Material-Specific Data*, cited 2018 November, Available from: https://www.epa.gov/facts-and-figures-about-materials-waste-and-recycling/textiles-material-specific-data.
21. B. Zamani, M. Svanström, G. Peters and T. Rydberg, *J. Ind. Ecol.*, 2014, **19**(4), 676.
22. G. Sandin and G. M. Peters, *J. Cleaner Prod.*, 2018, **184**, 353.
23. WRAP, *Textiles market situation report*, cited 2018 November, Available from: http://www.wrap.org.uk/sites/files/wrap/Textiles_Market_Situation_Report_2016.pdf.

24. F. Kuwonu, *Protectionist ban on imported used clothing*, cited 2018 November, Available from: https://www.un.org/africarenewal/magazine/december-2017-march-2018/protectionist-ban-imported-used-clothing.
25. S. Gray, *Mapping clothing impacts in Europe: the environmental cost*, WRAP, cited 2018 November, Available from: http://www.ecap.eu.com/wp-content/uploads/2018/07/Mapping-clothing-impacts-in-Europe.pdf.
26. EU Commission, *Closing the loop – An EU action plan for the Circular Economy*, cited 2018 November, Available from: https://eur-lex.europa.eu/resource.html?uri=cellar:8a8ef5e8-99a0-11e5-b3b7-01aa75ed71a1.0012.02/DOC_1&format=PDF.
27. Ş. Altun, *Fibres Text. East. Eur.*, 2012, **5**(94), 16.
28. J. M. Hawley, *Recycling in Textiles*, ed. Y. Wang, Woodhead Publishing Ltd, Cambridge, 2006, p. 7.
29. HM Sustainability Report 2014 Final Commitment 5, cited 2018 November, Available from: http://sustainability.hm.com/content/dam/hm/about/documents/masterlanguage/CSR/reports/HM_SustainabilityReport_2014_FINAL_Commitment_5.pdf.
30. Eileenfisher, *Eileenfisher*, cited 2018 November, Available from: https://www.eileenfisher.com/catalogsearch/result/?q=recycled+cotton.
31. Patagonia, *How to Recycle Patagonia Garments*, 2018, cited 2018 November, Available from: https://www.patagonia.com/recycling.html.
32. A. R. Köhler, L. M. Hilty and C. Bakker, *J. Ind. Ecol*, 2011, **15**(4), 496.
33. A. Köhler and L. Erdmann, *Hum. Ecol. Risk Assess.*, 2004, **10**(5), 831.
34. S. S. Muthu, Y. Li, J. Y. Hu and P. Y. Mok, *Ecol. Indic.*, 2012, **1**(13), 66.
35. N. Cherret, J. Barret, A. Clemett, M. Chadwick and M. J. Chadwick, *Ecological Footprint and Water Analysis of Cotton, Hemp and Polyester*, cited 2018 September, Available from: https://www.sei.org/mediamanager/documents/Publications/SEI-Report-EcologicalFootprintAndWaterAnalysisOfCottonHempAndPolyester-2005.pdf.
36. H. M. G. van der Werf and L. Turunen, *Ind. Crops Prod.*, 2008, **1**(27), 1.
37. Integrated Pollution Prevention and Control (IPPC) Reference Document on Best Available Techniques for the Textiles Industry, cited 2018 October, Available from: http://eippcb.jrc.ec.europa.eu/reference/BREF/txt_bref_0703.pdf.
38. USDA, United States Department of Agriculture, *The World And United States Cotton Outlook*, cited 2018 September, Available from: https://www.usda.gov/oce/forum/2018/commodities/Cotton.pdf.
39. ICAC and FAO, *Measuring Sustainability in Cotton Farming Systems Towards A Guidence Framework*, FAO and ICAC, Rome, cited 2018 September, Available from: https://www.icac.org/.Cotton./measuring-sustainability-cotton-farming-full-english.
40. Carbon Trust, *International Carbon Flows Cotton*, cited 2018 September, Available from: www.carbontrust.com/media/38354/ctc794-international-carbon-flows-cotton.pdf.

41. Cotton Incorporated, *Life Cycle Assessment of Cotton Fiber and Fabric*, cited 2018 November, Available from: http://resource.cottoninc.com/LCA/LCA-Full-Report.pdf.
42. WWF-India, *Cutting cotton carbon emissions Finding from Warangal*, India, 2013.
43. E. S. Laursen, J. Hansen, H. K. Knudsen, H. Wenzel, H. F. Larsen and F. M. Kristensen, *Ediptex-Environmental Assessment of Textiles*, Copenhagen, 2007.
44. ILO, *Child Labor in Cotton A Briefing*, Geneva, 2016.
45. K. Runde, P. Jurewicz, N. Responsible, *Cotton Sourcing Snapshot: A Survey of Corporate Practices to End Forced Labor*, cited 2018 September, Available from: https://www.sourcingnetwork.org/cotton-sourcing-snapshot/.
46. Yarns Ethically & Sustainably Sources, cited 2018 September, Available from: https://www.sourcingnetwork.org/yess.
47. OECD-FAO, *Lessons and Best Practices OECD-FAO Guidance for Responsible Agricultural Supply Chains*, cited 2018 September, Available from: https://mneguidelines.oecd.org/OECD-FAO-Guidance_Best-Practices.pdf.
48. PAN, Solidaridad, WWF. Sustainable Cotton Ranking 2017 Assesing Company Performance, cited 2018 September, Available from: https://www.sustainablecottonranking.org/.
49. J. Soth, C. Grasser and R. Salerno, *The Impact of Cotton on Fesh Water Resources and Ecosystems A Prelimanry Synthesis*, Gland, 1999.
50. A. K. Chapagain and A. Y. Hoekstra, *Water Resour. Manage.*, 2006, 35.
51. A. K. Chapagain, A. Y. Hoekstra, H. H. G. Savenije and R. Gautam, *Ecol. Econ.*, 2006, 186.
52. P. Nicolopoulou-Stamati, S. Maipas, C. Kotampasi, P. Stamatis and L. Hens, *Front. Public Health.*, 2016, 1.
53. F. Mancini, A. H. C. Van Brugen, J. L. S. Jiggins, A. C. Ambatipudi and H. Murphy, *Int. J. Occup. Environ. Health.*, 2005, 221–232.
54. M. I. Tariq, S. Afzal, I. Hussain and N. Sultana, *Environ. Int.*, 2007, 1107.
55. BCI, cited 2018 September, Available from: https://bettercotton.org/.
56. Textile Exchange, *Life Cycle Assessment (LCA) of Organic Cotton – A Global Average*, cited 2018 September, Available from: http://farmhub.textileexchange.org/upload/library/Farm%20reports/LCA_of_Organic_Cotton%20Fiber-Full_Report.pdf.
57. M. Pearson, *Geogr. J.*, 2006, 306.
58. A. E. Ziph and K. Rajasekaran, *J. New Seeds*, 2003, 115.
59. M. Marvier, C. McCreedy and J. Regetz, *Science*, 2007, 1475.
60. J. Kathage and M. Qaim, *Proc. Natl. Acad. Sci. U. S. A.*, 2012, 11652.
61. O. Van Cleemput and A. H. Samater, *Fert. Res.*, 1996, 81.
62. L. Dahllöf, A Case Study Concerning Fabric for a Sofa, Environmental Systems Analysis', Methodological Problems in the LCA Procedure for the Textile Sector. ESA-Report 2004:7, Chalmers University of Technology, Göteborg, 2004, Report No.: ISSN: 1404-8167.

63. S. Shahidi, A. Rashidi, M. Ghoranneviss, A. Anvari and J. Wiener, *Surf. Coat. Technol.*, 2010, **205**(1), 349.
64. R. Zhang and A. Wang, *J. Cleaner Prod.*, 2015, **87**, 961.
65. I. C. Gouveia, J. M. Fiadeiro and J. A. Queiroz, *Biochem. Eng. J.*, 2008, **41**, 157.
66. T. Bechtold, E. Burtscher and Y. T. Hung, *Handbook of Industrial and Hazardous Wastes Treatment*, ed. L. K. Wang, Y. T. Hung, H. H. Lo and C. Yapijakis, Marcel Dekker, New York. 2004, pp. 379–414.
67. S. Jose, S. Nachimuthu, S. Das and A. Kumar, *J. Text. Inst.*, 2018, **109**(2), 225.
68. Marks & Spencer, *ECP Minimum Standards Manufacturing Restricted Substances List (MRSL)*, vol. 2, cited 2018 October, Available from: https://corporate.marksandspencer.com/documents/plan-a-our-approach/ecp-modules-2016/mns-ecp-mrsl-and-rsl.pdf.
69. The European Commission, *Off. J. Eur. Union.*, 2014, **L174**, 45.
70. N. Bruce, N. Hartline, S. Karba, B. Ruff and S. Sonar, *Microfiber pollution and the apparel industry (Patagonia)*, cited 2018 December, Available from: http://www.esm.ucsb.edu/research/2016Group_Projects/documents/PataPlastFinalReport.pdf.
71. M. A. Browne, P. Crump, S. J. Niven, E. Teuten, A. Tonkin and T. Galoway, *et al.*, *Environ. Sci. Technol.*, 2011, **45**(21), 9175.
72. J. Talvitie, M. Heinonen, J. P. Pääkköne, E. Vahtera, A. Mikola and O. Setälä, *et al.*, *Water Sci. Technol.*, 2015, **72**(9), 1495.
73. K. A. V. Zubris and B. K. Richards, *Environ. Pollut.*, 2005, **138**, 201.
74. S. Palamutcu, *Energy*, 2010, **35**, 2945.
75. US EPA, cited 2018 October, Available from: https://www3.epa.gov/ttnchie1/ap42/ch09/final/c9s07.pdf.
76. WHO, cited 2018 October, Available from: http://www.who.int/occupational_health/publications/en/oehairbornedust.pdf?ua=1.
77. E. Koç and E. Çinçik, *Fibres Text. East. Eur.*, 2010, **18**(2 (79)), 14.
78. K. Schlüter, *Ullman's Encylopedia of Industrial Chemistry*, vol. 36: 2012, p. 75.
79. A. Sarkar, D. Sarkar, M. Gupta and C. Bhattacharjee, *Clean: Soil, Air, Water*, 2012, **40**(8), 830.
80. V. M. Correia, T. Stephenson and S. J. Judd, *Environ. Technol.*, 1994, **15**(10), 917.
81. B. Smith, Wastes From Textile Processing, in *Plastics and the Environment*, ed. A. L. Andrady, John Wiley & Sons Inc., 2005, pp. 243–309.
82. T. Stegmaier, W. Wunderlich, T. Hager, A. B. Siddique, J. Sarsour and H. Planck, *Clean: Soil, Air, Water*, 2008, **36**(3), 279.
83. US EPA, cited 2018 November, Available from: https://nepis.epa.gov/Exe/ZyPDF.cgi/50000HE9.PDF?Dockey=50000HE9.PDF.
84. E. Kalliala and P. Talvenma, *J. Cleaner Prod.*, 2000, **8**, 143.
85. X. Ren, *J. Cleaner Prod.*, 2000, **8**, 473.
86. R. B. Chavan, Environmentally friendly dyes, in *Handbook of Textile and Industrial Dyeing Principles, Processes and Types of Dyes*, ed. M. Clark Woodhead Publishing Ltd., 2011, pp. 515–561.

87. H. Ali, *Water, Air, Soil Pollut.*, 2010, **213**, 251.
88. C. O'Neill, F. R. Hawkes, D. L. Hawkes, N. D. Lourenco, H. M. Pinheiro and W. Delee, *J. Chem. Technol. Biotechnol.*, 1999, **74**(11), 1009.
89. E. Olsson, S. Posner, S. Roos and K. Wilson Kartläggning av kemikalieanvändning i kläder. Mölndal: 2009.
90. Greenpeace International, cited 2018 November, Available from: https://storage.googleapis.com/planet4-international-stateless/2012/03/0e8a0ec9-dirtylaundryreloaded.pdf.
91. AFIRM, *Restricted Substances List*, cited 2018 October, Available from: http://www.afirm-group.com/wp-content/uploads/2018/05/2018_AFIRM_RSL_EN.pdf.
92. Global Organic Textile Standart (GOTS) Version 5, cited 2018 October, Available from: https://www.global-standard.org/images/GOTS_Documents/GOTS_Standard_5.0_EN.pdf.
93. ECHA. *Reach Legislation*, 2018, Available from: https://echa.europa.eu/regulations/reach/legislation.
94. Swedish Chemical Agency, *Hazardous chemicals in textiles*, cited 2018 October, Available from: https://www.kemi.se/global/rapporter/2013/rapport-3-13-textiles.pdf.
95. A. Köhler, *Mater. Des.*, 2013, **51**, 51.
96. N. M. van der Velden, K. Kuusk and A. R. Köhler, *Mater. Des.*, 2015, **84**, 313.
97. T. Dawson, *Color. Technol*, 2011, **128**, 1.
98. Sustainable Apparel Coalition (SAC), *Higg Facility Tools*, cited 2018 December, Available from: https://apparelcoalition.org/higg-facility-tools/.
99. Nike, *Product & Materials Sustainability Indices*, cited 2018 December, Available from: https://sustainability.nike.com/product-material-sustainability-indices.
100. ZDHC, 2018, Available from: https://www.roadmaptozero.com/programme/.
101. European Environment Agency, *Life Cycle Assessment A guide to approaches, experiences and information sources*, 1997, Report No.: Environmental Issues Series no: 6.
102. ISO 14044:2006, *Environmental management—Life cycle assessment—Requirements and guidelines*, 2006-07.
103. Danish Ministry of Environment, EPA, *The Product, Functional Unit and Reference Flows in LCA*, Environmental News, 2004.
104. W. Klöpffer and B. Grahl, *Life Cycle Assessment (LCA): A Guide to Best Practice*, John Wiley & Sons, Weinheim, 2014.
105. S. Altun, A. Ruzgar, Z. I. Cin and F. Erdonmez, *Interdiscip. Environ. Rev.*, 2017, **18**(1), 28.
106. S. Altun, S. K. A. Solmaz, G. E. Üstün and İ. Kağnıcı, Life Cycle Assessment of Direct and Extrusion Spinning: A case study, *Fresenius Environ. Bull.*, 2018, **27**(2), 787–792.
107. T. L. van Winkle, J. Edeleanu, E. A. Prosser and C. A. Walker, *Am. Sci.*, 1978, **66**(3), 280.

108. G. G. Smith and R. H. Barker, *Resour., Conserv. Recycl.*, 1995, **14**, 233.
109. E. J. Kalliala and P. Nousiainen, *Autex Res. J.*, 1999, **1**(1), 8.
110. J. Madsen, B. Hartlin, S. Perumalpillai, S. Selby and S. Aumonier, *Mapping of Evidence on Sustainable Development Impacts That Occur in the Life Cycles of Clothing: A Report to the Department for Environment*, London, 2007.
111. J. S. Steinberger, D. Friot, O. Jolliet and S. Erkman, *Int. J. Life Cycle Assess.*, 2009, **14**(5), 443.
112. L. Dahllöf, *Licentiate of Engineering Thesis: Chalmers Unversity of Technology*, 2004.
113. L. Dahllöf ESA-Report 2003:9, Göteborg, 2003.
114. L. Shena, E. Worrell and M. K. Patel, *Resour., Conserv. Recycl.*, 2010, **55**, 260.
115. M. O. Tabone, J. J. Cregg and E. J. Beckman, *Environ. Sci. Technol.*, 2010, **44**(21), 8264.
116. N. M. van der Velden, M. K. Patel and J. G. Vogtländer, *Int. J. Life Cycle Assess*, 2014, **19**, 331.
117. M. De Saxce, S. Pesnel and A. Perwuelz, *J. Cleaner Prod.*, 2012, **37**, 221.
118. Y. Zhang, H. Kang, H. Hou, S. Shao, X. Sun and C. Qin, *et al.*, *Clean Technol. Environ. Policy*, 2018, **20**, 1355.
119. UK Environment Agency, *An updated lifecycle assessment study for disposable and reusable nappies*, cited 2018 December, Available from: http://randd.defra.gov.uk/Document.aspx?Document=WR0705_7589_FRP.pdf.
120. M. Cordella, I. Bauer, A. Lehmann, M. Schulz and O. Wolf, *J. Cleaner Prod.*, 2015, **95**, 322.
121. A. V. Weisbrod and G. Van Hoof, *Int. J. Life Cycle Assess.*, 2012, **17**, 145.
122. S. Roos, B. Zamani, G. Sandin, G. M. Peters and M. Svanström, *J. Cleaner Prod.*, 2016, **133**, 691.
123. C. Wang, L. Wang, X. Liu, C. Du, D. Ding and J. Jia, *et al.*, *J. Cleaner Prod.*, 2015, **108**, 464.
124. A. C. Woolridge, G. D. Ward, P. S. Phillips, M. Collins and S. Gandy, *Resour., Conserv. Recycl.*, 2006, **46**(1), 94.
125. L. Farrant, S. I. Olsen and A. Wangel, *Int. J. Life Cycle Assess.*, 2010, **15**(7), 726.
126. D. M. M. Yacout and M. S. Hassouna, *Environ. Monit. Assess.*, 2016, **188**, 444.
127. OECD, *Perfluorooctane Sulfonate (PFOS) and related chemical products*, cited 2018 October, Available from: http://www.oecd.org/chemicalsafety/risk-management/perfluorooctanesulfonatepfosandrelatedchemicalproducts.htm#5.
128. US Department of Labor-Occupational Safety and Health Administration, *Isocyanates*, cited 2018 November, Available from: https://www.osha.gov/SLTC/isocyanates/.
129. K. Fent, *Crit. Rev. Toxicol.*, 1996, **26**(1), 1.
130. E. Forgacs, T. Cserhati and G. Oros, *Environ. Int.*, 2004, **30**(7), 953.

131. A. K. R. Choudhury, *Sustainable Fibres and Textiles*, ed. M. Subramanian, Woodhead Publishing, 2017, pp. 267–322.
132. National Toxicology Program (June 2011), Report on Carcinogens, Twelfth Edition, cited 2018 October, Available from: http://ntp.niehs.nih.gov/go/roc12.
133. ZDHC, *Nonylphenol Ethoxylates (NPEOs) Guidence Sheet*, cited 2018 October, Available from: https://uploads-ssl.webflow.com/5c4065f2d6b53e08a1b03de7/5db6eb1066a7b24a82a22fe1_NPEO.pdf.
134. P. C. Nagajyoti, K. D. Lee and T. V. M. Sreekanth, *Environ. Chem. Lett.*, 2010, **8**(3), 199.
135. E. F. F. Mariussen, *Crit. Rev. Toxicol.*, 2006, **36**(3), 253.
136. C. Descorme, *Catal. Today*, 2017, **297**, 324.
137. H. I. Abdel-Shafy and M. S. M. Mansour, *Egypt. J. Pet.*, 2016, **25**(1), 107.
138. US EPA, *Volatile Organic Compounds' Impact on Indoor Air Quality*, 2018, cited 2018 November, Available from: https://www.epa.gov/indoor-air-quality-iaq/volatile-organic-compounds-impact-indoor-air-quality.

CHAPTER 9

Refining Made Circular: The Potential for Substituting Biogenic for Fossil Carbon

V. MOREAU,*[a] F. MEYLAN[b] AND S. ERKMAN[b]

[a] Ecole Polytechnique Fédérale de Lausanne, Laboratory of Environmental and Urban Economics, ENAC-IA-LEURE, Station 16, 1015 Lausanne, Switzerland; [b] University of Lausanne, Faculty of Geosciences and Environment, Institute of Earth Surface Dynamics, Geopolis, 1015 Lausanne, Switzerland
*Email: vincent.moreau@epfl.ch

9.1 Introduction

Carbon dioxide emissions from the combustion of fossil fuels must decline rapidly in order to mitigate climate change.[1] In response to the challenge of achieving global carbon neutrality by 2050, there have been many attempts to substitute bio based material and energy for fossil fuels. The concept of biorefinery, usually defined as "the sustainable processing of biomass into a spectrum of marketable products (food, feed, materials, chemicals) and energy (fuels, power, heat)" provides one solution to replace fossil with biogenic carbon.[2] Bio refineries are also heralded as one of the most effective negative emission technology when coupled with carbon capture and storage (CCS).[3] While negative emissions are required in most 1.5- or 2-degree (Celsius) global warming scenarios, the potential of bio energy with CCS to remove carbon from the atmosphere on that scale is doubtful.[4] Bio refineries might be more promising in displacing fossil fuel products and

Life Cycle Assessment: A Metric for the Circular Economy
Edited by Aiduan Borrion, Mairi J. Black and Onesmus Mwabonje

Published by the Royal Society of Chemistry, www.rsc.org

implementing a circular economy (CE) rather than removing anthropogenic carbon already in the atmosphere. From a CE perspective, substitution, removal and reuse of fossil carbon are relevant.[5] This chapter assesses the substitution potential of biorefinery products for chemicals and fuels in terms of total biogenic carbon available *vs.* current use of fossil carbon. In other words, biogenic carbon is accounted for in a life cycle perspective from sources to products.

Most research activities in the past have focused on energy, quantifying the global amount, in exajoules (EJ), that biomass might provide in the future, generally up to 2050 (see Chapter 14). Statistical and geographic information systems estimate the potential contribution of biomass from approximately 50 to 1200 EJ per year.[6,7] This broad range is due to multiple and uncertain factors (*e.g.* crops yields, climate change, world population, future diets *etc.*), which influence the future supply of biomass. Moreover, biomass has multiple uses with ordered preferences between food, feed, materials and energy. The sources of biomass have been the main distinction between generations of biorefineries: at first all crops, and particularly carbon-rich ones, were processed for fuels. The second generation improved upon the first by converting dedicated crops as well as waste food, agricultural and forestry products. The third generation relies on off-ground open ponds of highly productive microalgae. Second-generation biorefineries can produce a range of chemicals and fuels, substituting them for their fossil equivalents.[8]

Although a global energy and material potential from second-generation biorefineries is lacking, it is essential to estimate how fossil fuels and materials can be substituted with biogenic sources. Because biomass and carbon in particular can be found in very different forms and since all transformation processes are not suitable for any given source, a detailed investigation of biomass flows, in petagrams (Pg) of carbon, is required. Quantifying the carbon flows in products (*e.g.* polymers, liquid and gaseous chemicals) also provides a map of which and how much fossil carbon can be replaced. Therefore, this chapter evaluates the annual amount of biomass available globally in 2050, including dedicated crops, residues of wood and food industries as well as municipal organic wastes and organic matter contained in wastewater. Starch and sugar, vegetable oil, lignin, hemicellulose, cellulose and mixed wet materials are quantified separately. The average yields of appropriate transformation processes per type of biomass are then used to estimate the carbon that could be embedded in the final products made from biorefineries. The results are compared to the amount of fossil fuels currently used for equivalent products and applications to determine which could potentially be displaced.

In the next section, we explain the methodological approach. The results are two-fold, as presented in Sections 9.2.1 and 9.2.2. First, we show the estimated inputs, or biomass feedstock. Second, we evaluate the carbon that can be embodied in final products, as well as the comparison with the current consumption of fossil counterparts. These results and the methodology are then discussed, and a brief conclusion presents the main lessons and potential implications of bio based products.

9.2 Methodology

The problem at hand is essentially a mass balance between carbon-intensive products and applications currently sourced from fossil fuels and the potential substitutes made from biogenic carbon. Hence, the theoretical foundations underlying the results stem from substance flow analysis (SFA); in this case, carbon. This corresponds to the second step in life cycle assessment (LCA), or life cycle inventory (LCI), to quantify carbon flows at the product level on a global scale.[9] The use and end-of-life phases of the product life cycle are assumed equivalent for biogenic and fossil products alike, and can be ignored here. The reference year is 2050, when human activities should become carbon neutral to stand a chance of limiting global warming within 2 degrees.[10] Data availability, in petagrams of carbon per year (Pg C per year), and uncertainty will change under new climate conditions in 2050 and beyond. The data are converted from energy to mass units – in other words, from EJ to Pg of carbon – according to well-established recommendations.[11] The share of carbon in dry matter (DM) biomass is approximately 50% and we assume a gross calorific value of biomass of 18.3 MJ kg^{-1}. Therefore, one EJ of biomass corresponds to 0.027 Pg of carbon.

Turning biogenic carbon into fuels and chemicals involves a range of biological and chemical processes, such as fermentation and hydrolysis. We group them under biorefinery processes and characterize them according to their carbon yield, with inputs and outputs described below. Terrestrial biomass only is taken into account as potential inputs to biorefinery processes. As shown in Figure 9.1, five main sources of biomass lead to twice as many products and chemicals in output, both chemicals and fuels. Although the transformation of food, feed and direct utilization of wood can be associated with such processes, only their waste by-products are counted as inputs.

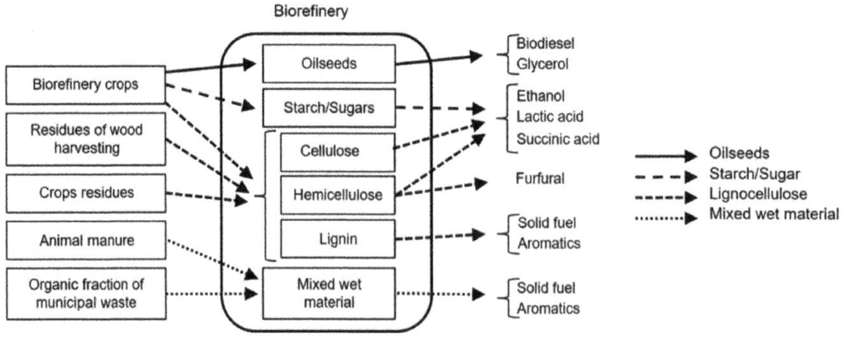

Figure 9.1 Biomass inputs and useful outputs of biorefineries.

Five sources of feedstock are accounted for, as shown in Figure 9.1, including dedicated crops, also known as energy crops, exclusively cultivated for biorefineries. Three categories of biorefinery crops are considered: oilseeds (*e.g.* rapeseed, palm), sugar and starch plants (*e.g.* corn, sugar cane) and lignocellulosic crops (LC) (*e.g.* switchgrass, miscanthus). The content of kernels in oilseeds after processing is ignored for lack of data. The wood residues of forestry are lignocellulosic products such as branches, bark and stumps. Crop residues include straw and other harvesting wastes. Animal manure and the organic fraction of municipal solid wastes are also included as potential inputs.

9.2.1 Inputs

Different molecules are extracted from the inputs described here: oilseeds, starch and sugars, lignocellulose and mixed wet materials. Lignocellulose consists of three bio-polymers (cellulose, lignin and hemicellulose), which are very abundant. They are produced from dedicated crops, as well as from forestry residues and crop cultivation. The "residue to product ratio" (RPR) indicates the amount of residues produced by a given crop. Roberts *et al.*[12] reported RPR ranging from 1.2 (oats) to 2.0 (soybeans) by averaging literature data. When accounting for the fraction of residues necessary for soil amendment, Vávrova *et al.*[13] reported lower estimates from 0.7 (spring barley) to 1.3 (triticale). A RPR equal to 1 remains on the conservative side given that residues might also be used elsewhere, for example as fuel in district heating networks with potentially higher value added, especially in the future.

To account for uncertainty in the future availability of biorefinery crops, three scenarios were derived based on the abundance of lignocellulose, constraints on sugar and starch crops, as well as the relative inefficiency of biodiesel production from oil seeds. Thus, the latter stays constant while scenarios 2 and 3 in Table 9.1 are double and triple the carbon in scenario 1, respectively. The first scenario of dedicated crops is representative of what has been achieved already.

9.2.1.1 Biorefinery Crops

The potential of biorefinery crops is highly uncertain, as it depends on parameters such as population growth, crop yields, diets and environmental

Table 9.1 Three scenarios of biorefinery crops.

Scenario	Lignocellulose (PgC per year)	Sugar/Starch (PgC per year)	Oil seeds (PgC per year)	Total (PgC per year)
1	0.856 (80.0%)	0.116 (10.8%)	0.098 (9.2%)	1.070
2	1.712 (83.8%)	0.232 (11.4%)	0.098 (4.8%)	2.042
3	2.568 (85.2%)	0.348 (11.5%)	0.098 (3.3%)	3.014

constraints. Hence, there is a large range covered by all three scenarios, as shown in Table 9.1.

In the first scenario, the production of biofuels is limited to 85 billion liters of ethanol (0.068 Pg) per year, as was the case in 2012 when the European Commission addressed concerns over indirect emissions of first-generation sources.[14] This corresponds to 0.035 Pg of carbon (PgC) per year, since ethanol is 52% carbon. Accounting for the carbon yield of ethanol production (59.9%; see Table 9.3), an estimated 0.058 PgC per year (0.035 PgC/0.599) is embodied in the sugar and starch feedstock of first-generation biofuels. If crop residues are also factored in with a RPR of 1, the total production comes to 0.116 PgC per year resulting from first-generation sugar crop cultivation (mainly corn and sugar cane).

For biodiesel, the annual production is equivalent to 0.018 Pg, which corresponds to 0.0139 Pg of carbon, with a carbon content in biodiesel of 77%. Taking into account the yield of biodiesel production (28.5%; see Table 9.3) and an identical RPR, an estimated 0.098 PgC per year is embodied into oilseeds cultivated for biodiesel production. Approximately the same amount of initial carbon serves as feedstock for biodiesel and bioethanol, although a larger volume of ethanol is produced annually. This shows that biodiesel production is relatively less efficient than ethanol. Popp *et al.*[14] reached the same conclusion, demonstrating that an identical surface area would produce four times more ethanol than biodiesel.

In theory, biorefinery crops can be selected according to the desired final product, for example, grains or sugarcane for ethanol production and oilseeds for biodiesel production). However, lignocellulosic crops (*e.g.* miscanthus, switch grass, poplar) are likely to become the main biorefinery crop in the future, since they can be cultivated on non-productive, abandoned or degraded land that is not suitable for food production *per se*. The residues of these crops can be entirely dedicated to biorefining, since the residues of conventional agricultural crops are already sufficient to cover alternative uses (feed, straw, *etc.*). The shares of the constituents of lignocellulosic material, both LC and residues, are based on average data from secondary sources, as shown in Table 9.2.

Table 9.2 Share of LC constituent in different residues and plants (% based on dry weight).

Sources	Cellulose	Hemicellulose	Lignin	References
Corn stover	35–40	21–25	11–19	56
Corn stover	38	23	20	41
Sugarcane bagasse	25–45	28–32	15–25	56
Sugarcane bagasse	39	25	23	41
Switchgrass	35–40	25–30	15–20	56
Switchgrass	31	24	18	41
Miscanthus	44	18	22	57
Sunflower stalks	39	34	18	57
Mean (geometric)	**38**	**25**	**19**	

Note that the amount of biogenic carbon available from lignocellulose is estimated at four times the combined content in starch, sugar and oilseed crops, in scenario 1. This grows to five and almost 6 times more carbon from lignocellulose in scenarios 2 and 3, respectively.

9.2.1.2 Wood Residues

Residues of wood processing are already widely used, for instance, in the form of chips and pellets for industrial and residential furnaces. Therefore, these residues have not been considered in this article as potential input for biorefineries. Residues from forestry and wood harvesting (such as branches, bark and stumps), however, are generally left on site. The amount available for biorefinery has been calculated by multiplying the total wood production by a factor of 0.15, which has been derived from the multiplication of the RPR of wood (0.6) by a recoverability factor (0.25), representing the share of wood residues recoverable in the current economic and technical conditions.[15] The total amount of wood harvested has been averaged based on the time series from 2004 to 2013.[16] The volumes reported in FAO statistics have been converted into mass of carbon (Pg of C) according to a conversion factor of 0.29 ton of C per m^3. This factor was derived from the averaged density of dry woods (0.58 tons m^{-3}), multiplied by the share of carbon in wood, or approximately 50%.[15,17] A slightly different share of lignocellulose constituents from that of crop residues has been assumed (40% cellulose, 25% hemicellulose and 25% lignin), since wood residues generally contain a larger share of lignin.

9.2.1.3 Crop Residues

Crop residues are the inedible part of plants, of which the unused fraction is approximately 33%.[18,19] All of it was assumed to be recoverable, because they are produced on easily accessible land. The share of lignocellulosic constituents is considered the same as for biorefinery crops (Table 9.2). A share of the residues should return to the soil to maintain some degree of fertility, which is accounted for in the RPR (see Section 9.2.1.1). In addition, some carbon may also return to the soil in a more stable form after biomass conversion, for example, the fraction of char from gasified or pyrolyzed biomass and solid residues of anaerobic digestion. Wastes and losses occurring during food transformation and consumption are considered part of municipal wastes, and this is accounted for in Section 9.2.1.5.

9.2.1.4 Animal Manure

The amount of recoverable manure is derived (mean) from secondary sources as well.[11,20] A fraction of 50% carbon is assumed for dry manure (dry matter or DM). The amount of recoverable manure depends on the size of global livestock, and consequently on current and future diets. We have not accounted for this parameter in our analysis, since we assume that the scenarios in Table 9.2 do not require significant dietary changes in order to cultivate biorefinery crops.

9.2.1.5 *Waste Treatment*

The organic fraction of municipal solid wastes (OFMSW) and the sewage sludge from wastewater treatment (WWT) can be used to produce biogas. The estimated recoverable amount of OFMSW in 2050 is estimated by averaging literature data.[11,21] For WWT, the amount of carbon is calculated using a production of 60–90 g of dry matter (DM) per day per capita.[22] A recoverability factor of 0.25 is used to calculate the input as a function of population.

From this baseline scenario for manure, OFMSW and WWT, which represents the current conditions, we assume that the amount of carbon in mixed wet materials increases 20% and 40% in scenarios 2 and 3, respectively.

9.2.2 Outputs

For each of the feedstock above, at least one process for the production of chemicals and another for the production of fuels were analyzed. For example, the conversion of cellulose into lactic acid and/or into ethanol are both taken into account. Processes with the highest yields were selected as proxies to the technological progress that might occur ahead of 2050. The effective carbon yield CY_{ef} is calculated as the product of an efficiency rate η, specific to each conversion process, and the theoretical carbon yield CY_{th}, as shown in eqn (9.1).

$$CY_{ef} = \eta \times CY_{th} \tag{9.1}$$

The efficiency rates (*eta*, in %) were estimated from secondary data available in the literature and summarized in Table 9.3. They correspond to the ratio of products effectively obtained over the maximal amount that could be theoretically produced according to the chemical reaction in ideal conditions. The theoretical carbon yield (CY_{th}) is calculated on the basis of the chemical stoichiometry of the reactions. It corresponds to the share of carbon embodied in the final product that is neither emitted as CO_2 nor transferred in by-products. Based on the data compiled from the literature, the uncertainty in the efficiency of biorefinery processes is within 5–10%, which means a similar uncertainty in effective carbon yields, as shown by the minimum and maximum values in Table 9.3.

9.2.2.1 *Oilseeds*

Oilseeds must be processed to extract triglycerides through trituration. The amount of oil extracted depends on the type of feedstock. The efficiency reported in Table 9.3 is the mean based on data reported for different feedstocks.[23–26] In general, potential losses in the process, outside of

Table 9.3 Carbon yield of biorefinery processes.

Feedstock	Products	Processes	Value	CY_{th} (%)	η (%)	CY_{ef} (%)
Oilseeds	Biodiesel	Extraction and transesterification	**Min**	95.0	20.0	19.0
			Mean	95.0	30.0	28.5
			Max	95.0	40.0	38.0
Oilseeds	Biodiesel	Extraction and transesterification	**Min**	5.0	20.0	1.0
			Mean	5.0	30.0	1.5
			Max	5.0	40.0	2.0
Starch and sugars						
Starch and sugars	Ethanol	Fermentation (+hydrolysis if required)	**Min**	66.6	85.0	56.6
			Mean	66.6	90.0	59.9
			Max	66.6	95.0	63.3
Starch and sugars	Lactic acid	Fermentation (+hydrolysis if required)	**Min**	100.0	85.0	85.0
			Mean	100.0	90.0	90.0
			Max	100.0	95.0	95.0
Starch and sugars	Succinic acid	Fermentation (+hydrolysis if required)	**Min**	114.3	70.0	80.0
			Mean	114.3	80.0	91.4
			Max	114.3	90.0	102.9
Cellulose						
Cellulose	Ethanol	Hydrolysis and fermentation	**Min**	66.6	75.0	50.0
			Mean	66.6	80.0	53.3
			Max	66.6	85.0	56.6
Cellulose	Lactic acid	Hydrolysis and fermentation	**Min**	100.0	83.0	83.0
			Mean	100.0	88.0	88.0
			Max	100.0	93.0	93.0
Cellulose	Succinic acid	Hydrolysis and fermentation	**Min**	114.3	65.0	74.3
			Mean	114.3	75.0	85.7
			Max	114.3	85.0	97.2
Hemicellulose						
Hemicellulose	Ethanol	Hydrolysis and fermentation	**Min**	66.6	60.0	40.0
			Mean	66.6	70.0	46.6
			Max	66.6	80.0	53.3
Hemicellulose	Lactic acid	Hydrolysis and fermentation	**Min**	100.0	65.0	65.0
			Mean	100.0	75.0	75.0
			Max	100.0	85.0	85.0
Hemicellulose	Succinic acid	Hydrolysis and fermentation	**Min**	114.3	35.0	40.0
			Mean	114.3	45.0	51.4
			Max	114.3	55.0	62.9
Hemicellulose	Furfural	Hydrolysis and dehydration	**Min**	100.0	45.0	45.0
			Mean	100.0	55.0	55.0
			Max	100.0	65.0	65.0
Lignin						
Lignin	Heat and power	Substitution (coal)	**Min**	100.0	50.0	50.0
			Mean	100.0	60.0	60.0
			Max	100.0	70.0	70.0
Lignin	Aromatic compounds	Extraction	**Min**	60.0	3.8	2.3
			Mean	60.0	10.0	6.0
			Max	60.0	20.0	12.0
Mixed wet materials						
Mixed wet materials	Biomethane	Anaerobic fermentation	**Min**	70.0	55.0	38.5
			Mean	75.0	60.0	45.0
			Max	80.0	65.0	52.0
Mixed wet materials	Digestate	Anaerobic fermentation	**Min**	20.0	100.0	20.0
			Mean	25.0	100.0	25.0
			Max	30.0	100.0	30.0

products and residues, are assumed to be negligible and ignored in this case. Thus, the theoretical yields CY_{th} in g of products divided by g of reactants can be applied directly to calculate the effective carbon yields CY_{ef} in Table 9.3. After trituration, the triglycerides react with an alcohol to produce fatty methyl ester and glycerol; a process called transesterification. The residues of trituration are oil meals, usually used for animal fodder, thanks to their high content in proteins and cellulose.[24] An alternative might be to produce lignocellulosic ethanol, which could serve as a substitute for the alcohol (usually methanol) used in transesterification.[5] The amount of ethanol produced from these residues would be sufficient for transesterification and the process may even increase the concentration of proteins in the oil meals.[27] A valuable co-product of transesterification is glycerol, which can be the precursor to a variety of chemicals or products such as epoxy resins, ethylene glycol and 1,3 propanediol.[26] The theoretical carbon yield CY_{th} is based on formula and reaction Scheme 9.1.

9.2.2.2 *Starch and Sugars*

Starch and sugars (glucose) are easily fermentable to produce chemicals and fuels. The yield of ethanolic fermentation is almost identical for starch and sugars and slightly higher than that of cellulose or hemicellulose, since no pretreatment (hydrolysis) is required. A similar statement can be made for lactic and succinic acid production, which were selected as synthesizable chemicals in this chapter. The theoretical carbon yields were calculated based on the reaction Scheme 9.2.

Succinic acid has attracted much attention for carbon capture purposes since the theoretical carbon yield, at 114%,[28] exceeds 100%. Some CO_2 is reacted and absorbed in succinic acid production, either from external sources or as carbonates in the medium.[29] Hence the high potential of succinic acid in displacing fossil carbon sources and sinks. Succinic acid has been selected as

$CH_2OCOO(CH_2)_{17}CH_3$
|
$CHOCOO(CH_2)_{17}CH_3$ + 3 CH_3CH_2OH → 3 $CH_3CH_2OCOO(CH_2)_{17}CH_3$ + $OHCH_2\text{-}CHOH\text{-}CH_2OH$
|
$CH_2OCOO(CH_2)_{17}CH_3$

Scheme 9.1

(Ethanol)	$C_6H_{12}O_6 \rightarrow 2\ C_2H_6O + 2\ CO_2$
(Lactic acid)	$C_6H_{12}O_6 \rightarrow 2\ C_3H_6O_3$
(Succinic acid)	$7\ C_6H_{12}O_6 + 6\ CO_2 \rightarrow 12\ C_4H_6O_4 + 6\ H_2O$

Scheme 9.2

one of the top 12 value added chemicals derived from biomass.[30] The acid can be easily transformed into butanediol (BDO), tetrahydrofuran (THF) and gamma-butyrolactone. These molecules can be used as solvents in green chemistry, or further transformed into precursors of elastic fibbers and polymers such as polyurethane. In addition, succinic acid and its derivatives can be used as intermediates for a number of chemicals (*e.g.* dyes and perfumes) and pharmaceutical compounds (*e.g.* sedative and anti-cancer drugs).

9.2.2.3 *Cellulose*

Cellulose is a polycarbonate composed of C6 sugars, which can be hydrolyzed and fermented. As for starch and sugars, ethanol, lactic and succinic acids are among the most useful outputs. Compared to a simple sugar fermentation, an additional step is required to break down the cellulose in monomeric fermentable entities. The theoretical yield of the hydrolysis is approximated at 90%. The yield for ethanol production is then calculated by multiplying the individual steps; hydrolysis (90%), fermentation (93%) and purification (98%).[31] Similar yields as for glucose fermentation apply for lactic and succinic acid production, adjusted for the hydrolysis step. For comparison purposes, the yields described in the literature are also indicated in Table 9.4. Again, yields for succinic acid production exceed 100%, this time due to hydrogen as an additional reactant in the fermentation process, which changes the stoichiometry of the reaction.[32] In terms of mass, the ratio of products over feedstock must be corrected for a 1.12 g of succinic acid per gram of cellulose at maximum yield of 100%. Table 9.4 shows an excerpt of Table 9.3, with ranges in efficiency and effective yields which were narrowed to approximately 80% for CY_{ef} in the production of lactic and succinic acid.

9.2.2.4 *Mixed Wet Materials*

Anaerobic digestion (AD) is an effective way to provide energy from mixed inputs such as animal manure and the organic fraction of municipal wastes. In addition to biogas, AD also provides solid and liquid residues (digestate) containing a large fraction of valuable soil nutrients such as nitrogen, phosphorous and potassium compounds. A fraction of the initial carbon is also present in the digestate under a stabilized form. Therefore, using this digestate as soil amendment has a great potential to displace fertilizers and increase carbon sequestration in soils.[33] Biogas itself contains 60% methane and 40% carbon dioxide.[33,34]

Table 9.4 Carbon yield of cellulose transformation.

Feedstock	Products	Processes	CY_{th} (%)	η (%)	CY_{ef} (%)	References
Cellulose	Ethanol	Fermentation	67	82	55	31
Cellulose	Lactic acid	SSF[a]	100	93	93	44
Lignocellulosic material	Succinic acid	Fermentation	114	4–96	5–109	32

[a]Simultaneous saccharification and fermentation.

9.2.2.5 *Lignin*

Except for vanillin, very few chemicals produced from lignin are commercially viable.[35] As Ragauskas *et al.*[36] claim: "One can make anything from lignin except money". Approximately 40% of the global supply of lignin is sufficient to provide the heat and power necessary for operating biorefineries. The remainder can be burned in power plants *in lieu* of coal, in which case 1 PgC of lignin is more or less equivalent to 1 PgC of coal.[37]

Lignin has great potential as material substitutes, given the promising results for carbon fiber synthesis.[36] In addition, lignin is a valuable and abundant feedstock to produce aromatic chemicals such as benzene, toluene, xylene and phenolic compounds.[38,39] Yields are uncertain, since very few applications have reached commercial stages. Yields are typically given in terms of g of (mixed) products per g of lignin and are higher for products such as phenolic compounds isolation, since no oxygen atoms need to be removed. To account for unavoidable losses due to separation and purification of crude mixture to pure products, the lower ends of the ranges for yields are reported in Table 9.5. Calculation of the carbon yields is based on a theoretical depolymerization and hydrogenolysis reaction from coniferyl alcohol, which is considered as a model of lignin in this particular case. The efficiency of processes is equally theoretical and shown in Scheme 9.3.[40]

In the scenarios of material and chemical production, earmarking 40% of the initial feedstock to operate the biorefineries is not necessary as the by-products of extraction can largely cover the energy requirement.

9.2.2.6 *Hemicellulose*

Hemicellulose is a polycarbonate containing mainly C5 sugars (xylose and arabinose) and a small share of C6 sugars (mannose and galactose). Except for softwood, the percentage of C5 sugars generally exceed 80%.[28,41,42]

Table 9.5 Carbon transfer during lignin valorization.

Feedstock	Products	Processes	CY_{th} (%)	η (g^{out}/g^{in})	η (%)	CY_{ef} (%)	References
Lignin	Benzene/toluene/p-xylene	DPO[a] + hydro-deoxygenation	60	0.02	4	2.40	39
Lignin	Phenolic compounds	DPO[a] + hydro-genolysis	60	0.10	20	12.00	38
Lignin	Heat and power	Coal substitution	100	N.A.	60.00[b]	60.00	Own assumption
Lignin	Aromatic compounds	"Extraction"	60	0.02	4.00	2.40	Own assumption

[a]Depolymerization.
[b]40% of the feedstock is reserved to cover energy requirement of biorefineries.

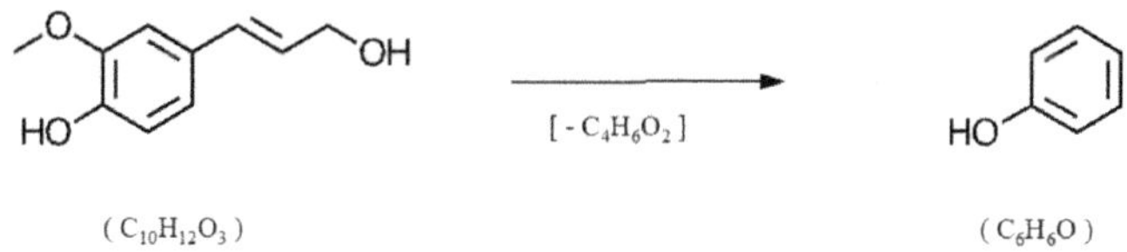

$$C_{10}H_{12}O_3 \rightarrow C_6H_6O\ \text{(phenol)} + C_4H_6O_2$$

Scheme 9.3

(Ethanol)	$3\ C_5H_{10}O_5 \rightarrow 5\ C_2H_6O + 5\ CO_2$
(Lactic acid)	$C_5H_{10}O_5 \rightarrow 5\ C_3H_6O_3$
(Succinic acid)	$7\ C_5H_{10}O_5 + 5\ CO_2 \rightarrow 10\ C_4H_6O_4 + 5\ H_2O$
(Furfural)	$C_5H_{10}O_5 \rightarrow C_5H_4O_2 + 3\ H_2O$

Scheme 9.4

Table 9.6 Carbon transfer during hemicellulose transformation.

Feedstock	Products	Processes	CY_{th} (%)	η (%)	CY_{ef} (%)	References
Hemicellulose	Ethanol	Hydrolysis + fermentation	67	71	48	31
Xylose	Furfural	Dehydration	100	85	85	58
Hemicellulose	Furfural	Hydrolysis + dehydration	100	56	56	59
Xylose	Lactic acid	Fermentation	100	86	0.86	44
Xylose	Succinic acid	Fermentation	114	49–61	56–70	Calculated from[43]
Hemicellulose	Lactic acid	Hydrolysis + fermentation	100	70	70	Own assumption
Hemicellulose	Succinic acid	Hydrolysis + dehydration	114	49	56	Own assumption

A hydrolysis step is necessary to decompose hemicellulose into its constituents. The yield of hemicellulose hydrolysis, as well as that of C5 sugars fermentation to ethanol (85%) are slightly lower than in the case of cellulose.[31] This is also true for xylose fermentation to succinic acid, whose yields are lower than for C6 sugars such as glucose.[43] While current research focuses primarily on simple sugar fermentation, improvements can be expected in the near future. This is similar for lactic acid production, although to a smaller extent. In this case, instead of the phosphoketolase pathway producing one lactate (C3) and one acetate (C2) per molecule of xylose (C5), another synthesis route based on the pentose phosphate pathway has been developed.[44] Furfurals were selected as a chemical valorization of hemicellulose, albeit with a 50% efficiency. Interestingly, this molecule is exclusively biosynthesized by dehydration of xylose. The reaction Scheme 9.4 was used to derive the theoretical carbon yields, as reported in Table 9.6.

All these chemical and fuel production pathways were evaluated simultaneously and compared per products and per state (solid, liquid and gaseous) with current consumption from fossil origin.

9.3 Results and Discussion

The mass balance exercise described here, leads to the following results, where we discuss the substitution of biogenic carbon for fossil carbon in fuels and chemicals. More emphasis is placed on the latter as the potential is greater than for the former.

9.3.1 Biomass Inputs

The total amount of terrestrial biomass available for second-generation biorefineries in 2050 is reported in Figure 9.2, according to the three different scenarios in Table 9.2.

As show in Figure 9.2, the potential inputs for biorefinery processes are very dependent on the availability of dedicated crops. In all three scenarios, lignocellulosic materials represent the main contribution, between 65% and 74% of the total. Therefore, much can be achieved by optimizing the supply chain of these inputs, for example, by increasing the recovery factor of wood and crop residues, without necessarily changing the residue to product ratio (RPR). This is valid for all crops including oilseeds and starch and sugar crops. By contrast, forestry residues have low potential to supply biorefineries with biogenic carbon, between 1% and 4% depending on the scenario.

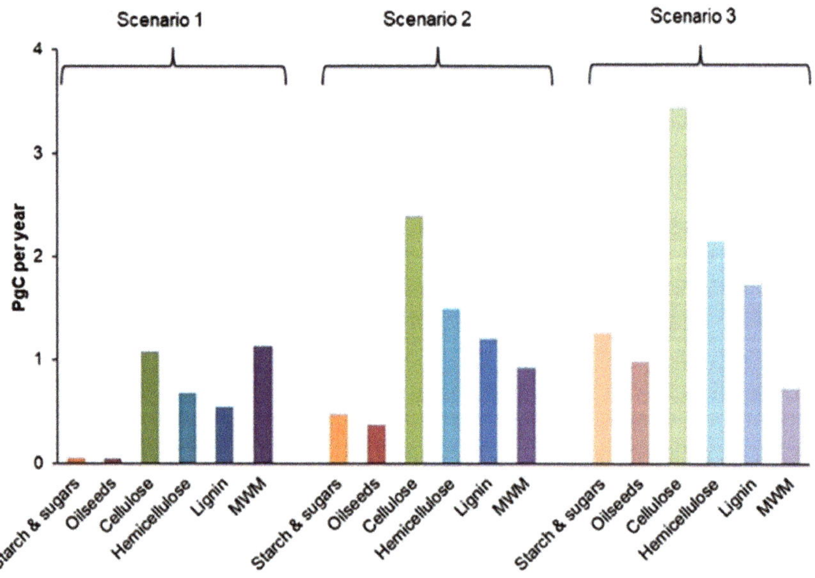

Figure 9.2 Potential inputs to biorefineries.

Two reasons explain this low availability: (1) this is a conservative estimate for the recoverability of wood residues at 25%; and (2), more importantly, residues from forestry, sawmills and pulp and paper industries are already well utilized for industrial, commercial and residential heating. The largest contributor to mixed wet materials (MWM) is animal manure. Although the potential availability decreases from scenario 1 to 3, in absolute terms, the potential remains high for biogas production.

This inventory of biogenic carbon potentially available for biorefinery processes might seem large, ranging from approximately 3.5 to 10.4 Pg of carbon or an average of 100 to 280 PJ per year. In comparison, Germany uses approximately 9000 PJ of final energy per year, 20% of which is electricity which is partly decarbonized.[45] In terms of carbon dioxide, Germany emits about 780 Mt of fossil CO2 per year from domestic activities,[45] whereas the biogenic carbon potential could displace up to 38 Gt or almost 50 times more. This clearly show the importance of mapping biogenic carbon potential to where the displacement of fossil sources is highest.

9.3.2 Fuel and Chemical Outputs

What can be achieved with the biogenic carbon potential above is quite remarkable even if the potential remains small on a global scale. If on average 0.027 Pg of carbon leads to 1 EJ, for energy purposes the conversion can be much more efficient in specific cases. Biorefinery processes themselves can be improved in light of the theoretical carbon yields, which exceed 100% in the case of succinic acid. Carbon yield should be optimized, either by improving the share of carbon conversion in existing pathways, or by developing innovative pathways able to incorporate additional carbon dioxide. Similarly, biorefinery processes can be tailored to specific products with high efficiencies.

9.3.2.1 Fuels

The results show that liquid biofuels, ethanol and biodiesel could substitute biogenic carbon for approximately 1.8 to 5.4 Pg of carbon in fossil fuels every year, depending on the chosen scenario. As shown in Figure 9.3, in scenario 1, the production of ethanol from starch and sugar crops alone is approximately 0.037 PgC per year out of 0.952 PgC of ethanol from all dedicated crops. Given a 60% yield, this is consistent with the production indicated by Popp *et al.*[14] at 85 billion liters (0.067 Pg of carbon assuming a density of 789 kg m^{-3}). This result demonstrates that the parameters used in scenario 1 were well calibrated. Moreover, the results for scenario 1 allow for more cultivation of LC crops but exclude the utilization of additional productive lands to grow starch and sugar or oilseed crops.

Solid fuels, namely lignin, ranges from 0.3 to 1 Pg C, and gaseous fuel from 0.3 to 0.5 Pg C per year. In comparison, the carbon contained in current fossil fuel combustion can be easily estimated from the amount of C in CO_2

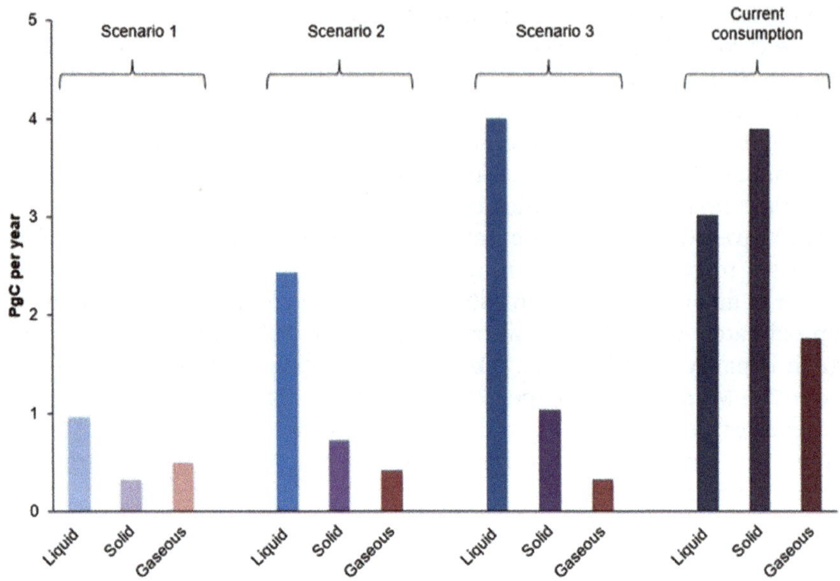

Figure 9.3 Potential of carbon substitution in fuels by biorefineries.

emissions, since the carbon content in fuels varies significantly and is difficult to estimate. An estimated 2.95 Pg of C in oil, 4.04 Pg of C in coal and 1.74 Pg of C in natural gas were burned in 2013.[46] Therefore, biomass conversion in liquid, solid and gaseous fuels may only displace a fraction of the current fossil fuel consumption. In the most optimistic scenario 3, however, liquid fuels could be entirely covered from biogenic carbon.

9.3.2.2 *Chemicals*

From a circular economy perspective, the potential of biogenic carbon to displace carbon of fossil origin in products, materials and chemicals, is equally, if not more relevant. The maximal amount of biogenic carbon that can be transformed into chemicals is shown in Figure 9.4. Values are reported in Pg of C per year, except for the smaller production volumes of glycerol and aromatics, which are in Tg of C per year. These results correspond to the maximum possible production in each scenario. For instance, if all carbohydrates (starch and sugars, cellulose and hemicellulose) were fermented in succinic acid, a maximal amount of 1.326, 3.255 or 5.212 Pg of C could be embodied in products every year, in scenarios 1, 2 and 3, respectively. The results can now be compared to the global fossil carbon contained in the main chemicals produced annually.[40,47] Approximately 220–240 Tg of carbon are embodied in chemicals every year, which means that biogenic sources can supply more than 5 times the carbon even in the most pessimistic scenario.

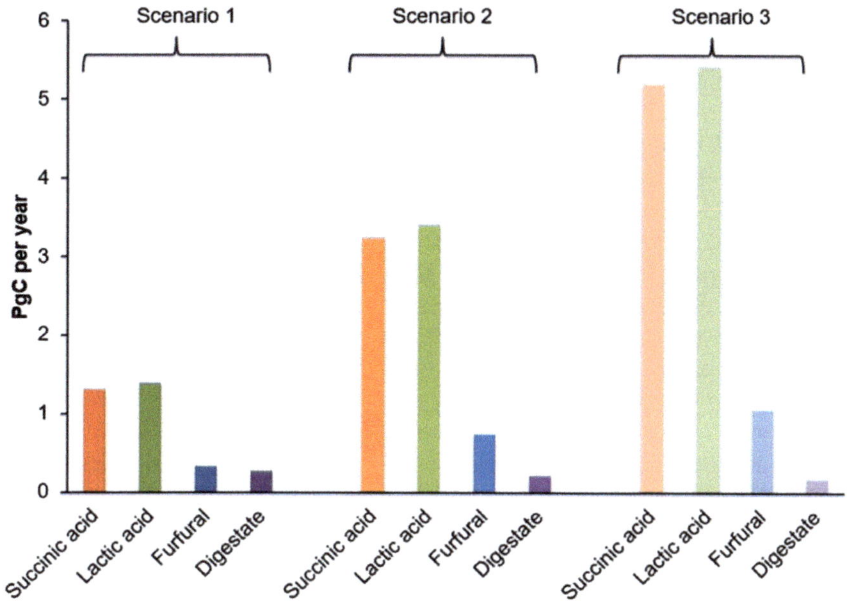

Figure 9.4 Potential of carbon substitution in fuels by biorefineries.

The carbon that is potentially convertible in chemicals by fermentation is superior than that for ethanol, since the carbon yield for the production of lactic and succinic acids is higher. Even in the pessimistic scenario, the potential is much more important than the current production of these two chemicals. The annual production of succinic acid is approximately 40 kt, but could reach 700 kt in 2020.[48] This represents 0.3 Tg of C annually immobilized in succinic acid (41% of C by weight). Biorefineries have the potential to completely substitute the fossil-derived production of succinic acid. The case of lactic acid is slightly different since almost all the current production is bio based.[49] The global production is approximately 400 kt, corresponding to 0.160 Tg of C.[48] Biomass supply is also largely sufficient to cover this supply, especially with cellulose as precursor, thanks to the relatively high carbon yield of its transformation.

Given that lactic and succinic acids can be easily transformed in polymers, the comparison with plastic production is also very relevant from a carbon sequestration and/or circular economy standpoint. In 2014, the global plastic production was approximately 311 Mt.[50] Assuming a carbon share of 75%, this represents 0.233 Pg of carbon. The main challenge consists of developing bio-plastics with similar or better proprieties than fossil ones, in order to completely replace fossil polymers with bio based equivalents in the near future. Furfural is also interesting, because it is synthesized from biomass sources only. The global production is approximately 200 kt per year,[48] corresponding to 0.126 Tg of C (63% of C by weight). Hemicellulose can provide enough feedstock in any of the scenarios above. Therefore, developing new chemical routes based on precursors such as furfural would be highly recommended to fully harness the potential of biorefineries.

Production of benzene derivatives from lignin remains a challenge. Using conservative assumptions, we estimate that approximately 5 to 17 Tg of carbon could be extracted from lignin, which is below the current market volume of these chemicals. The production of benzene and p-xylene is approximately 75 Mt and that of phenol is 10 Mt annually,[47] which corresponds to 67.5 and 7.7 Tg of carbon embodied in benzene and phenolic compounds assuming a carbon content of 90% and 77%, respectively. Therefore, the extraction of chemicals from lignin has to be optimized.

A growing production of biodiesel means that more uses of the co-products, in particular glycerol, will be needed. Indeed, the petrochemical synthesis of glycerol has been largely displaced. The total production is estimated at 3 Tg per year, with biodiesel contributing over 65% or 1.95 Tg.[51] In carbon terms, it corresponds to 0.761 Tg of C (39% of C by weight) close to the 0.940 Tg of C calculated in scenario 1. If all biorefineries currently sell their glycerol, new uses will be needed to absorb future production, especially in scenarios 2 and 3. Finally, the use of the digestate as soil amendment can store small amounts of carbon in soils, approximately 0.2 Pg of C. Although this is not a formal substitution of fossil carbon (except for small amount contained in certain fertilizers), this practice may avoid indirect emissions due to fertilizer production as well as land use change, which are significant.[52]

In terms of fossil carbon displacement, the results are promising and if the technology exists, bringing down the costs is the next challenge. Fortunately, some fuels and/or chemicals are co-products of one another, which illustrates the main avenue for cost reduction: process integration, including with pulp and paper mills, for example. Cost estimates are seldom found and are highly context-specific. Moreover, they can be positive and negative depending on the prices of the main and co-products. One example for furans and furfurals from lignocellulosic biomass shows that at approximately USD 1500 per ton, the production of these chemicals can compete with fossil equivalents.[53] In addition, as biorefinery processes scale up, costs should go down, compounded by the increasing costs of the incumbent fossil carbon processes.

9.4 Conclusion

Multiple solutions will be needed to move away from fossil carbon, not only to mitigate the impacts of climate change but also to close material cycles. The results presented here show that the production of chemicals and other materials based on coal, oil and gas can be replaced by comparable bio based products on a global scale. More importantly, this potential for substituting biogenic with fossil carbon is high in all scenarios, today and in the future. This is despite the rather conservative parameter estimates taken into account here, including for biomass sources and food wastes.

The fossil carbon in fuels, however, can only partly be offset by biogenic sources of carbon, except in the most optimistic scenario. Power-to-gas and the electrification of transportation will also reduce the need for fossil carbon such that biofuels, mostly ethanol, will displace more fossil carbon in the future. Biorefineries and bio based chemicals and fuels, are at the core of a larger integrated industrial ecosystem to use and reuse carbon for multiple applications.[54,55]

This bodes well for biogenic carbon and all processes necessary to convert it into useful products. The integration of these processes are an active field of research, bringing down the costs and making sure co- and by-products are reused to their full potential. In many ways, a biorefinery is a circular economy in itself, which contributes to closing the cycles of materials and products used everywhere else.

References

1. C. Figueres, H. J. Schellnhuber, G. Whiteman, J. Rockström, A. Hobley and S. Rahmstorf, *Nat. News*, 2017, **546**, 593.
2. A. Sonnenberg, J. Baars and P. Hendrickx, *Biorefinery*, International Energy Agency, Netherlands, 2007.
3. D. L. Sanchez and D. M. Kammen, *Nat. Energy*, 2016, **1**, 15002.
4. K. Anderson and G. Peters, *Science*, 2016, **354**, 182.
5. F. D. Meylan, V. Moreau and S. Erkman, *J. CO2 Util.*, 2015, **12**, 101–108.
6. H. Long, X. Li, H. Wang and J. Jia, *Renewable Sustainable Energy Rev.*, 2013, **26**, 344–352.
7. R. Slade, A. Bauen and R. Gross, *Nat. Clim. Change*, 2014, **4**, 99–105.
8. M. K. Patel, A. Bechu, J. D. Villegas, M. Bergez-Lacoste, K. Yeung, R. Murphy, J. Woods, O. N. Mwabonje, Y. Ni, A. D. Patel, J. Gallagher and D. Bryant, *Biofuels, Bioprod. Biorefin.*, 2018, **12**, 426–441.
9. V. Moreau and G. Massard, *Oxford Research Encyclopedia of Environmental Science*, 2017.
10. IPCC, *Global warming of 1.5°*, Special report, 2018.
11. H. Haberl, T. Beringer, S. C. Bhattacharya, K.-H. Erb and M. Hoogwijk, *Curr. Opin. Environ. Sustain.*, 2010, **2**, 394–403.
12. J. J. Roberts, A. M. Cassula, P. Osvaldo Prado, R. A. Dias and J. A. P. Balestieri, *Renewable Sustainable Energy Rev.*, 2015, **41**, 568–583.
13. K. Vávrová, J. Knápek and J. Weger, *Renewable Sustainable Energy Rev.*, 2014, **35**, 436–444.
14. J. Popp, Z. Lakner, M. Harangi-Rákos and M. Fári, *Renewable Sustainable Energy Rev.*, 2014, **32**, 559–578.
15. E. M. W. Smeets and A. P. C. Faaij, *Clim. Change*, 2007, **81**, 353–390.
16. FAO, http://faostat3.fao.org/faostat-gateway/go/to/home/E.
17. FAO, *Global Forest Products Facts and Figures*, Food and Agriculture Organization of the United Nation, 2012.
18. F. Krausmann, K.-H. Erb, S. Gingrich, C. Lauk and H. Haberl, *Ecol. Econ.*, 2008, **65**, 471–487.

19. N. S. Bentsen, C. Felby and B. J. Thorsen, *Prog. Energy Combust. Sci.*, 2014, **40**, 59–73.
20. S. Cornelissen, M. Koper and Y. Y. Deng, *Biomass Bioenergy*, 2012, **41**, 21–33.
21. J. Koornneef, P. van Breevoort, P. Noothout, C. Hendriks, L. Luning and A. Camps, *GHGT-11*, 2013, **37**, 6043–6052.
22. L. Appels, J. Lauwers, J. Degrève, L. Helsen, B. Lievens, K. Willems, J. Van Impe and R. Dewil, *Renewable Sustainable Energy Rev.*, 2011, **15**, 4295–4301.
23. L. Panichelli, A. Dauriat and E. Gnansounou, *Int. J. Life Cycle Assess.*, 2009, **14**, 144–159.
24. J. Schmidt, *Int. J. Life Cycle Assess.*, 2010, **15**, 183–197.
25. G. Pehnelt and C. Vietze, *Uncertainties about the GHG emissions saving of rapeseed biodiesel*, Univ. u.a., Jena, 2012.
26. S. J. Gerssen-Gondelach, D. Saygin, B. Wicke, M. K. Patel and A. P. C. Faaij, *Renewable Sustainable Energy Rev.*, 2014, **40**, 964–998.
27. E. M. Visser, D. O. Filho, M. A. Martins and B. L. Steward, *Biomass Bioenergy*, 2011, **35**, 489–494.
28. F. Cherubini and A. H. Strømman, *Energy Fuels*, 2010, **24**, 2657–2666.
29. K.-K. Cheng, X.-B. Zhao, J. Zeng and J.-A. Zhang, *Biofuels, Bioprod. Biorefin.*, 2012, **6**, 302–318.
30. T. Werpy, G. Peterson, A. Aden, J. Bozell, J. Holladay, J. White and A. Manheim, *Top Value Added Chemicals From Biomass: Results of Screening for Potential Candidates from Sugars and Synthesis Gas*, NREL, 2004.
31. F. Cherubini and G. Jungmeier, *Int. J. Life Cycle Assess.*, 2010, **15**, 53–66.
32. J. Akhtar, A. Idris and R. Abd Aziz, *Appl. Microbiol. Biotechnol.*, 2014, **98**, 987–1000.
33. J. Smith, A. Abegaz, R. B. Matthews, M. Subedi, E. R. Orskov, V. Tumwesige and P. Smith, *Biomass Bioenergy*, 2014, **70**, 58–72.
34. D. Tonini, G. Dorini and T. F. Astrup, *Appl. Energy*, 2014, **121**, 64–78.
35. M. P. Pandey and C. S. Kim, *Chem. Eng. Technol.*, 2011, **34**, 29–41.
36. A. J. Ragauskas, G. T. Beckham, M. J. Biddy, R. Chandra, F. Chen, M. F. Davis, B. H. Davison, R. A. Dixon, P. Gilna, M. Keller, P. Langan, A. K. Naskar, J. N. Saddler, T. J. Tschaplinski, G. A. Tuskan and C. E. Wyman, *Science*, 2014, DOI: 10.1126/science.1246843.
37. J. McKechnie, Y. Zhang, A. Ogino, B. Saville, S. Sleep, M. Turner, R. Pontius and H. L. MacLean, *Biofuels, Bioprod. Biorefin.*, 2011, **5**, 279–292.
38. R. J. A. Gosselink, W. Teunissen, J. E. G. van Dam, E. de Jong, G. Gellerstedt, E. L. Scott and J. P. M. Sanders, *Bioresour. Technol.*, 2012, **106**, 173–177.
39. A. L. Jongerius, P. C. A. Bruijnincx and B. M. Weckhuysen, *Green Chem.*, 2013, **15**, 3049–3056.
40. F. Cherubini and A. H. Strømman, *Biofuels, Bioprod. Biorefin.*, 2011, **5**, 548–561.
41. US DOE, *Biomass Feedstock Composition and Property Database*, US Department of Energy, 2004.

42. C. N. Hamelinck, G. van Hooijdonk and A. P. Faaij, *Biomass Bioenergy*, 2005, **28**, 384–410.
43. M. Bradfield and W. Nicol, *Bioprocess Biosyst. Eng.*, 2015, 1–12.
44. M. S. Ou, L. O. Ingram and K. T. Shanmugam, *J. Ind. Microbiol. Biotechnol.*, 2011, **38**, 599–605.
45. BMWi, *Energiedaten*, Bundesministerium für Wirtschaft une Energie, Berlin, 2018.
46. IEA, *CO_2 Emissions from Fuel Combustion, Highlights*, International Energy Agency, 2015.
47. CIEC, *The Essential Chemical Industry - Basic chemicals*, http://www.essentialchemicalindustry.org/chemicals/, (accessed 23 December 2015).
48. S. Choi, C. W. Song, J. H. Shin and S. Y. Lee, *Metab. Eng.*, 2015, **28**, 223–239.
49. M. A. Abdel-Rahman, Y. Tashiro and K. Sonomoto, *Bioenergy Biorefinery Biomass Innov. Technol. Dev*, 2013, **31**, 877–902.
50. Plastics Europe, *Plastics – the Facts 2015*, 2015.
51. M. Anitha, S. K. Kamarudin and N. T. Kofli, *Chem. Eng. J.*, 2016, **295**, 119–130.
52. C. Liptow, M. Janssen and A.-M. Tillman, *Int. J. Life Cycle Assess.*, 2018, **23**, 2110–2125.
53. A. J. J. E. Eerhart, M. K. Patel and A. P. C. Faaij, *Biofuels, Bioprod. Biorefin.*, 2015, **9**, 307–325.
54. V. Codina Gironès, S. Moret, E. Peduzzi, M. Nasato and F. Maréchal, *Energy*, 2017, **137**, 789–797.
55. R. Suciu, L. Girardin and F. Maréchal, *Energy*, 2018, **157**, 830–842.
56. V. Menon and M. Rao, *Prog. Energy Combust. Sci.*, 2012, **38**, 522–550.
57. ECN, *Phyllis2, database for biomass and waste, Energy research Centre of the Netherlands*, 2010.
58. R. Weingarten, J. Cho, W. C. Conner and G. W. Huber, *Green Chem.*, 2010, **12**, 1423–1429.
59. R. Sahu and P. L. Dhepe, *ChemSusChem*, 2012, **5**, 751–761.

CHAPTER 10

Case Study – Packaging

M. NIERO*[a] AND A. MANZARDO[b]

[a] Aalborg University, Department of Planning, A.C. Meyers Vænge 15, 2450, Copenhagen, Denmark; [b] University of Padova, Department of Industrial Engineering, Via Marzolo 9, 351531, Padova, Italy
*Email: monian@plan.aau.dk

10.1 Introduction

10.1.1 Packaging: Evil or Necessary Good?

In modern society, packaging plays a central role that brings in controversial management issues.[1] According to the Cambridge dictionary, packaging is defined as the material in which objects are wrapped before being sold.[2] Packaging fulfils four main functions for the marketing of goods and specifically for non-durable consumer products:[3]

- Containment, being used to contain goods such as liquids and solids;
- Protection, being used to preserve the characteristics and performance of the product contained; as such, packaging allows for the transportation of goods over a large distance and guarantees their quality until they reach their final destination: the consumer;
- Convenience, provided in the carriage of the product from one place to another, in stocking and in consuming;
- Communication, being a means of marketing: as such, packaging allows for the identification of the product and its brand and therefore directly supports the selling of the goods it contains.

Life Cycle Assessment: A Metric for the Circular Economy
Edited by Aiduan Borrion, Mairi J. Black and Onesmus Mwabonje

Published by the Royal Society of Chemistry, www.rsc.org

All of these functions can be performed at different levels, according to the following classifications:[3]

- Primary packaging, which is in direct contact with the contained product that is usually directly sold to the final customer and supports all of the above-mentioned functions;
- Secondary packaging, which is used to contain a number of primary packages; for example, a corrugated case or box that is used for facilitating transportation of goods and is sometimes used for containing and displaying the product in the retail setting;
- Tertiary packaging, which is used to contain a number of secondary packages and mainly used to secure transportation of goods in space; this is usually a stretch-wrapped pallet of corrugated cases.

Considering all these aspects, packaging is fundamental to respond to the needs of a growing globalized market (local products sold all over the world) and changing consumer habits (smaller portions, outdoor consumption). The packaging sector generates approximately 2% of the gross national product in developed countries, and approximately half of this packaging concerns the food sector.[3]

Despite being a necessary good, packaging can have a negative influence on the natural environment for several reasons.[4] First of all, the actual global packaging market is based on materials extracted from limited primary resources such as oil and ores. In 2017 the market share relied on plastics, *i.e.* flexible packaging (39%) and rigid plastics (28%), followed by metals (11%), glass (9%), paper-based containers (7%) and, finally, poly-laminated cartons (used for packaging liquids, 6%).[5] Secondly, once used these resources have to be managed as waste, and packaging is a major contributor to municipal solid waste. For example, in 2015, European citizens generated 84.5 million tons of packaging waste, 38% of which were paper, 21% plastic, 20% glass, 15% wood and 6% metals.[6] In the same year, considering a *per capita* packaging waste generation of 166.3 kg, 131.9 kg were recovered and only 109 kg were actually recycled.[6]

10.1.2 Packaging in the Circular Economy from a Policy Perspective

The Circular Economy (CE) both offers new directions and supports current practice in the packaging sector, *e.g.* in terms of preventing the production of packaging waste, and promoting the reuse, recycling and other forms of recovering of packaging waste, instead of its final disposal. In the context of the EU Action Plan for Circular Economy,[7] the Directive (EU) 2018/852 of the European Parliament and of the Council of 30 May 2018 amending Directive 94/62/EC on packaging and packaging waste[8] has identified a set of measures aiming at:

- Preventing the generation of packaging waste and minimizing the environmental impact of packaging, *e.g.* through extended producer responsibility schemes;

- Increasing the share of reusable packaging put on the market and of systems to reuse packaging without compromising food safety, *e.g.* through deposit-return schemes;
- Meeting the specific recycling targets for each material (for 2025 and 2030, respectively): plastic (50%, 55%); wood (25%, 30%); ferrous metals (70%, 80%), aluminium (50%, 60%), glass (70%, 75%), paper and cardboard (75%, 85%), and overall packaging (65%, 70%).

The Directive (EU) 2018/852 applied since 4 July 2018 and had to become law in the EU countries by 5 July 2020. The essential requirements that the packaging placed on the market by EU countries must meet are:[8] (1) to limit the weight and volume of packaging to a minimum in order to guarantee the required level of safety, hygiene and acceptability for consumers; (2) to reduce the content of hazardous substances and materials in the packaging material and its components; and (3) to design reusable or recoverable packaging.

Packaging recovery systems should be set up to provide for the return and/or collection of used packaging and/or packaging waste, as well as the reuse or recovery including recycling of the packaging and/or packaging waste collected. Producer responsibility schemes for the return and/or collection of used packaging and/or packaging waste and its channelling to the most appropriate waste management option, as well as for reuse or recycling of the collected packaging and packaging waste, should be established by 2025 for all packaging.[8]

10.1.3 Overview of CE Strategies Relevant for Packaging

CE strategies are linked to three main actions for waste and reduction of virgin resource consumption within the 3Rs principles: reduce, reuse and recycle.[9,10] The 3Rs are embedded in the well-known CE illustration developed by the Ellen MacArthur Foundation (EMF), *i.e.* the so-called "butterfly diagram",[10] (see Figure 10.1). The distinction between technical and biological cycles was derived by the Cradle to Cradle® (C2C) design framework.[12] According to the C2C terminology, the technical metabolism is defined as a closed-loop system of manufacture, recovery and reuse, maintaining its highest value through many product life cycles.[13] In the technical cycle, "technical nutrients" act as "products of service", *i.e.* materials are used as part of durable goods that provide a service to customers or, as stated in the EMF diagram, to "users", which are at the centre of this cycle, see Figure 10.1. Opposed to products of service are in the biological cycle the so-called "products of consumption", which are made of "biological nutrients", *i.e.* materials that are meant to be returned to the biosphere by composting or anaerobic digestion through several cascades, by targeting those who in the EMF diagram are called "consumers". The technical metabolism illustrates how the value of technical finite materials should be kept through continuous loops based on strategies such

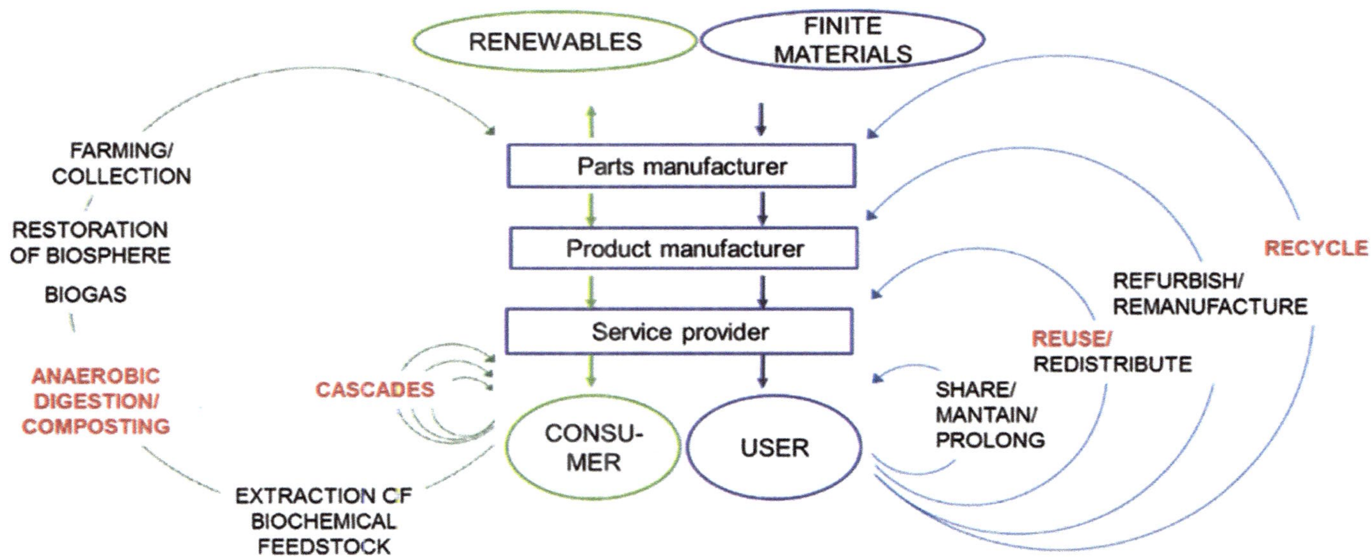

Figure 10.1 Circular economy (CE) system diagram where the CE strategies relevant for packaging are highlighted.

as maintain (*i.e.* inner loop based on reduce), reuse/redistribute, and refurbish/remanufacture (*i.e.* middle loops based on reuse), and recycle (*i.e.* outer loop),[14] meanwhile, the biological metabolism illustrates a regenerative process which is targeting renewable materials. Figure 10.1 illustrates the above-mentioned cycles of the CE and highlights the CE strategies that are relevant for packaging, namely reuse (*e.g.* in the case of glass bottles, steel kegs, wooden or plastic pallets) and recycling (relevant for all materials) for the technical cycle, and biodegradability/compostability for the biological cycle, which is relevant for biodegradable materials such as paper, wood or biodegradable plastics.

Pauer *et al.*[15] performed an overview of circularity indicators relevant for packaging, *i.e.* those that inform about material flows at the product level and the use of renewable energy. Their classification includes three distinct categories, *i.e.* indicators referring either to: (1) material inputs; (2) material outputs; or (3) energy flows. Input-related indicators refer to the materials used to produce packaging, *i.e.* recycled content (RC) being the proportion, by mass, of a recycled material in goods or packaging (for the the technical cycle) or the renewable content, which informs about the use of biological sources. Output-related indicators refer to the fate of the materials after disposal of packaging, namely recyclability or reusability for the technical cycle and compostability for the biological cycle. Finally, the indicator "share of renewable energy" informs about the use of renewable energy for the production, use and disposal of packaging. These indicators can be combined and two of the most used metrics for measuring circularity (including in the packaging sector) are the Material Circularity Indicators (MCI) and the Material Reutilization Score (MRS). The MCI, developed by the EMF and Granta Design,[16] provides an indication of how much a product's materials circulate, in order to allow companies to understand how far they are in the transition from "linear" to "circular". It is calculated considering the following parameters: (1) destination after use, distinguishing among the percentages of recycling collection rate (RR) and reuse rate (ReR); (2) percentage of recycled feedstock (RC); (3) efficiency of the recycling process, *i.e.* the yield during recycling, and (4) utility during use stage, *i.e.* use intensity with regard to an average product in the market. The MRS is the metric used to quantify material reutilization, *i.e.* the criterion included in the C2C certification program addressing the recycling value of the materials,[17] which quantifies the recyclability or compostability potential of a product, in the technical or biological cycle, respectively. In the case of the technical cycle, the MRS considers two variables: the intrinsic recyclability (IR) of the product, *i.e.* the percentage of the product that can be recycled at least once after its initial use stage and the percentage of RC. Since each metric only includes a selection of features that are relevant for assessing the performances of different packaging, the use of material circularity indicators (such as MRS and MCI) provide only a partial picture on the overall sustainability performance of packaging and needs to be complemented with other indicators based on life cycle thinking.[18–20]

10.2 Life Cycle Assessment of Packaging

10.2.1 Focus at the Product Level

The global trend in terms of increased packaging consumption and the CE policy measures including targets on recycling put further pressure on companies in the packaging sector, specifically in the food and beverage industry, to take action for quantifying the environmental impacts of their activities. One of the methodologies that has been widely adopted to perform such task is Life Cycle Assessment (LCA).[21,22] Beverage packaging was the first product category where the LCA methodology was applied to compare the environmental performances of different alternatives.[23]

Vignali[24] presented a literature review of 172 LCA papers related to food packaging systems published in scientific journals with impact factors, from 1990 to June 2015. The analysis highlighted the interest of the packaging industry for LCA results and specifically the intention of the studies to quantify environmental impacts related to the use and end-of-life management of packaging materials, which are also key aspects in a CE perspective. Other relevant topics identified by the same author are the comparison among different packaging in order to identify more sustainable solutions for the systems understudy and the introduction of new bio-based materials, as well as smart and functional packaging to reduce the related environmental impacts. Most of the investigated studies agreed that both from a CE perspective and from an environmental impacts reduction perspective it is relevant to work on the shelf life of packaging (along with its influence on food waste) and its recyclability at the end of life.

LCA has, throughout the years, been applied to quantify the potential environmental impacts of specific packaging solutions, mainly referring to primary packaging, such as glass bottles, aluminium cans, plastic bottles such as polyethylene terephthalate (PET) or high-density polyethylene (HDPE) bottles or aseptic cartons, as reported in Table 10.1 (which includes a non-exhaustive list of studies). Only a limited number of cases of LCA specifically applied to secondary packaging, such as cardboard boxes (*e.g.* Koskela *et al.*[25]) or tertiary packaging, such as pallets (*e.g.* Niero *et al.*[26]), are available. However, as packaging is ubiquitous in everyday life and fulfils the four above-mentioned main functions (containment, protection, convenience and communication), it is necessary for a wide range of not only consumer products, such as food and beverages, personal hygiene products or household products, but also for durable goods, such as electronics.

The life cycle of a packaging includes the following stages: raw materials extraction and production, filling and packing, distribution, use, end of life (EoL), as represented in Figure 10.2, which includes primary, secondary and tertiary packaging.

The hotspot in the packaging life cycle, *i.e.* the phase responsible for most of the environmental impact, is in most of the cases the raw materials extraction, as summarized in Table 10.1. Raw materials extraction refers to

Table 10.1 Non-exhaustive list of product Life Cycle Assessment studies in the beverage and food sector including packaging made of different materials. P = primary, S = secondary, T = tertiary packaging.

Material category	Material	Packaging	Product	Hotspot	Source
Plastic	PET	Bottle (P)	Olive oil	Raw materials extraction, distribution	Accorsi *et al.*[96]
	PET	Bottle (P)	Carbonated drinks	Raw materials extraction	Amienyo *et al.*[51]
	PET	Bottle (P)	Water	Raw materials extraction	Pasqualino *et al.*[34]
	HDPE	Bottle (P)	Milk	Raw materials extraction	Meneses *et al.*[97]
	HDPE	Bottle (P)	Milk	Raw materials extraction	Scipioni *et al.*[52]
	PET	Tray (P)	Sliced meat	Raw materials extraction	Toniolo *et al.*[35]
	PP	Box (P) (reusable)	Food container	Use (washing)	Gallego-Schmid *et al.*[32]
	PP	Pot (P)	Baby food	Raw materials extraction	Humbert *et al.*[98]
	PP	Crates (P), reusable	Fruit & vegetable	Raw materials extraction	Albrecht *et al.*[99]
	PS	Tray (P)	Poultry product	Raw materials extraction	Zampori & Dotelli[100]
	HDPE	Crates (*S*) reusable	Bread delivery	Distribution	Koskela *et al.*[25]
Wood	Wooden	Pallet (T)	n.a.	Raw materials extraction	Niero *et al.*[26]
	Wooden	Box (P)	Fruit & vegetable	Distribution	Albrecht *et al.*[99]
Ferrous metals	Steel	Keg (P)	Beer	Distribution	Cordella *et al.*[30]
Aluminium	Aluminium	Can (P)	Carbonated drinks	Raw materials extraction	Amienyo *et al.*[51]
	Aluminium	Can (P)	Beer	Raw materials extraction	Pasqualino *et al.*[34]
	Aluminium	Can (P)	Beer	Raw materials extraction	Detzel and Möncker[31]
	Aluminium	Can (P)	Beer	Raw materials extraction	Niero *et al.*[27]
	Aluminium	Tray (P)	Poultry product	Raw materials extraction	Zampori & Dotelli[100]
Glass	Glass	Bottle (P)	Olive oil	Raw materials extraction, distribution	Accorsi *et al.*[96]
	Glass	Bottle (P)	Carbonated drinks	Raw materials extraction	Amienyo *et al.*[51]
	Glass	Bottle (P)	Beer	Raw materials extraction	Cordella *et al.*[30]
	Glass	Bottle (P) – refillable	Beer	Filling	Detzel and Möncker[31]
	Glass	Box (P) (reusable)	Food container	Use (washing)	Gallego-Schmid *et al.*[32]

Table 10.1 (*Continued*)

Material category	Material	Packaging	Product	Hotspot	Source
	Glass	Jar (P)	Tomato sauce	Raw materials extraction	Manzardo *et al.*[81 b]
	Glass	Jar (P)	Baby food	Raw materials extraction	Humbert *et al.*[98]
Paper and cardboard	Aseptic	Carton (P)[a]	Milk	Raw materials extraction	Pasqualino *et al.*[34]
	Aseptic	Carton (P)[a]	Milk	Raw materials extraction	Meneses *et al.*[97]
	Aseptic	Carton (P)[a]	Milk	Raw materials extraction	Scipioni *et al.*[52]
	Aseptic	Jar (P)[a]	Tomato sauce	End of life	Manzardo *et al.*[81 b]
	Cardboard	Box (P)	Fruit & vegetable transport	Raw materials extraction	Albrecht *et al.*[99]
	Cardboard	Box (P)	Bread delivery	Raw materials extraction	Koskela *et al.*[25]

[a]Refers to a multimaterial.
[b]Only covers water footprint.

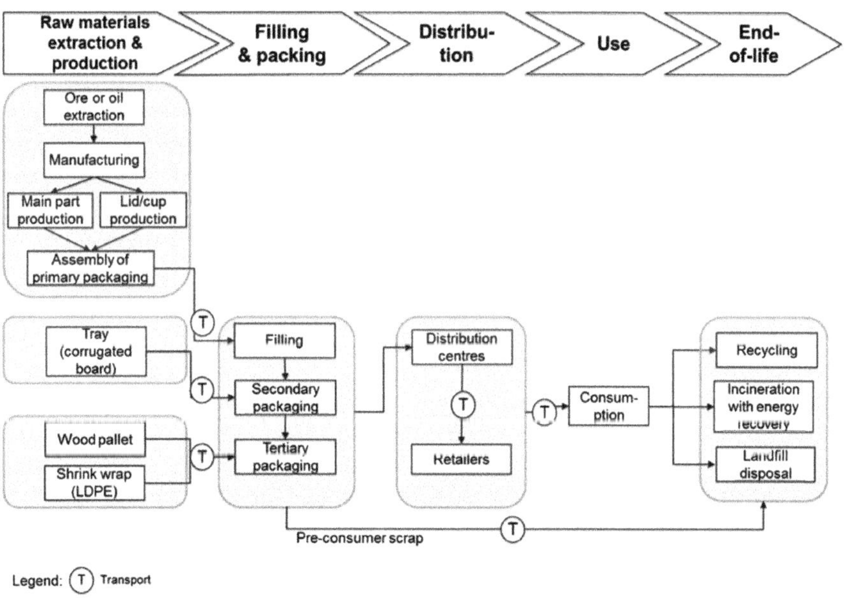

Figure 10.2 Exemplification of the life cycle of packaging.
Adapted from ref. 33 with permission from Elsevier, Copyright 2016.

primary production from bauxite for aluminium,[27] crude oil extraction for polymer production for the different plastics types, silicon dioxide and alkaline and earth-alkaline metal oxides,[28] cellulose fibres for cartons and lignin for wood.[29] The only exceptions, where raw materials extraction is not the most impacting life cycle stage, refer to cases of reusable packaging, where either the distribution or use phase are the hotspots. Distribution emerged as the hotspot in the case of reusable HDPE crates for bread delivery,[25] refillable steel kegs[30] or refillable glass bottles[31] for beer. The use phase emerged as the hotspot in the case of reusable glass and plastic food containers,[32] due to impacts of washing.

A critical aspect for the quantification of the environmental impacts of packaging refers to the EoL, as the selection of the final disposal (recycling, incineration with energy recovery and disposal in landfill) can greatly influence the results of the LCA, as demonstrated in the case of aluminium cans, *e.g.* Niero *et al.*,[33] laminated cartons, *e.g.* Pasqualino *et al.*[34] or plastic packaging, such as trays in Toniolo *et al.*[35] In particular, the LCA results can be affected by the choice of the Life Cycle Inventory (LCI) modelling option for recycling and the multitude of available methods complicates the choice of an appropriate method for the LCA practitioner.[36] Given the relevance of recycling as a key CE strategy, this aspect is particularly challenging and will therefore be discussed in detail in Section 10.3.3.

10.2.2 Organizational Life Cycle Assessment in the Packaging Sector

Along with the quantification and analysis of the environmental impacts of products, a growing number of companies worldwide are showing an increasing interest in the application of the life cycle perspective at organizational level.[37] The reasons for this interest are manifold. In some countries such as France and Italy, the management of environmental impacts is considered to be strategic and supports the success of an organization.[38] Moreover, together with the increased attention of the international community towards environmental issues such as climate change, the behaviour of consumers has also changed: they have started to include the environmental impact of products and services in their decision-making process.[39]

To support this attention a variety of organizations has invested in research and harmonization towards the definition of Organizational Life Cycle Assessment (OLCA).[38] ISO/TS 14072,[40] a technical specification focused on methodological aspects, defines OLCA as the "compilation and evaluation of the inputs, outputs and potential environmental impacts of the activities associated with the organization as a whole or portion thereof adopting a life cycle perspective". This document includes requirements and guidelines for the application of LCA to organizations. The UNEP Life Cycle Initiative has also worked on the refinement of the methodology and to support its application in industries. In 2015, a guidance document on OLCA[38,41] was published, presenting also a first range of embryonal applications of the methodology. Moreover, another initiative called Organization Environmental Footprint (OEF), launched by the European Commission,[42] presents a method to address environmental impacts of organizations from a life cycle perspective. Even if OEF can be seen as a type of OLCA, some of the principles and requirements have been identified as not being in complete alignment with the constitutional documents of LCA, such as in the case of comparative assertion related to organizations, weighting and reporting.[43]

Aligning with product LCA, the general structure of an OLCA study is based on four phases: goal and scope definition, life cycle inventory analysis (LCI), life cycle impact assessment (LCIA) and life cycle interpretation. Most of these phases are consistent with principles and requirements of product LCA standards,[21,22] however, considering the specificity of organizations, new aspects are introduced in the scoping phase of LCA, as well as in the inventory analysis.

With specific reference to the goal and scope definition, the following aspects significantly influence and affect all the phases of an OLCA study:[41] definition of the organizational boundaries, reporting unit and system boundaries.

The definition of organizational boundaries is a fundamental step to determine who the organization understudy is. This is particularly relevant with big companies that usually have complex organizational structures, with financial shares in other companies and several divisions that can release very different products. In the case of packaging companies, an example could be of an organization committed in the production of both packaging material and filling machines (Figure 10.3) or also a company with production taking place in several countries and/or an affiliated or subsidiary company. ISO/TS 14072[40] introduces two potential approaches to deal with organizational boundaries definition, namely: the financial/operational control and the equity share approach. The first one, usually adopted by small companies, supports the quantification and monitoring of impacts of the activities directly under the financial or operational control of the organization understudy. The second one supports the quantification

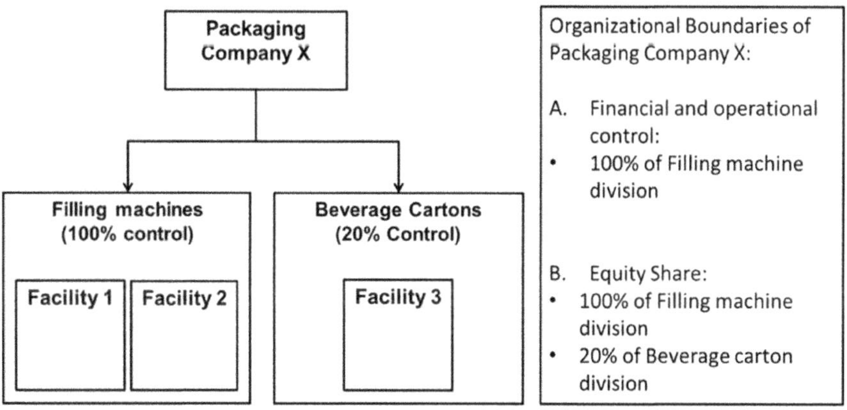

Figure 10.3 Illustration of organizational boundaries definition in the case of (A) financial and operational control and (B) equity share.

and monitoring of impacts of all the activities on which the company has a share.

Figure 10.3 represents an example of organizational boundaries adopting the two consolidation approaches; the packaging Company X fully owns the activities of division Filling machines and has a share of 20% of division Beverage Carton, but no financial or operational control on its facilities. If the company decides to adopt the operational and financial approach method it should consider all of the activities of the division it fully owns. In the case the equity share approach is selected, the Packaging Company X should also take care of the 100% of the Filling machine division and the 20% of impacts generated from division Beverage Carton.

The definition of the reporting unit is introduced instead of the product LCA functional unit concept. This reflects the complexity of organizations that can deliver a variety of products (also known as product portfolio).[38] The reporting unit is used as a reference for the quantification of the potential environmental impacts of the activities of these organizations. In the case of a packaging company, the reporting unit could be the overall production in a reference period (*e.g.* a year) or the revenues associated with the selling of products. The reporting unit was designed to support the flexibility of OLCA methodology to meet the goal of the study. A recent study by Manzardo *et al.*[44] proved that the definition of the reporting unit can be a challenging issue, specifically when the products delivered by the organizations change a lot over time. To overcome this issue, the concept of the "activity portfolio" is introduced and defined as the set of activities performed by the organization as declared in its company statute and specified in job descriptions.[44] Such activities are in fact transversal to the products of the organization, and generally stable over time. Due to its recent introduction, no case studies on the application of the activity portfolio can be identified in the packaging sector. Nevertheless, some studies have already elaborated on the reporting unit for packaging companies. Scipioni *et al.*[45] presented one of the first applications of LCA to organizations and defined the reporting unit of a beverage carton company as the overall number of packs equivalent produced over a solar year period. Pack equivalent was used as a reference unit by the company representing a standard surface of beverage carton material. In this case, the definition of the reporting unit was an easy task, as the products sold by the company had exactly the same function and differed only in the size of the surface of the carton. Manzardo *et al.*[39] presented a case study of a beverage company where both glass and plastic bottles for containing mineral water are formed and filled; in this study, the reporting unit was referred to the content of the packaging integrated by the definition of a product portfolio reporting on the number of formats used to contain the mineral water.

With reference to system boundaries, considering the complexity of organizations, flexibility is allowed if correctly justified,[41] allowing the organization understudy to focus on either the organization as a whole or its sections such as business divisions, brands, regions or facilities.[38] ISO/TS

14072[40] introduces different perspectives when defining system boundaries ranging from the product perspective (where the overall production of a single product is considered), to the facility/division perspective (where all of the activities of such facility/division are considered), to the more comprehensive organizational perspective. It is important to point out that the system boundary depends directly on the organizational boundary definition, the reporting unit and the reference period chosen for the assessment. Pelletier *et al.*[46] highlight that often the resource use and emissions linked to upstream processes (*e.g.* raw material extraction, processes of material production and supply transport processes) and/or downstream processes (*e.g.* distribution and end-of-life processes) can be as or more determinant of the overall environmental profile of the reporting organization. Scipioni *et al.*[45] and Scipioni *et al.*[47] confirm this, where 95% of final impacts on climate change of a beverage carton company takes place on raw material extraction and end-of-life management. The correct definition of system boundaries proved to be a fundamental aspect in several studies due to its direct influence on the capacity of the organization to have a correct view of the impact of its activities and therefore to set improvement strategies. Such critical aspects will be discussed in depth in Section 10.3.2.

With specific reference to inventory analysis, ISO/TS 14072[40] introduces three data collection and calculation procedures. The first one is called the top-down approach, where data are collected, looking at the whole organization without referring to a specific product. This choice is usually preferred when no specific data are available; however, it limits the capacity to investigate the causes of environmental impacts. The second approach is called the bottom-up approach: this is the case when data on specific products are available. This procedure can be more complex because the details of the data are greater; however, it allows a better understanding of the contribution that each product and process brings to the overall impacts related to the organization understudy. Finally, the hybrid approach is a combination of the previous two.[41] Case studies of OLCA of packaging companies usually adopt this last approach: the top-down approach is considered for energy and other resource use,[45] while the bottom-up approach is for raw materials and packaging.[48]

Requirements and methodological aspects of product LCA impact assessment and interpretation phases are directly applicable without modification of OLCA.

Results of OLCA applications to packaging companies answer to a variety of goals and applications. In Scipioni *et al.*[45] the goal of the analysis was the overall monitoring of the carbon footprint performance of a beverage carton company with two sites located in Italy. The study included all of the activities of the company in a life cycle perspective and was related to the quantification of a baseline to set and verify emissions reduction objectives. Most of the impacts were shown to be located outside of the direct control of the organization, specifically on raw material production and end of life.

In Manzardo *et al.*,[39] the goal of the study was to quantify the environmental impacts of a water bottling division (glass and PET bottle) of a beverage company located in the north east of Italy. The intended applications of the OLCA were to provide a system for tracking its environmental-performance activities and to develop a pilot model to be extended in the future to the other divisions and sites of the company. The inventory analysis was performed, adopting a bottom-up approach that was checked at the end of the analysis, matching the results with data collected adopting a top-down approach. This allowed verification of the consistency of the collected data and information. The adopted inventory analysis approach allowed the company to verify impacts at different levels such as a single product, single production line, overall lines per packaging (Glass and PET) and the division. Impact assessment confirmed the importance of the supply chain, in particular the relevance of the impacts of raw materials extraction.

10.3 Methodological Challenges for LCA of Packaging in the CE

10.3.1 Functional Unit Definition

The functional unit (FU) is "the quantified performance of a product system for use as a reference unit",[21] which defines the qualitative aspects and quantifies the quantitative aspects of the function, by answering the questions "what?", "how much?", "for how long/how many times?", "where" and "how well?".[49] For food and beverage packaging, the FU is typically defined as carrying (and protecting) a certain amount of beverage or food until the expiry date. In the case of aluminium cans, either the single packaging can be taken as the starting point to quantify the FU, *e.g.* 1000 cans of 33 cl size with an overall volume of 330 l, as done by Stichling and Nguyen-Ngoc,[50] or the total volume contained by the packaging, *e.g.* 100 l (=1 hl) in the case of Detzel and Mönckert.[31] The selection of the containment of a fixed volume of beverage, *i.e.* 1 l or 1 hl, allows for the quantitative comparison of alternative ways of providing the function by different packaging for the same product category, *e.g.* carbonated soft drink,[51] milk[52] or beer.[19,31]

However, given the CE ambition of using materials in continuous loops, the function of the product is not only to carry (and protect) its content, but also to supply secondary resources for the next uses of the packaging. The idea of multiple co-functions is introduced in the ILCD Handbook, Annex C[53] with an example of an aluminium beverage can, which has a first co-function (carrying and protecting the beverage it contains), and as a second co-function (the aluminium scrap it provides as secondary resource for subsequent product systems). Based on this definition, Niero and Olsen[54] defined a so-called "Circular economy-inspired FU" for aluminium cans as "the containment of 1 hl of beer until the expiry date and supply of resource after its use stage for 30 loops". The choice of the 30 loops was arbitrary to take into account a temporal span of 5 years, as one recycling loop for aluminium cans corresponds to 60 days.[55]

10.3.2 System Boundaries Definition

According to the European Environment Agency[56] "when using an LCA for determining the environmental impacts of improved product circularity, the methodology must be adapted to assess the impacts over different life-cycles instead of the linear concept of from cradle to grave. This means that extra care needs to be taken in establishing the system boundaries, as these may seriously affect the outcome of an LCA." To overcome this challenge and to be consistent with the "circular economy inspired-FU" introduced in Section 10.3.1, Niero and Olsen[54] suggest shifting the focus from modelling a pure aluminium (Al) flow,[57] which is common practice when modelling 1 life cycle, to modelling the actual alloy composition, *i.e.* including the production and change in composition of the alloying elements and considering that aluminium scraps are used for manufacturing the body of the can and not the lid, as represented in Figure 10.4.

According to the ILCD Handbook Annex C,[53] to provide the first co-function (containment of 1 hl of beer) the can has to be produced, meanwhile to provide the second function in the form of scrap (supply of secondary resource flow), it is sufficient if the aluminium grade the can is made of is produced, *i.e.* considering the different alloy composition of the body and lid. All other steps, such as the transport of the aluminium to the can plant and making the can, are indeed not attributable to the provision of

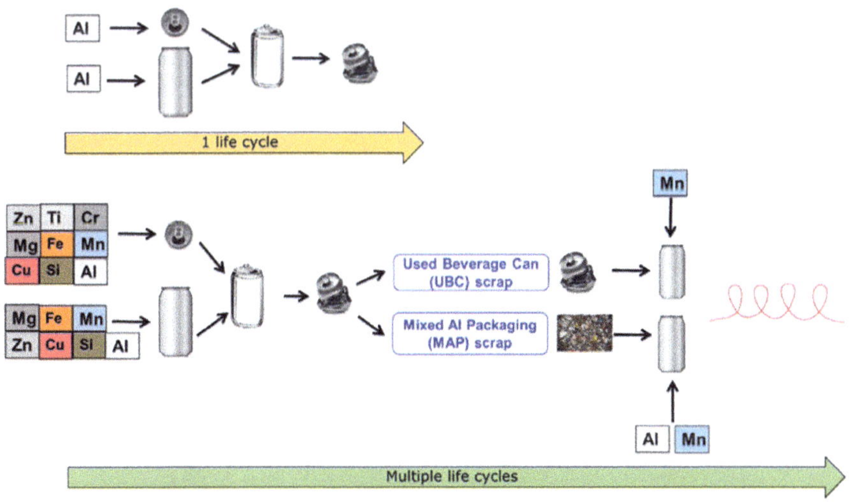

Figure 10.4 Difference between modelling one life cycle of aluminium cans based on pure Al flow (on the top) and multiple life cycle considering the actual alloy composition of the body and lid/tab and supply of alternative aluminium scrap (on the bottom).

the scrap. Therefore, the system boundaries have to be enlarged, considering the actual alloy composition of both the lid and the can body and differentiating between the first production loop and the subsequent recycling and production loops. Apart from the can body and lid manufacturing, the actual can production including lacquering, filling of the can, and then the recycling with pre-processing, remelting and alloy adjustment stages, which are all included in the system boundaries.[54] Excluded from the system boundaries are the distribution system of the can, its use, as well as all the transports of raw materials from suppliers and waste scrap to the EoL treatment. As shown in Figure 10.5a, in the first loop, both the lid and the can body are manufactured from primary aluminium and virgin alloying elements, including Mn, Fe, Si, Cu, Cr, Ti, Zn and Mg, according to the

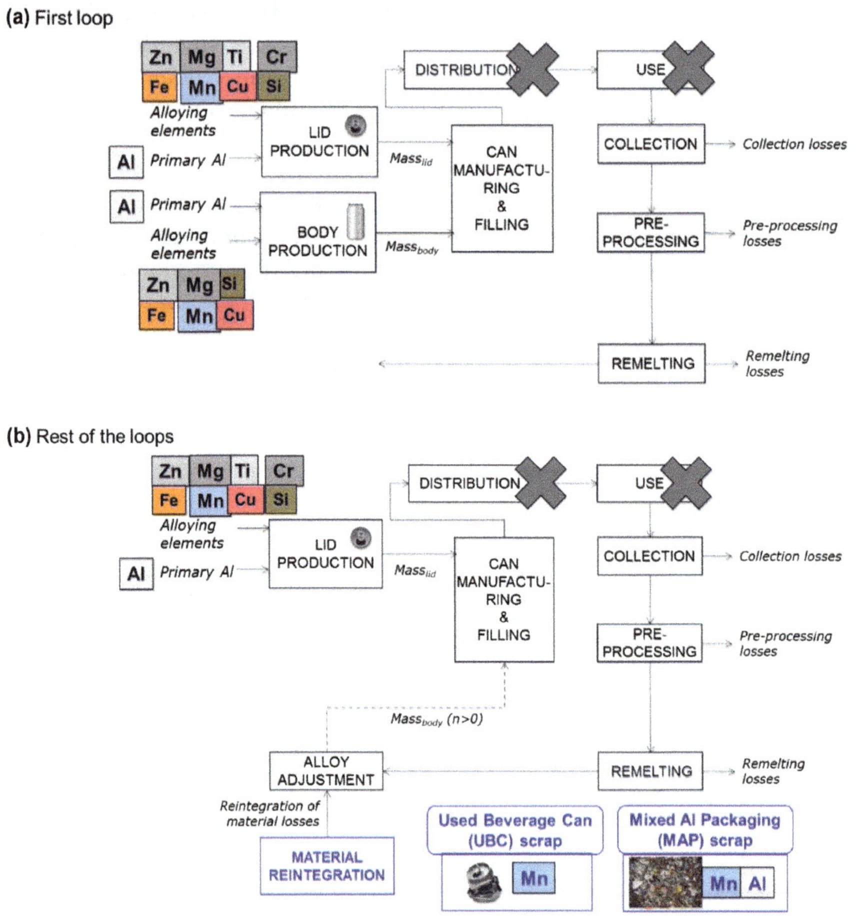

Figure 10.5 System boundaries definition in the case of multi-loop recycling for aluminium cans, distinguishing between (a) first loop and (b) rest of the loops.
Adapted from ref. 54 with permission from Elsevier, Copyright 2016.

specific alloy composition, *i.e.* AA3004 or AA3104 (with higher manganese content) for the body and AA5182 (with higher magnesium content) for the lid.[58] Meanwhile, in the following loops (see Figure 10.5b) the body production originates from the combination of recycled material from the previous loop and the amount of material (either virgin materials or post-consumer packaging scrap) that needs to be reintegrated to close the mass balance. The composition of the reintegrated material changes according to the scrap source and the percentage of recycling rate (RR). Two cases of post-consumer packaging scrap supply and three levels of RR are considered in Niero and Olsen,[54] namely Used Beverage (UBC) scrap and Mixed Aluminium Packaging (MAP) scrap.

As already mentioned in Section 10.2.2, the definition of system boundaries is also a relevant aspect in OLCA applications. This definition directly depends on three aspects:

1. The organizational boundaries definition (to determine who the organization is);
2. The specific goal of the study (to determine which part of the organization will be included); and
3. The nature of the activities of the organization (to determine which life cycle stages should be included in term of relevance, *e.g.* packaging companies should usually take care of raw material production and end-of-life management of the packaging material).

Most of the OLCA applications published so far identified the definition of the system boundaries to be particularly critical and sometimes difficult to implement.[37] These challenges are usually related to the nature and dimension of the organization understudy and can limit the capability of companies in identifying and managing environmental hotspots. To overcome the issue, Manzardo *et al.*[44] introduced the control and influence approach adapted from ISO 14001.[59] Control is the leadership and responsibility of the organization over the activities and their environmental impacts (*e.g.* the company owns the process); therefore, it is regarded as the capacity of the company to take direct action against the impacts generated by those processes or products. Manzardo *et al.*[39] recommend at least including such activities in an OLCA study. Influence is represented by the indirect effects of a company's choices, *e.g.*, those related to products and services used by the organization and those related to products and services provided by the organization.[59] The control and influence approach has been successfully applied in OLCA in the construction sector, but no experiences can be identified in the packaging industry, thus this approach should be further investigated in the future.

A recent application of OLCA of a beverage company by Manzardo *et al.*[48] confirmed the importance of system boundaries setting in OLCA and highlighted the issue of so-called "production allocation burden". This can take place when a company with two or more sites perform OLCA of only part of its facilities and decide to act on allocation of production to improve its environmental performances. It is therefore recommended to perform a general OLCA of the company adopting a top-down approach in order to verify that actions on one or more sites guarantee an overall improvement of the environmental performances of the organization understudy.

10.3.3 End-of-life Modelling

As reported in Figure 10.2, the main EoL options for packaging are recycling, incineration with energy recovery and landfill disposal. Recycling is one of the key CE strategies (see outer loops of the CE diagram by the EMF in Figure 10.1), which should be promoted instead of disposal through incineration or landfill. From a methodological point of view, recycling is a case of multi-functionality in LCA, with the product to be recycled having two functions: firstly, the function(s) the product was primarily made for, and secondly the function of providing secondary resources for use in subsequent life cycles/systems.[53] As the recycling process cannot only be considered as a waste management process, but also as a production process for material, it is shared between two product systems, *i.e.* the one producing the recycled material and one using the recycled material.[36] How to model recycling in LCA is one of the most debated issues in the LCA scientific community, see *e.g.* Frischknecht,[60] Allacker *et al.*,[61] Schrijvers *et al.*[62]

According to the ISO standards 14040-44,[21,22] two types of recycling can be used:

- Closed-loop recycling, which applies to closed-loop product systems and to open-loop product systems with no changes in inherent properties of recycled material, where allocation is avoided since the use of secondary material displaces the use of virgin (primary) materials;
- Open-loop recycling, which applies to open-loop product systems where the material is recycled into other product systems and the material undergoes a change to its inherent properties; in this case allocation should be performed based on: (1) physical properties; (2) economic value, *e.g.*, market value of the scrap material or recycled material in relation to market value of primary material; or (3) the number of subsequent uses of the recycled material.

Van der Harst *et al.*[36] evaluated six widely used methods for LCA modelling of recycling: three substitution methods (*i.e.* substitution based on equal quality, a correction factor, and alternative material), allocation based on the number of recycling loops, the recycled-content method, and the equal-share method. The three substitution methods (also known as the avoided burden approach or "0 : 100" approach) stimulate the recyclability of the product and assign credits for the obtained recycled material. The choice to either apply a correction factor, or to account for alternative substituted material is encouraged over simply ignoring quality loss.[36]

The allocation-on-number-of-recycling-loops method focusses on the life expectancy of material itself, rather than on a specific separate product. The recycled-content method (also known as the cut-off approach or "100 : 0" approach) stimulates the use of recycled material, *i.e.* credits the use of recycled material in products and ignores the recyclability of the products. The equal-share method (also known as the "50 : 50 approach") is a compromise between the substitution methods and the recycled-content method.

Van der Harst *et al.*[36] applied all six methods to the case of an aluminium can, outlining the underlying assumptions to each methods. The results are significantly influenced by the selection of the method chosen to perform the LCA modelling of recycling, but also by other factors such as the percentage of RC and the number of recycling loops.[36] The influence of %RR and %RC is confirmed by other LCAs of aluminium cans in a CE context, *e.g.* Stotz *et al.*[63] and Niero *et al.*[33] The choice of the method to model recycling depends on the specific situation that has to be modelled. Closed-loop recycling is suited when the used aluminium beverage scrap is used to produce other aluminium cans or products which are made of the same aluminium alloys as the body of the can.[54] Meanwhile, a case of open-loop recycling is suited when UBC scrap is mixed with other aluminium scrap to produce products made of cast alloy, this leading to the so-called "down-cycling" phenomenon.[64]

Down-cycling takes place when the inherent technical properties of the secondary material (*i.e.* the material obtained from recycling) downgrade during the recycling process. Such qualitative degradation during the use and recycling stages may limit the number of cycles that such materials can afford. In other cases, a higher amount of the recycled material is necessary compared with virgin material to provide the same functionality. The main consequences are that the secondary material can replace the primary material only to a limited extent and the quality of recycled materials is often lower compared with the corresponding primary materials.[29] All materials that are used for packaging production experience to a higher or lesser extent the down-cycling phenomenon, even the so-called "permanent materials", *i.e.* those materials whose inherent properties do not change during use and through solid–liquid transformation, and which can revert to their initial state.[65] Grosso *et al.*[65] highlighted the challenges that affect the recycling of end-of-life for aluminium (*i.e.* product contamination by alloying elements), steel (concentration of tramp elements) and glass (*e.g.* the presence of elements affecting the cullet quality, such as iron and chromium that affect the colour, and organics such as residues of sticking papers, glue, oil and grease on the glass fragments, which act as a reducing agent).

Rigamonti *et al.*[29] described the down-cycling phenomenon for most of the materials used for packaging production, *i.e.* paper, plastics, metals (aluminium and steel) and wood. While modelling recycling in waste management-oriented LCA studies, they recommended taking into account the quality of the waste-derived material, *i.e.* to use a quality factor that considers to what extent the inherent properties of the material change during recycling. The same recommendation emerged from the work of Schrijvers *et al.*,[62] which encouraged the use of a quality-correction factor that "represents to what extent the inherent properties of the material are lost, using the limiting factor as quality parameter", therefore allowing to incorporate down-cycling by combining the end-of-life recycling rate with such quality correction-factor. The main implication is that the common assumption of a 1 : 1 substitution ratio of recycled to virgin materials leads to an overestimation of the benefits of recycling since the down-cycling phenomena are neglected.[66]

The only method able to take into account of the different approaches to model recycling in LCA (favouring both the use of recycled content and the recyclability at end of life) and the down-cycling issue, is the equal share method with quality correction.[36] This is the method recommended in the Product Environmental Footprints (PEF) guide[42] indicated as the most suitable for considering the multiple uses of resources in continuous loops,[61] and upgraded to the Circular Footprint Formula.[67]

It should be noted that, although most of the packaging materials belong to the technical cycle (see Figure 10.1), some materials, such as paper and plastics, can fit the biological cycle, too. Paper is in fact a renewable and biodegradable material, which makes it suitable for a number of different recovery pathways, not limited to material recycling, but including renewable energy recovery, both *via* combustion or biological anaerobic degradation, as well as its degradation during biological aerobic processes. Whatever the disposal pathway is for paper, it will contribute in terms of material or energy recovery, or at least it will degrade with limited or negligible impact on the environment, with the exception of uncontrolled landfilling.[68] Plastics can also fit the biological cycle in the case of biodegradable plastics. However, many unresolved significant issues relate to the choice of bioplastics as the most suitable solution to replace conventional materials in the packaging sector, especially when it comes to their role in waste management and to their end-of-life, as discussed by Calabró and Grosso.[69]

10.3.4 Modelling of Reuse

Reuse refers to any operation by which products or components are used again for the same purpose for which they were conceived.[70] Reuse plays a central role in the CE, as a waste prevention activity, being positioned in an inner loop of the EMF diagram scheme compared to recycling (see Figure 10.1). A key parameter in the modelling of reuse is the number of times a product can be reused, and packaging is no exception to this. In the case of comparative LCAs between different packaging alternatives for food savers,[32] the glass option showed 12–64% higher impacts than the plastic alternative. Moreover, it was found that glass food savers should have up to 3.5 times greater lifespan to match the environmental footprint of plastic containers.

Rigamonti *et al.*[71] propose a methodology to collect data on packaging reuse that can be applied in different contexts to make them comparable for legislative compliance and statistical purposes. They consider the following qualitative characteristics: the material, the business application, the sector of use, the main basic characteristics (such as average size and/or weight), the possible reconditioning process applied, and the type of service on which it is managed (*e.g.*, rental). Biganzoli *et al.*[72] assessed the impacts associated with the life cycle of reusable Intermediate Bulk Containers (IBCs), which are reusable industrial containers designed for the transport and storage of bulk liquid and granulated substances, such as chemicals, food ingredients, solvents and pharmaceuticals. Results showed that the impact of the reconditioning process is modest compared to that associated with the IBCs production. Its contribution to the overall impacts increases with the number of rotations, but it is always below 20% for most of the considered indicators. The only exceptions are in the ozone depletion and climate change impact categories, where the reconditioning process contributes for a maximum of 40%. Reconditioning and reusing IBCs is thus preferable to a situation where IBCs are used only once and then sent to recycling/disposal. In addition, the benefits associated with the practice of reuse increase with the number of rotations.

10.3.5 Impact Assessment and Footprinting

Impact assessment is the third phase of both product LCA[21] and OLCA[40] and consists of the characterization of inventory data into potential environmental impacts, their classification and eventually their normalization and weighting. The first decision related to impact assessment consists of the selection of impact assessment categories that take place in the goal and scope definition of the study.

With reference to impact assessment of packaging, a key reference is the publication of a Product Category Rule (PCR) under the Environmental Product Declaration (EPD) International Programme Operator.[73] Other examples of reports that provide recommendations on impact categories and indicators are UNEP/SETAC Global Guidance for Life Cycle Impact Assessment Indicators[74] and the ILCD Recommendations for life cycle impact assessment in the European context,[75] though this is not specific for packaging.

In general the following impact assessment categories are usually assessed in LCA studies of packaging system and packaging material:[24] global warming potential (GWP), natural resource depletion (including fossil fuels and water), stratospheric ozone depletion (despite being a marginal issue in packaging production), acidification, photochemical ozone creation, eutrophication, human toxicity and aquatic toxicity.

Following an increasing trend in LCA studies, a growing number of applications have focused on the quantification of footprints. A footprint is defined as metric(s) used to report life cycle assessment results addressing an area of concern.[76] As such, a footprint is not as comprehensive as a full LCA but, according to the goal and scope of the study, allow investigating specific performances related to an environmental subject defined by the interest of society. Footprints, being limited in scope, cannot be used in comparative assertion to be disclosed to the public, but it was shown to be particularly relevant in terms of communication effectiveness.[76] For this specific reason, the number of footprint studies in the packaging sector has increased a lot in the past few years.

Most investigated footprints refer to the so-called carbon footprint,[77] a metric that quantifies the potential impact of a product on climate change, and the water footprint, defined as a metric to quantify the potential impact of a product related to water.[78] The former, expressed in CO_2 eq, is usually assessed both in single studies and within more comprehensive LCA. Scipioni *et al.*[47] presented an approach for the monitoring of product carbon footprint over time. The study focused on a beverage carton material and highlighted the importance of raw material selection and end-of-life management. With reference to the specific product understudy, the quantification of carbon footprint allowed the company to push for the separate collection of beverage carton in Italy therefore supporting a more circular management of poly-laminated packaging. Cimini and Moresi[79] presented a carbon footprint study of a beer product and its related PET and glass bottles. Results of the analysis identified raw material production and end-of-life management to be the main hotspot of the two packaging with better overall performances for the glass one. The authors presented a sensitivity analysis, proving that higher recycled content of the packaging resulted in significant carbon footprint reduction. However, even if GWP has proven to be a good predictor for other impact categories in the case of packaging, see *e.g.* Amienyo *et al.*,[51] Niero *et al.*,[33] Scipioni *et al.*,[52] preference should be given to the inclusion of a broad set of impact categories.[80]

With reference to water footprint assessment of packaging, due to the recent introduction of this concept in LCA,[78] only a very limited number of case studies are available in the scientific literature. Manzardo *et al.*[81] investigated the impacts of different packaging solutions of a tomato sauce produced in the United States. A background analysis of the regions where the different materials of the sauce are produced, suggested that it would be necessary to investigate several impact assessment categories: water availability, eutrophication, acidification, human toxicity and eco-toxicity. A beverage carton was compared to a glass jar to support the company in the decision of which container could guarantee better environmental performances related to water. The glass bottle scored worse performances in terms of acidification, human toxicity and eco-toxicity due to the energy used for its production and recycling. The packaging was shown to be equivalent when other impact categories were addressed.

10.4 Implications for the Packaging Sector and its use of LCA Towards CE

10.4.1 Trends and Challenges for CE Implementation in the Packaging Sector

The packaging industry and the food and beverage industry, as part of the Fast Moving Consumer Goods (FMCG) sector, are given high priority for CE implementation.[11,82] As shown by Stewart and Niero[83] in their systematic review of 46 corporate sustainability reports from 2016 in the FMCG, CE started to be integrated into the corporate sustainability agenda of companies in the food and beverage and packaging sectors. Most reported activities aiming at applying the CE concept are oriented toward the main product and packaging, focusing on end-of-life management (*i.e.* recycling) and sourcing strategies (*i.e.* of recycled material and renewable feedstock as raw materials). However, their analysis also revealed that the breadth of CE-related activities remains to be explored and the systemic dimension of CE is rarely present in companies' narratives about CE, as well as poorly rooted in CE-related activities, considering the limited focus on circular product design, consumer engagement, and circular business models. Most of the FMCG companies referring to CE in their corporate sustainability reports in 2016 have reported the use of LCA (and other footprint methodologies) to assess and/or monitor their environmental performances, but with no explicit link to CE activities.[83] Therefore, there is a huge potential for LCA to be used to contribute to the environmental sustainability assessment of CE strategies in the packaging sector. Moreover, based on the categorization of CE strategies introduced by Bocken *et al.*,[84] it is important to expand on the traditional understanding of the CE concept for packaging, therefore, not only aiming at narrowing resource loops (*e.g.* lightweight) or closing resource loops (recycling), but also slowing resource loops, *e.g.* through reuse.

10.4.2 The Role of Design and Business Models

Product design plays a key role to overcome the challenges for the implementation of CE strategies. This is especially true for packaging, which has to fit both product and its use environment and to take into account the increasingly complex packaging technology and business ecosystems.[85] The definition of the optimal beverage packaging life scenario, specifically addressing product design, is the first of the 4 steps in the decision support framework developed by Niero *et al.*,[85] which combines LCA and C2C certification program for the development of continuous loop packaging systems. The distinction between the technical cycle and biological cycle can help in identifying the best use of the packaging, which in most cases depends on the material choice. LCA can then be used to assess the potential environmental impacts of different scenarios aiming at continuous loop recycling of packaging inspired by the C2C design framework.[85] In the C2C certification program, the potential for "design for recyclability" or "design for compostability" is assessed by a specific material circularity indicator, *i.e.* MRS.[17] With regard to the technical cycle, the MRS quantifies the recyclability potential of a product considering two variables: the intrinsic recyclability of the product, *i.e.* the percentage of the product that can be recycled at least once after its initial use stage and the %RC. The recyclability of the packaging material, depends on both its technical recyclability, *i.e.* the ease with which it can be reprocessed and used to manufacture new products, and on the availability of facilities to collect, sort, and reprocess the material.[86] However, Niero and Kalbar[19] showed that the exclusive use of indicators addressing material circularity to assess CE at the product level can be misleading in the case of packaging and therefore recommended the combination of such indicators with LCA-based indicators. The fourth step in the above-mentioned decision-support framework by Niero *et al.*[85] refers to the assessment of the benefits of closing the loop, not only from an environmental perspective, but also incorporating the economic perspectives *via* formulation of a business model. Stewart *et al.*[87] explored the possibilities for the business implementation of a closed-loop supply of aluminium beverage cans and therefore proposed a business model-inspired framework. Such a framework was derived from the basic three-pillar-based understanding of business models: building the value network, engaging network actors and generating combined environmental and economic value.[87] The combination of LCA with business model assessment in the context of CE should be further investigated,[88] as well as the role of the Life Cycle Sustainability Assessment (LCSA) framework in implementing CE strategies.[20]

10.4.3 The Role of Food-packaging Systems and Consumer Perceptions

Wikstrom *et al.*[89] highlighted the importance of exploring the potential of packaging systems to reduce the overall environmental impact of food-packaging systems, *i.e.* the combined impact from the food and the packaging. Packaging can indeed be used as a means to reduce or entirely prevent food waste, but potential trade-offs need to be investigated by means of LCA.

Boesen *et al.*[90] investigated how well-educated young consumers living in Denmark understand the environmental sustainability of five different kinds of packaging for liquid food (milk, beer, soft drink, olive oil and skinned tomatoes). Their results showed that the consumers included in the sample only to a limited extent consider that packaging prevents food waste. Only 36% of around 200 respondents indicated that it was important that the packaging was formed in such a way that no product was wasted and only 3% stated the view that small portion sizes increased the environmental sustainability of the packaging. Moreover, their results showed that the consumers involved in the study base the perception of the environmental sustainability of liquid food packaging on the material type and on what they can personally do at the end-of-life stage. Amongst the investigated packaging types, bio-based types and glass emerged to be perceived as the most environmentally sustainable ones, and plastic in general as the least favourable, with laminated cartons receiving a mixed perception. The results of the consumer research were compared with LCA results (in the case of beer and soft drinks) and validated by means of comparative LCAs of the five product categories published in scientific literature. From the quantitative environmental sustainability assessment it emerged that that plastic and especially laminated cartons can be environmentally preferable solutions, even though they may be difficult to recycle. The results of the LCA performed on beer and soft drink showed that there is a significant difference in environmental performance between one-way glass and refillable glass, but consumers seem not to be aware of this difference. This gap between consumers' perception of environmental sustainability of packaging and LCA results was found also in other studies, see *e.g.* Steenis *et al.*[91] in the case of tomato sauce, but requires further investigation. Understanding the motivations that lead consumers to change their behaviour is imperative for the success of CE strategies[92] and use of LCA[93] in policy making.

10.4.4 Food Safety as a Constraint for CE Development in the Packaging Sector

Given the fundamental role that recycling has been playing in the discussion around implementation of the CE in the packaging sector, then the safety issues connected with recycling of food packaging waste into new food packaging should not be neglected. Recycling and the consequent use of recycled food packaging may indeed increase the levels of potentially hazardous chemicals in the packaging, as discussed by Geueke *et al.*[20] In their review, they first described recycling processes of commonly used food packaging materials, including plastics, paper and board, aluminium, steel and multimaterial multilayers (*e.g.*, beverage cartons). Then, they provided an overview of typical migrants from all types of recycled food packaging materials, and summarized approaches to reduce chemical contamination.

With the increased use of recycled materials, not only there is a risk of increase of the levels of potentially hazardous chemicals in the packaging,[94] but – after migration – also in the food.[95] Therefore, given the potential of chemicals to migrate to food, the findings of Ernstoff *et al.*[95] suggested that consumer exposure to chemicals in food packaging could be an important pathway to consider in LCAs of food packaging materials, and thus should be further explored.

10.5 Conclusions

Packaging plays a key role in everyday life, being such a visible aspect of human consumption, not only in terms of goods' functions fulfilment (containment, protection, convenience and communication), but also in terms of waste generation. The packaging sector has been at the centre of discussions on sustainability for a long time and the environmental impacts of packaging have been investigated since the early development of the LCA methodology. In this context, the CE both offers new directions and supports current practice in the packaging industry. The truth is that there is no evil or good packaging in absolute terms, but it is the way that packaging is managed that significantly changes its impacts on the environment. The use of LCA proved to be an effective tool in guiding decisions related to packaging design and management confirming that CE practices applied to packaging act exactly where most of the environmental impacts of packaging materials are located: raw material extraction and end-of-life treatment. This is the evidence that "closing the loop" can result in a double advantage, not only in terms of material re-circulation, but also from an environmental perspective. However, there is a huge potential in exploring the implementation of CE strategies based on slowing the resource loop, *e.g.* by prolonging the use of the packaging material. Consumers play a central role in determining the success of CE strategies, therefore it is key to investigate and understand the consumption dynamics around packaging and improve the LCA modelling of the use stage of packaging. There is room to improve the integration of LCA both in the product design phase as well as at the strategic organizational level, *e.g.* in the shift of business model or through the use of OLCA. Moreover, despite the significant efforts of research on packaging several trade-offs should be investigated and solved in the future. One of the main constraints related to the reduction of packaging material impacts is related to the food sector, where the food contact regulations either limit or avoid the use of recycled material in primary packaging.

References

1. Y. Wyser and D. Shires, *Packag. Technol. Sci.*, 2019, **32**, 3–5.
2. Cambridge University Press, Packaging definition, https://dictionary.cambridge.org/dictionary/english/packaging, last accessed 2020.02.13.
3. G. L. Robertson, *Food Packaging Principles and Practice*, CRC Press Taylor & Francis Group, Boca Rotan, FL, 3rd edn, 2012.
4. E. von Falkenstein, F. Wellenreuther and A. Detzel, *Int. J. Life Cycle Assess.*, 2010, **15**, 938–945.
5. Flexible Packaging Association, Share of the global packaging market in 2017, by packaging type, https://www.statista.com/statistics/258962/share-of-the-global-packaging-market-by-packaging-type/, (accessed 3 January 2019).

6. Eurostat, Packaging & Packaging Waste Statistics 2007–2016, https://ec.europa.eu/eurostat/statistics-explained/index.php/Packaging_waste_statistics#Time_series_of_packaging_waste_generation_and_treatment, (accessed 3 January 2019).
7. EC, *COM (2015) 614 Communication from the Commission to the European Parliament, the Council, the European Economic and Social Committee and the Committee of the Regions. Closing the loop - An EU action plan for the Circular Economy*, 2015.
8. European Comission, DIRECTIVE (EU) 2018/852 OF THE EUROPEAN PARLIAMENT AND OF THE COUNCIL of 30 May 2018 amending Directive 94/62/EC on packaging and packaging waste (Text with EEA relevance), Brussels, 2018.
9. P. Ghisellini, C. Cialani and S. Ulgiati, *J. Cleaner Prod.*, 2016, **114**, 11–32.
10. J. Kirchherr, D. Reike and M. Hekkert, *Resour., Conserv. Recycl.*, 2017, **127**, 221–232.
11. EMF, *Towards the circular economy. Opportunities for the Consumer Goods Sector*, Ellen MacArthur Foundation, 2013.
12. W. McDonough and M. Braungart, *Cradle to Cradle*, North Point Press, New York, 2002.
13. M. Braungart, W. McDonough and A. Bollinger, *J. Cleaner Prod.*, 2007, **15**, 1337–1348.
14. L. L. Kjaer, D. C. A. Pigosso, M. Niero, N. M. Bech and T. C. Mcaloone, *J. Ind. Ecol.*, 2018, **23**(1), 22–35.
15. E. Pauer, B. Wohner, V. Heinrich and M. Tacker, *Sustainability*, 2019, **11**, 925.
16. EMF and Granta Design, *Circularity Indicators - An Approach to Measuring Circularity - Project overview*, 2015.
17. Cradle to Cradle Products Innovation Institute, *Cradle to Cradle Certified Product Standard Version 3.1*, 2016.
18. M. Niero and M. Z. Hauschild, in *Proceedings of the 11th Italian LCA Network Conference (Atti del XI Convegno della Rete Italiana LCA: Resource Efficiency and Sustainable Development Goals: the role of Life Cycle Thinking), Siena, 22-23 giugno 2017*, ed. B. Valentina, A. Niccolucci and C. Arianna, 2017, pp. 11–19.
19. M. Niero and P. Kalbar, *Resour., Conserv. Recycl.*, 2019, **140**, 305–312.
20. M. Niero and M. Z. Hauschild, *Procedia CIRP*, 2017, **61**, 685–690.
21. ISO, *ISO 14040: Environmental management — Life cycle assessment — Principles and framework*, Geneva, Switzerland, 2006.
22. ISO, *ISO 14044: Environmental management. Life cycle assessment. Requirements and guidelines*, Geneva, Switzerland, vol. 44, 2006.
23. R. G. Hunt and W. E. Franklin, *Int. J. Life Cycle Assess.*, 1996, **1**, 4–7.
24. G. Vignali, in *Environmental Footprints of Packaging*, ed. S. S. Muthu, Springer Science + Business Media Singapore, 2016, pp. 1–22.
25. S. Koskela, H. Dahlbo, J. Judl and M. Korhonen, *J. Cleaner Prod.*, 2014, **69**, 83–90.

26. M. Niero, F. Di Felice, J. Ren, A. Manzardo and A. Scipioni, *Int. J. Life Cycle Assess.*, 2014, **19**, 901–918.
27. M. Niero, S. I. Olsen and A. Laurent, *J. Ind. Ecol.*, 2018, **22**, 760–772.
28. B. Geueke, K. Groh and J. Muncke, *J. Cleaner Prod.*, 2018, **193**, 491–505.
29. L. Rigamonti, M. Niero, M. Haupt, M. Grosso and J. Judl, *Waste Manage.*, 2018, **76**, 261–265.
30. M. Cordella, A. Tugnoli, G. Spadoni, F. Santarelli and T. Zangrando, *Int. J. Life Cycle Assess.*, 2008, **13**, 133–139.
31. A. Detzel and J. Mönckert, *Int. J. Life Cycle Assess.*, 2009, **14**, 70–79.
32. A. Gallego-Schmid, J. M. F. Mendoza and A. Azapagic, *Sci. Total Environ*, 2018, **628–629**, 979–989.
33. M. Niero, A. J. Negrelli, S. B. Hoffmeyer, S. I. Olsen and M. Birkved, *J. Cleaner Prod.*, 2016, **126**, 352–362.
34. J. Pasqualino, M. Meneses and F. Castells, *J. Food Eng.*, 2011, **103**, 357–365.
35. S. Toniolo, A. Mazzi, M. Niero, F. Zuliani and A. Scipioni, *Resour., Conserv. Recycl.*, 2013, **77**, 61–68.
36. E. van der Harst, J. Potting and C. Kroeze, *Waste Manag.*, 2016, **48**, 565–583.
37. UN Environment, *Road testing Organizational Life Cycle Assessment all over the world: applications, experiences and lessons learned*, 2017.
38. UNEP, *Guidance on Organizational Life Cycle Assessment*, 2015.
39. A. Manzardo, A. Loss, A. Mazzi and A. Scipioni, in *Environmental Footprints of Packaging*, ed. S. S. Muthu, Springer Science + Business Media Singapore, 2016, 2016.
40. ISO, *ISO/TS 14072: Environmental management – Life cycle assessment – Requirements and guidelines for organizational life cycle assessment*, Geneva, Switzerland, 2014.
41. J. Martínez-Blanco, A. Inaba and M. Finkbeiner, *Int. J. Life Cycle Assess.*, 2015, **20**, 829–841.
42. EC, *Commission Recommendation of 9 April 2013 on the use of common methods to measure and communicate the life cycle environmental performance of products and organisations*, European Commission, Brussels, 2013.
43. M. Finkbeiner, *Int. J. Life Cycle Assess.*, 2014, **19**, 266–271.
44. A. Manzardo, A. Loss, R. Jingzheng, F. Zuliani and A. Scipioni, *J. Cleaner Prod.*, 2018, **184**, 264–273.
45. A. Scipioni, M. Mastrobuono, A. Mazzi and A. Manzardo, *J. Cleaner Prod.*, 2010, **18**, 299–306.
46. N. Pelletier, K. Allacker, R. Pant and S. Manfredi, *Int. J. Life Cycle Assess.*, 2014, **19**, 387–404.
47. A. Scipioni, A. Manzardo, A. Mazzi and M. Mastrobuono, *J. Cleaner Prod.*, 2012, **36**, 94–101.
48. A. Manzardo, A. Loss, M. Niero, C. Vianello and A. Scipioni, *Procedia CIRP*, 2018, **69**, 429–434.

49. A. Bjørn, M. Owsianiak, A. Laurent, S. I. Olsen, A. Corona and M. Z. Hauschild, in *Life Cycle Assessment: Theory and Practice*, ed. M. Z. Hauschild, R. K. Rosenbaum and S. I. Olsen, Springer International Publishing AG, 2018, pp. 75–116.
50. J. Stichling and D. Nguyen-Ngoc, *Life Cycle Inventory and Impact Analysis for Beverage Cans. Final Report*, PE International, 2009.
51. D. Amienyo, H. Gujba, H. Stichnothe and A. Azapagic, *Int. J. Life Cycle Assess.*, 2013, **18**, 77–92.
52. A. Scipioni, M. Niero, A. Mazzi, A. Manzardo and S. Piubello, *Int. J. Life Cycle Assess.*, 2013, **18**, 673–682.
53. EC-JRC-IES, *ILCD Handbook - General Guide on LCA – Detailed Guidance*, European Union, 1st edn, 2010.
54. M. Niero and S. I. Olsen, *Resour., Conserv. Recycl.*, 2016, **114**, 18–31.
55. EAA, Press Release. Recycling rate for cans at new record level, 2015, http://european-aluminium.eu/media/1285/201503_recycling-rate-for-cans-at-new-record-level_europeanaluminium.pdf, (accessed 14 June 2016).
56. EEA, *Circular by design: Products in the circular economy*, 2017.
57. EAA, *Environmental Profile Report for the European Aluminium Industry April 2013- Data for the year 2010 Life Cycle Inventory data for aluminium production and transformation processes in Europe*, European Allumi-nium Association, 2013.
58. The University of Liverpool, Aluselect, http://aluminium.matter.org.uk/aluselect/06_wrought_comp.asp, (accessed 15 September 2015).
59. ISO, *ISO 14001:2015 Environmental management systems – Requirements*, Geneva, Switzerland, 2015.
60. R. Frischknecht, *Int. J. Life Cycle Assess.*, 2010, **15**, 666–671.
61. K. Allacker, F. Mathieux, S. Manfredi, N. Pelletier, C. De Camillis, F. Ardente and R. Pant, *Resour., Conserv. Recycl.*, 2014, **88**, 1–12.
62. D. L. Schrijvers, P. Loubet and G. Sonnemann, *Int. J. Life Cycle Assess.*, 2016, **21**, 994–1008.
63. P. M. Stotz, M. Niero, N. Bey and D. Paraskevas, *Resour., Conserv. Recycl.*, 2017, **127**, 96–106.
64. D. Paraskevas, K. Kellens, W. Dewulf and J. R. Duflou, *J. Cleaner Prod.*, 2015, **105**, 357–370.
65. M. Grosso, M. Niero and L. Rigamonti, *Waste Manage. Res.*, 2017, **35**, 793–794.
66. L. Rigamonti, M. Grosso and M. C. Sunseri, *Int. J. Life Cycle Assess.*, 2009, **14**, 411–419.
67. V. Bach, A. Lehmann, M. Görmer and M. Finkbeiner, *Sustainability*, 2018, **10**(8), 2898.
68. M. Grosso, M. Niero and L. Rigamonti, *Waste Manage. Res.*, 2017, **35**(8), 793–794.
69. P. S. Calabrò and M. Grosso, *Waste Manage.*, 2018, **78**, 800–801.
70. European Union, *Directive 2008/98/EC of the European Parliament and of the Council of 19 November 2008 on waste and repealing certain Directives*, 2008.

71. L. Rigamonti, L. Biganzoli and M. Grosso, *J. Mater. Cycles Waste Manage.*, 2018, **21**, 1–9.
72. L. Biganzoli, L. Rigamonti and M. Grosso, *Procedia CIRP*, 2018, **69**, 827–832.
73. Environdec, *Packaging Product Category Classification: Multiple CPC. PCR 2019:13 version 1.0 valid until 2023-11-08*, 2019.
74. UNEP, *Global Guidance for Life Cycle Impact Assessment Indicators Volume 1*, 2016.
75. EC-JRC-IES, *International Reference Life Cycle Data System (ILCD) Handbook – Recommendations for Life Cycle Impact Assessment in the European context*, EUR 24571 EN, Publications Office of the European Union, Luxemburg, 1st edn, 2011.
76. ISO, *ISO 14026:2017. Environmental labels and declarations – Principles, requirements and guidelines for communication of footprint information*, Geneva, Switzerland, 2017.
77. ISO, *ISO 14067:2018 Greenhouse gases – Carbon footprint of products – Requirements and guidelines for quantification and communication*, 2018.
78. ISO, *ISO 14046 Environmental Management – Water Footprint – Principles, Requirements and Guidelines*, 2014.
79. A. Cimini and M. Moresi, *J. Food Eng.*, 2018, **236**, 1–8.
80. A. Laurent, S. Olsen and M. Hauschild, *Environ. Sci. Technol.*, 2012, **46**, 4100–4108.
81. A. Manzardo, A. Mazzi, A. Loss, M. Butler, A. Williamson and A. Scipioni, *J. Cleaner Prod.*, 2016, **112**, 4657–4666.
82. European Commission, *Closing the loop – An EU action plan for the Circular Economy*, Brussels, 2015.
83. R. Stewart and M. Niero, *Bus. Strategy Environ.*, 2018, **27**, 1005–1022.
84. N. M. P. Bocken, I. de Pauw, C. Bakker and B. van der Grinten, *J. Ind. Prod. Eng.*, 2016, **33**, 308–320.
85. M. Niero, M. Z. Hauschild, S. B. Hoffmeyer and S. I. Olsen, *J. Ind. Ecol.*, 2017, **21**, 742–753.
86. K. Verghese, H. Lewis and L. Fitzpatrick, *Packaging for Sustainability*, Springer, 1st edn, 2012.
87. R. Stewart, M. Niero, K. Murdock and S. I. Olsen, *Procedia CIRP*, 2018, **69**, 810–815.
88. M. Niero and X. C. Schmidt-Rivera, *Procedia CIRP*, 2018, **69**, 793–798.
89. F. Wikström, H. Williams, K. Verghese and S. Clune, *J. Cleaner Prod.*, 2014, **73**, 100–108.
90. S. Boesen, N. Bey and M. Niero, *J. Cleaner Prod.*, 2019, **210**, 1193–1206.
91. N. D. Steenis, E. van Herpen, I. A. van der Lans, T. N. Ligthart and H. C. M. van Trijp, *J. Cleaner Prod.*, 2017, **162**, 286–298.
92. J. Camacho-Otero, C. Boks and I. N. Pettersen, *Sustainability*, 2018, **10**(8), 2758.
93. E. Polizzi di Sorrentino, E. Woelbert and S. Sala, *Int. J. Life Cycle Assess.*, 2016, **21**, 237–251.
94. J. Lee, A. B. Pedersen and M. Thomsen, *Environ. Int.*, 2014, **73**, 312–322.

95. A. S. Ernstoff, M. Niero, J. Muncke, X. Trier, R. K. Rosenbaum, M. Z. Hauschild and P. Fantke, *Int. J. Life Cycle Assess.*, 2019, **24**, 543–552.
96. R. Accorsi, L. Versari and R. Manzini, *Sustainability*, 2015, **7**, 2818–2840.
97. M. Meneses, J. Pasqualino and F. Castells, *J. Environ. Manage.*, 2012, **107**, 76–83.
98. S. Humbert, V. Rossi, M. Margni, O. Jolliet and Y. Loerincik, *Int. J. Life Cycle Assess.*, 2009, **14**, 95–106.
99. S. Albrecht, P. Brandstetter, T. Beck, P. Fullana-i-Palmer, K. Grönman, M. Baitz, S. Deimling, J. Sandilands and M. Fischer, *Int. J. Life Cycle Assess.*, 2013, **18**, 1549–1567.
100. L. Zampori and G. Dotelli, *Int. J. Life Cycle Assess.*, 2014, **19**, 206–217.

CHAPTER 11

Case Study: Agricultural Crop Production

T. L. OLDFIELD, W. CHEN, F. MURPHY AND N. M. HOLDEN*

University College Dublin, School of Biosystems and Food Engineering, Belfield, Dublin 4, Ireland
*Email: Nick.holden@ucd.ie

11.1 Introduction

The food we eat today (2019) is unsustainable for two reasons: (1) the food production system causes unacceptable environmental impacts; and (2) it depletes non-renewable resources.[1] Therefore, our food can be regarded as 'fossil food' because its production relies on fossil fuel, non-renewable mineral resources, depletion of groundwater reserves and excessive soil loss. Garnett[2] conceptualised three perspectives on how food security and, ultimately, the global food sustainability challenge can be addressed: (perspective 1) the production challenge: the need to improve efficiency; (perspective 2) the consumption challenge: the need for demand restraint; (perspective 3) the socio-economic challenge: the necessity to improve governance. The adoption of a circular economy falls within the improve efficiency perspective (perspective 1). Circular economy thinking has emerged as a leading theory as a way to create a *Circular Bioeconomy*, which focuses on the valorisation of biological waste at the end of a product's life, in order to create looped systems. Ideally, these would be closed-loop systems, but more commonly for biological waste valorisation it creates extended linearisation systems and open-loop systems.[3]

Crop production systems are inherently wasteful, as only a small percentage of what is grown is consumed as food. This results in a significant

Life Cycle Assessment: A Metric for the Circular Economy
Edited by Aiduan Borrion, Mairi J. Black and Onesmus Mwabonje

Published by the Royal Society of Chemistry, www.rsc.org

percentage of biomass being underutilised and conventionally thought of as waste. A universally accepted definition of what is a *Waste* in agriculture has still not been agreed.[4] For example, Oldfield *et al.*[5] suggested using *Waste* to strictly describe those materials that are not utilisable and are disposed of in the biosphere sink (waste should be prevented to minimum as possible); *Residue*, these materials that are unavoidable but not consumable (can be regarded as a bioresource – prepare for reuse, recycle, compost or energy recovery); *Wasted food/product*, these are materials that have been mismanaged and should never end up in secondary processing (prevent to a practical minimum, then prepare for reuse, recycle, compost or energy recover). The H2020 project AgroCycle[6] suggested grouping all these types of materials together and have named them Agricultural Waste Co-products and By-products (AWCB). For simplicity, we use the term AWCB in this chapter. In Europe, it is estimated that, annually, 700 million tonnes of AWCB are produced. An in-depth analysis of the different types, and annual amounts of AWCB produced in Europe can be found on the AgroCycle MarketPlace,[7] a web-based AWCB exchange software platform.

An analysis carried out by the AgroCycle project[8] endeavoured to understand which ACWB should be prioritised for valorisation in a Circular Bioeconomy. The AgroCycle framework is simple, using two criteria (availability *versus* avoidability) to evaluate whether an AWCB should be valorised in a Circular Bioeconomy, or whether it is already valorised or the focus for the AWCB should remain as prevention, which would be in line with the Waste Hierarchy.[9] The AgroCycle framework and approach (Figure 11.1) is not exhaustive but gives a good indication on what type of AWCB should be prioritised by those working in AWCB valorisation, and how we should be thinking when it comes to considering using technology to add value to AWCB. At a high level it can be surmised that AWCBs that are available and

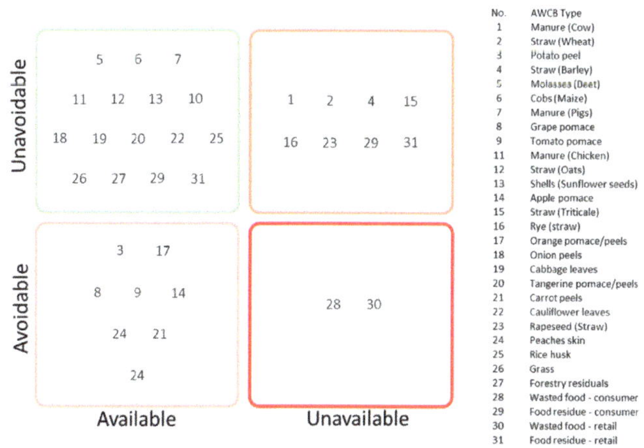

Figure 11.1 Availability *vs.* avoidability of AWCB. (Adapted from White and Oldfield.)[8]

un-avoidable should be the priority for AWCB valorisation in a Circular Bioeconomy.

First-generation AWCB valorisation technologies include anaerobic digestion (methane – energy production; digestate – organic fertiliser), composting (compost – organic fertiliser), incineration (energy) and animal feed.[10,11] No complete list of second-generation technologies is available; the following list has been taken from a number of sources:[6,11]

- Pyrolysis: biochar – energy/soil amendment/carbon capture; gas – energy production; oil – energy production
- Microbial fuel cell: energy production
- Biosolarization: pesticide technology
- Fermentation: energy production
- Biocomposite production: material production
- Metamorphosis *via* Black Soldier Fly: feed and oil production.

As the Circular Bioeconomy gains traction and novel AWCB valorisation technologies progress up the technology readiness (TRL) scale, from laboratory to industrial, it is imperative that these technologies are interrogated from an environmental sustainability perspective. It is suggested that, when carrying out the environmental analysis of a whole system, that the use of life cycle assessment (LCA) allows for a more holistic analysis of multiple impacts per a defined unit,[12] which enables the averting of environmental burden shifting.

Life cycle assessment has been used to analyse AWCB valorisation (Table 11.1). These were predominantly first-generation technologies, but more recently, second-generation technologies are being investigated using LCA.

A major challenge in carrying out an LCA of AWCB management/valorisation is the choice of system boundary, and the majority of those

Table 11.1 Published LCAs of AWCB valorisation technologies.

Nr	Country	Technology	AWCB type	Reference
1	Denmark	C	FW	13
2	Belgium	AD/I/L	MSW (FW)	14
3	Sweden	AD/C/I	FW	15
4	UK	AD/I/L	FW	16
5	Denmark	AD/I	FW	17
6	Italy	AD/I	FW	18
7	Singapore	AD/I/C	FW	19
8	South Korea	I/C/FP	FW	20
9	Spain	C	FW	21
10	US	C	FW	22
11	Ireland	C/L	MSW (FW)	23
12	United States of America	B	TP	24
13	Spain/Italy/Belgium	P	FW/OR/WR/M	25
14	Ireland	I/LF/AD/C	WF/FR	5
15	Cuba	FP/AD	SCB	26

studies listed in Table 11.1 chose to follow the zero-burden assumption, which gives AWCB a zero embedded impact. The tenability of the zero-burden assumption has been discussed,[27–29] and it has been suggested when an AWCB is being valorised in a circular bioeconomy, rather than being managed in a linear economy, the upstream *embedded* impact should be included in the system boundary, and a downstream functional unit (FU), of the value-added product used. In a linear system, the system boundary usually starts (the Cradle) at the point of material extraction and ends (the Grave) when the product is disposed of. In a circular system, theoretically, there should be no cradle and no grave, just a continuous loop, like a perpetual motion machine.[30] If we apply this to the agri-food sector and, in particular, crop production this becomes the ideal.

11.2 Agricultural Systems – Linear *vs.* Circular

The conventional model of the agri-food system has been the 'take-make-use-waste' concept that has underpinned the global economy since the industrial revolution.[31] This approach is 'linear' (Figure 11.2). The farmer uses resources, such as energy, phosphorus, agro-chemicals, water, soil and land (the take) to grow a primary crop, such as a cereal (the make), which is then processed to a greater or lesser extent into a food product for human consumption (the use) and any bi- or co-products and unwanted food are discarded (the waste). This type of food system is recognised as being inherently unsustainable because it is depleting non-renewable resources *i.e.* we consume fossil food.[1]

To make the resources that have been consumed go further, we have the option to do more with them. For instance, it is possible to generate energy from by-products and wasted food, it is possible to utilise co-products more efficiently and it is possible to valorise wasted food to create new materials as inputs to other value chains.[6] While this extended linear system is perhaps more eco-efficient, depending on what the newly valorised products displace in the market, it does not necessarily fix the problem of depleting non-renewable resources, and thus we are still consuming fossil food, even if its impact is reduced.[32]

A further development of the system is to try to return as many of the resources as possible after the use stage back to the make stage, which reduces the waste stage to the smallest possible amount of activity and also decreases the activity in the take stage.[33] In an ideal world, the nutrition extracted from

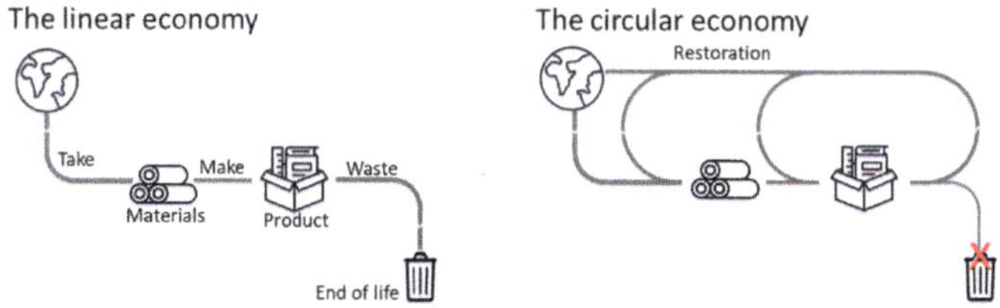

Figure 11.2 Linear economy verses circular economy.

food would occur at the same, or a slower rate than its renewable production (through photosynthesis to capture carbon) and biological decay and geological weathering to release nutrients. In this situation, the circulating mass would then remain more or less the same, supporting the food system, with the only significant extraction being the growth of humans, who ultimately die and return their small mass to the system as well.

The distinction between these system designs is important because it determines the rate at which humans deplete natural resources, and whether humanity is operating within the limits of the planetary boundary. Presently, we exceed some of these limits; therefore, our food systems are not sustainable.[1]

11.3 Methodology to Compare Linear and Circular Agricultural Systems

To compare the sustainability of linear and circular agricultural systems with LCA, a comprehensive, consistent and effective evaluation framework is required. The evaluation aspects of the framework should cover environmental, social and economic dimensions. The maturity of a technology is defined following the Technology Readiness Level (TRL).[34] Where an AWCB valorisation technology is positioned on the TRL would influence the availability of data and data quality, which will determine the types of investigated impact categories in LCA and the accuracy of LCA results.[35] Therefore, a tiered life cycle sustainability assessment (LCSA) framework was proposed for AWCB valorisation systems. The tiered LCSA framework has three tiers, and each tier has different impact categories, specified characterisations for life cycle inventory assessment and input data requirement. Figure 11.3 illustrates the investigated impact categories that are

Environmental issue	Indicator	Unit	Social issue	Indicator	Unit	Economic issue	Indicator	Unit
Global warming	Global warming potential	kg CO_2-eq	Fair wage	Cumulative risk of fair wage in supply chain	Euro	Production Cost	Labour cost	€
Acidification	Acidification potential	Kg SO_2-eq	Health and safety	Cumulative fatality rate (injury rate) in supply chain	Number of cases	Production Cost	Transportation cost	€
Eutrophication	Aquatic eutrophication potential	kg PO_4^{3-}-eq or kg NO_3-eq	Working time	Cumulative working time in supply chain	Second	Production Cost	Equipment cost	€
	Terrestrial eutrophication potential	m^2 UES	Public living condition	Avoided environmental cost on human health in supply chain	Euro	Production Cost	Electricity and fuel cost	€
Water use	Blue water consumption	m^3	Provision of employment	Created employment opportunity	Number of jobs	Production Cost	waste collection and material cost	€
Land use	Land occupation	m^2-years	Child labour	Cumulative child labour in supply chain	Labour unit	Production Cost	Maintenance cost	€
Mineral resource depletion	Mineral extraction	MJ extra	Forced labour	Cumulative forced labour in supply chain	Labour unit	Production Cost	Soft cost (design fee, permit fee…)	€
Human toxicity	Human toxicity, non-carcinogens	kg C_2H_3Cl-eq	Contribution to economy	Total added value to economy in macro scale	Euro	Profitability	Net income per function unit	€
	Human toxicity, carcinogens	kg C_2H_3Cl-eq	Technology development	Change of production efficiency	%	Efficiency of value creation	Net value added per kg waste	€/kg
Ozone layer depletion	Ozone layer depletion potential	kg CFC-11-eq	Promoting social responsibility	Waste reused or recycled	Kg waste			
Eco-toxicity	Fresh (Marine) water eco-toxicity potential	kg TEG-eq	Resource and energy security	Avoided resource and energy	kg resource or MJ energy			
	Terrestrial eco-toxicity potential	kg TEG-eq						
Photochemical smog	Photochemical oxidation potential	m2*ppm*hours or kgC_2H_4-eq						

Figure 11.3 Tiered life cycle sustainability assessment framework for AWCB system.

proposed in the LCSA framework for AWCB valorisation systems. The indicators in yellow, green and blue boxes represent the investigated impacts in tier I, II and III, respectively. The number and types of indicators in tier III are not fixed. It should be adjusted based on the characteristics of AWCB system in study.

Tier I of the LCSA framework was designed for carrying out a streamlined LCA. It provides a fundamental profile of the sustainability performance of the system as an accessible and entry level LCSA. It is focused on impacts of greatest public and policy concern (*e.g.* climate change, employment, production cost). It has low demand for data quality. Tier II is a 'business-as-usual' LCSA. It aims to reduce uncertainty by increasing the specification of data quality and includes more impact indicators with a greater data demand. Tier III has a high temporal and spatial resolution for LCA results. It includes a greater set of indicators, which may not be always reported, and all elementary flows within investigated system should be included. This type of LCA is used for advancing the LCA technique of experienced LCA practitioners or research outcome. The tiered LCSA framework can specify the comparison criteria and provide robust sustainability information of AWCB valorisation systems, based on the TRL of associated technologies, and the quantity and quality of available data. The tiered LCSA framework will also meet the different demands of sustainability information in AWCB valorisation. It can not only encourage stakeholders to adopt LCSA by providing an easy entry route, but also offer a framework for experienced LCA professions and researchers to improve LCA methodology and data collection schemes in AWCB valorisation systems and achieve a comprehensive and accurate information of sustainability performance. According to Chen and Holden,[35] the Tier II LCSA can be used for managing the sustainability of matured AWCB valorisation system. Tier I has merit as it has a low demand on time and effort and includes information on key environmental and social-economic indicators (*e.g.* climate change, cost, employment). Tier III would be more suitable for internal evaluation with advanced sustainability goals.

11.4 Examples of Crop System ACWB Valorisation

Five case studies are presented to illustrate how LCA can be used to understand the implications of valorising AWCB from crop production systems. These includes: general crop AWCB valorised for animal feed; rice AWCB valorised by composting; olive AWCB valorised by pyrolysis; potato AWCB valorised as an input to composite material production; and tomato AWCB valorised as a pesticide by biosolarization.

11.4.1 Crop Production (Numerous) → Residue (Examples) → Animal Feed (Business-as-usual)

Research[20,24] identified that AWCBs are currently valorised as a feed for different livestock systems. Feed is defined as a first-generation AWCB valorisation pathway.[10] In order to understand the implications of diverting an AWCB from this pathway to a second-generation technology, the business-as-usual (BaU) pathway must first be established as the benchmark. This first example illustrates how this can be done, allowing the LCA practitioner to evaluate the current impact of AWCB valorisation.

For this first example, it is hard to justify calling the system *circular* as the AWCB is exported from a farm and valorised as input into another farm system. This is unless the system boundary is extended and the manure from the animal, which consumes the feed is included and is circulated back to the crop where the AWCB was derived from. Therefore, AWCB valorisation for feed is best described as an *extended-linear* system, when the system boundary does not include the whole animal production system (Figure 11.4) and the manure (*i.e.* the nutrients taken from the soil) is not imported back to the farm where the AWCB originates from.

In this example, the goal of the study would be to calculate the environmental benefit of valorising the AWCB generated from the *Food Production System* (Figure 11.4). In this case, as a feed supplement for cattle we chose tomato pulp, which is a by-product from the tomato processing industry and is produced in large amounts in California, see Oldfield *et al.*[24] The system boundary would need to include the upstream production of the tomato system (Food Production System – Figure 11.4), it would also include the processing of the tomato to produce the primary product (assumed to be tomato paste). These aggregated impacts would have to be allocated between the primary product: (1) tomato paste; and (2) the valorised AWCB (by-product), the tomato pulp used as a feed. The system boundary could be curtailed at the feed (tomato pulp) entering the *Animal Production System*. If the system was extended to include the animal production system, the feed efficiency should be considered.

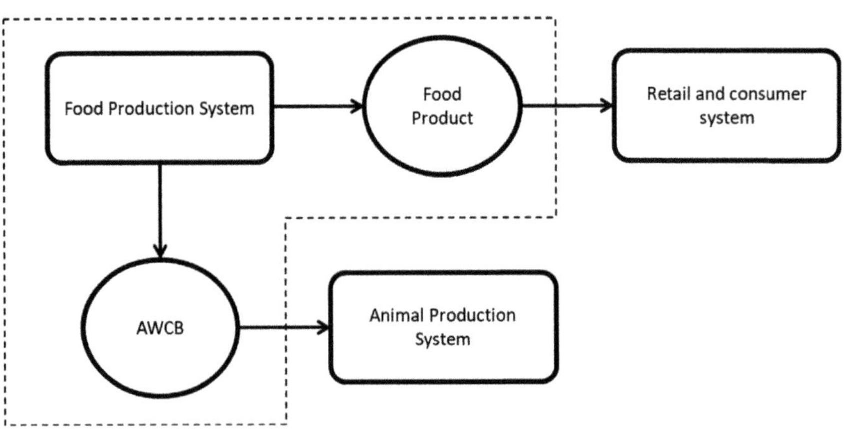

Figure 11.4 System diagram, with boundary, for AWCB valorisation as livestock feed.

Table 11.2 Example stakeholder analysis looking as differing perspectives of system.

Stakeholder	Stakeholder interest	Stage of interest	Influence	Function (perspective)
Farmer – Tomato production	High yield; best quality product; low cost; protection of soil	Production	High	kg crop per hectare; quality (protein/fat *etc.*) of crop per hectare; € of crop per hectare
Tomato Processor	Lowest cost for processing tomato; highest price for tomato and AWCB	Processing	Medium	Lowest/highest € per kg tomato paste/pulp processed/produced
Retailer	Lowest cost for quality of product, to get best profit	Retail	Medium	Quality (protein/fat *etc.*) of tomato per € purchased
Customer	Lowest price for best product	Consumer	Low	Quality (protein/fat *etc.*) of rice per € sold
Regulator	Provide product without exceeding environmental limitations	All	High	kg of product
Farmer – Cattle	Lowest price for feed; best-quality feed	Production	Low	kg of feed; € per kg of feed

The functional unit of such a system depends on the perspective chosen; when systems have multiple perspectives, it has been suggested that a stakeholder analysis is undertaken,[27] an example of a stakeholder analysis is presented in Table 11.2. For this study, the perspective chosen was the *cattle farmer*, and therefore per kg of feed (*i.e.* per kg of tomato pulp), in order to establish the BaU benchmark.

The use of AWCB as a feed for animal production is well established and can therefore be described as a technology that is high on the TRL scale. This also means that there are a lot of data available that enables a more detailed LCA evaluation of the system. Table 11.3 gives an indicative idea of the data requirements for carrying out such an analysis. The table is broken down into the different sub-systems and processes involved.

The results will differ depending on the AWCB value chain chosen, the valorisation technology, and the location and distance of where the AWCB derived feed is distributed. An example of what the results could look like (based on previous analysis carried out) are presented in Figure 11.5 for three indicators: Climate Change (CC), Acidification Potential (AP) and Eutrophication Potential (EP).

This allows a LCA practitioner to establish a baseline model for the AWCB valorisation pathway and includes a number of indicators where trade-offs can be assessed. The interpretation that could be given from these results would be a comparison of process contribution per impact, and an analysis

Table 11.3 Life Cycle Inventory (LCI) data for carrying out AWCB valorisation for livestock feed.

Agricultural Operations – Tomato Production				
In	Machinery fuel consumption	Litres	Primary activity data – litres consumed	Database for emission factors
	Fertiliser (production and emissions)	kg	Primary activity data – amount and type	Database for emission factors
	Pesticide (production and emissions)	kg	Primary activity data – amount and type	Database for emission factors
	etc.			
Out	Tomato	kg	Primary activity data	—
Food Processing				
In	Energy consumption	kWh	Primary activity data – amount consumed	Database for emission factors
	Other consumable materials	$kg\,L^{-1}$	Primary activity data – amount consumed	Database for emission factors
	Machinery	kg	Primary activity data – amount required	Database for emission factors
	etc.			
Out	Tomato paste	kg	Primary activity data	—
	Tomato pulp	kg	Primary activity data	—
Transport of Product (*e.g.* Tomato paste)				
—	Type/distance	$t\,km^{-1}$	Primary activity data – distance/type	Database for emission factors
AWCB Processing/Valorisation				
In	Machinery for drying	kg	Primary activity data – amount required	Database for emission factors
	Energy consumption for drying/processing	kWh	Primary activity data – amount consumed	Database for emission factors
	Other consumable materials	$kg\,L^{-1}$	Primary activity data – amount consumed	Database for emission factors
Out	Dried tomato pulp (feed)	kg	Primary activity data	—

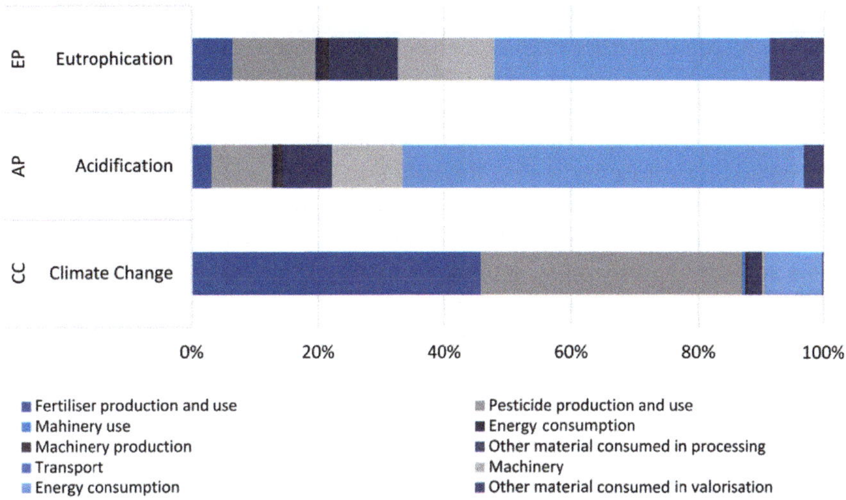

Figure 11.5 Contribution analysis – percentage impact per parameter for climate change; acidification; eutrophication.

of whether this is the same across impact categories. It also provides the practitioner with an impact (Figure 11.5) for three impact categories per functional unit chosen, which would allow for a comparison with other cattle feed, assuming the system boundary is the same.

11.4.2 Crop Production (Rice) → Residue → Composting → Crop Production

This example also presents how to analyse an AWCB that has already been valorised. In this scenario, the BaU for the AWCB is its valorisation as a feed for livestock (Example 1 approach). The alternative AWCB valorisation pathway chosen was another first-generation technology: composting. Composting technology can be installed at a farm level, thus creating a Circular Bioeconomy system at a farm level, whereby nutrients [nitrogen (N); phosphorus (P); and potassium (K)], carbon and organic matter are circularised. The crop system chosen in this case is rice production, which is based on research being carried out by the AgroCycle[6] project. The production of white rice produces two by-products: bran and husk. Husk is commonly used for parboiling;[36] whilst bran is valorised for animal feed production due to its high fat content.[37] If the bran and husk are then diverted from this pathway, the implications must be considered, *i.e.* what will be the marginal feed that will replace the bran, and what source of energy would replace the heat produced *via* husk combustion used in parboiling.

Composting is one of the most circular agricultural technologies that is commercially viable and widely used today. It takes an AWCB and transforms it into a value-added product that returns nutrients (NPK) back to the soil.

The percentage of nutrients returned is by no means 100% of those consumed, for example, Oldfield *et al.*[25] estimated that when composting wasted food and food residue approximately, 3% of N; 5% of P and 0.3% of K is actually recycled, which means the remainder is either consumed or lost somewhere else in the system. AWCB valorisation *via* composting as a means of circularising nutrients is best described as circular (Figure 11.6) when the systems boundary is placed around the farm and the only nutrients that are exported out are those embedded in the primary product, *i.e.* the rice.

The goal of the study is to calculate the environmental benefit of diverting an AWCB from its current valorisation pathway (feed) to an alternative, more circular pathway (composting). When circulating an AWCB there maybe limitations to how much of the system can be converted from a linear system to a circular system. This limitation was found in this case study, where only 33% of the system could be converted to a circular approach, *i.e.* AWCB from 100 ha^{-1}, was enough to convert 33 ha^{-1} to a compost system and the remaining 67 ha^{-1} would still remain as a linear fossil-based system. As composting technology is high on the TRL scale and in line with the AgroCycle Protocol Rules for AWCB valorisation we suggest including multiple impact categories. This can be supplemented with other metric analysis, such as *Circularity of Resource Use*[30] or *Carbon Return on Investment.*[27]

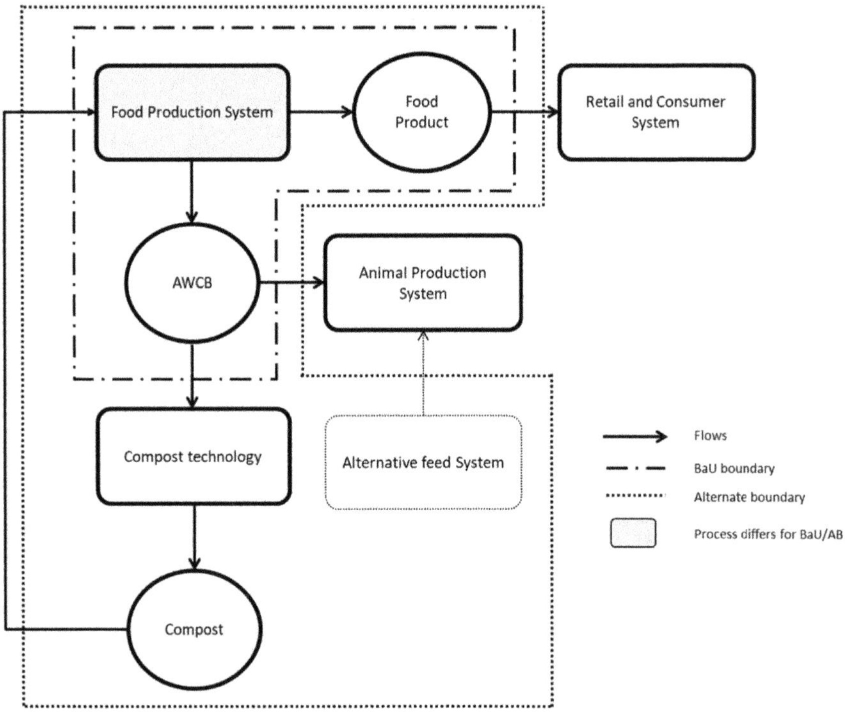

Figure 11.6 System diagram, with system boundary for AWCB valorised *via* competing pathways.

The system boundary in this example must include a number of sub-systems that are inter-related, which form the overall system under analysis. The different sub-systems are listed in the system diagram (Figure 11.6) and the example LCI table (Table 11.4) but include a feed sub-system; a rice production sub-system (agriculture operations and processing). As per the previous example, the functional unit of such a system depends on the perspective chosen (an example stakeholder analysis is included in Example 1). For this study, the perspective chosen was that of the farmers, and therefore a functional unit of per kg of processed white rice would be used.

The valorisation of an AWCB *via* composting is well established and is a technology that is high on the TRL scale. This also means that there are a lot of data available that can enable a more detailed LCA evaluation of the system. However, it is worth noting though that composting is highly variable, depending on many parameters such as feedstock, technology, therefore specific data may have to be substituted with more generic composting data. Table 11.4 gives an indicative idea of the data requirements for carrying out such an analysis. The table is broken down into the different sub-systems and processes involved.

The results for such an analysis will differ depending on the AWCB value chain chosen, the valorisation technology, the location and distance of where the AWCB derived feed is distributed, *etc.* Figure 11.7a and b give an indication of what the results may look like. The two options should be presented, showing their total impact and the contribution for each stage of the valorisation system.

Results such as those that are illustrated in Figure 11.7a and b would result in the conclusion that the circulation of AWCB would provide a minor improvement for carbon impact/abatement. The circulation does, however, result in nutrient, carbon and organic matter being returned to the soil, which were not captured in the analysis, but would lead to more carbon saving.

11.4.3 Crop Production (Olives) → Residue → Pyrolysis → Crop Production (+Energy)

This example presents the case of olive production with olive pomace produced as a co-product. In this scenario, the BaU for olive pomace is its valorisation as a feed for livestock (similar to the first example). The alternative AWCB valorisation pathway is pyrolysis, a second-generation technology suitable for treatment of agricultural residues, forestry residues, wood and animal manures.[38] Pyrolysis is a process which involves heating biomass materials so that they produce useful products, without complete combustion in the absence of oxygen.[39] The pyrolysis process is one of the simplest and oldest methods of processing one source of biomass to form three primary products; liquid fuel or bio-oil, gas and biochar.[38] The biochar produced from pyrolysis of olive pomace is a promising soil amendment on old olive plantations as it can increase the ability of sandy soils to retain

Table 11.4 Life Cycle Inventory (LCI) data for carrying out AWCB valorisation *via* composting.

Agricultural Operations – Rice Production with Mineral Fertiliser[a]				
In	Machinery fuel consumption	Litres	Primary activity data – litres consumed	Database for emission factors
	Fertiliser (production and emissions)	kg	Primary activity data – amount and type	Database for emission factors
	Pesticide (production and emissions)	kg	Primary activity data – amount and type	Database for emission factors
	Irrigation – fuel consumption	Litres	Primary activity data – litres consumed	Database for emission factors
	etc.			
Out	Whole rice	kg	Primary activity data	—
Food Processing – Milling/Parboiling[a]				
In	Whole rice	kg	Primary activity data	—
	Energy consumption	kWh	Primary activity data – amount consumed	Database for emission factors
	Other consumable materials	$kg\,L^{-1}$	Primary activity data – amount consumed	Database for emission factors
	etc.			
Out	White rice	kg	Primary activity data	—
	Rice husk	kg	Primary activity data	—
	Rice bran	kg		
Transport of Product (*e.g.* Rice)[a]				
—	Type/distance	$t\,km^{-1}$	Primary activity data – distance/type	Database for emission factors
AWCB Processing/Valorisation – Composting[b]				
In	Rice/bran husk in	kg	Primary activity data – amount required	Database for emission factors
	Energy consumption for composting	kWh	Primary activity data – amount consumed	Database for emission factors
	Other consumable materials	$kg\,L^{-1}$	Primary activity data – amount consumed	Database for emission factors
Out	Compost	kg	Primary activity data	—
Agricultural Operations – Rice Production with Compost[b]				
In	Machinery fuel consumption	Litres	Primary activity data – litres consumed	Database for emission factors
	Compost (production and emissions)	kg	Primary activity data – amount and type	Database for emission factors
	Irrigation – fuel consumption	Litres	Primary activity data – litres consumed	Database for emission factors
	etc.			
Out	Whole rice	kg	Primary activity data	—
Alternative Feed System[b]				
—	Type based on nutritional composition	kg	Primary activity data – distance/type	Emission factor from previous study

[a]Data for BaU.
[b]Data for new valorisation pathway.

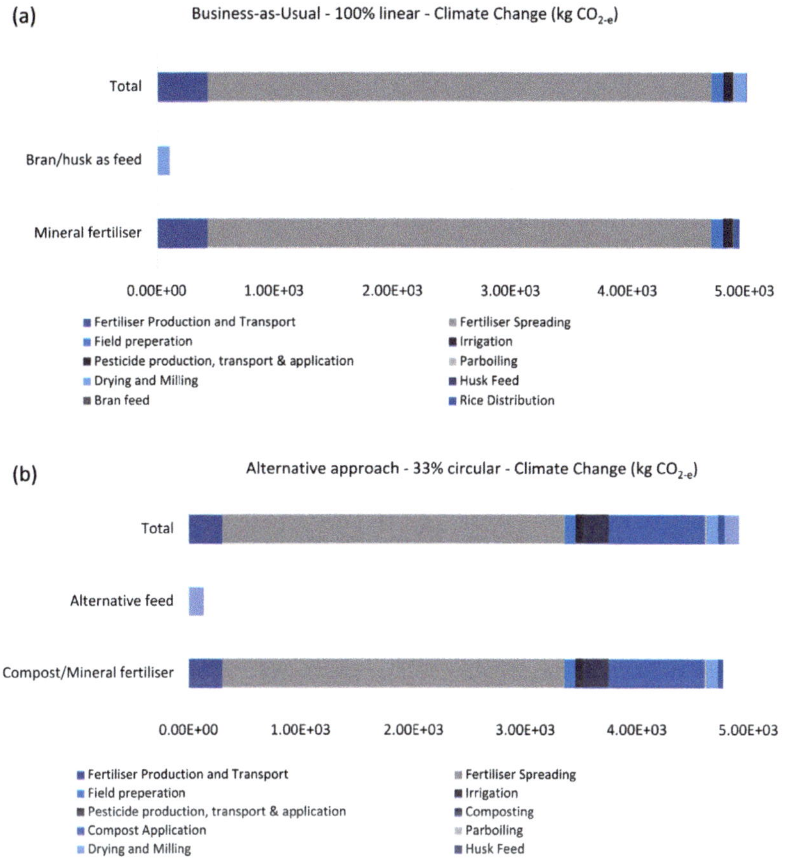

Figure 11.7 (**a**) Contribution and total impact of Business-as-Usual, 100% linear, (**b**) Contribution and total impact of AA, 33% circular and 67% linear.

water and nutrients and can help to offset the acidifying effects of N fertilisers.[40] The biochar can also have a role in carbon sequestration,[11] *i.e.* a process by which carbon dioxide is removed from the atmosphere and held in solid or liquid form. Thus, when applied at the farm level, biochar produced from pyrolysis is returned to the same land that provided the feedstock, thus resulting in a Circular Bioeconomy system at a farm level. The other products of pyrolysis; liquid fuel or bio-oil and gas, can displace fossil fuels, thus an extended linear system expands beyond the farm level.

The production of olive oil produces two by-products in the olive mill; pomace and wastewater. It is assumed that the wastewater is treated in a conventional manner; whilst the pomace is valorised in the pyrolysis process. If the pomace is diverted away from its conventional use as animal feed, production of the marginal feed that will replace the pomace will need to be considered. This approach encroaches into the realms of Consequential LCA, whereby the marginal product displaced must be identified and modelled.

As the olive pomace is wet, the pyrolysis process must begin with drying of the biomass as wet biomass generates water in the bio-oil.[41] This is an energy-intensive process, thus attention must be paid to sourcing sustainable fuels for heating, minimising heat loss and potential use of natural/solar drying in a greenhouse.[40] It is then necessary to grind the biomass to below a few millimetres in size (1–3 mm).[41] Heat is then supplied externally to the pyrolysis vessel and the process proceeds in a carefully controlled air supply.[39]

The goal of the study is to calculate the environmental benefit of diverting an AWCB from its current valorisation pathway (feed) to an alternative, more circular pathway (pyrolysis).

The system boundary in this example includes several sub-systems that are inter-related, which form the overall system under analysis. The different sub-systems are listed in the system diagram (Figure 11.8) and the example LCI table (Table 11.5), including an olive oil production sub-system (agriculture operations and processing); a feed sub-system; and a fossil fuel sub-system. For this study, the perspective chosen was that of the farmers, as the pyrolysis system represents a partial circular bioeconomy at

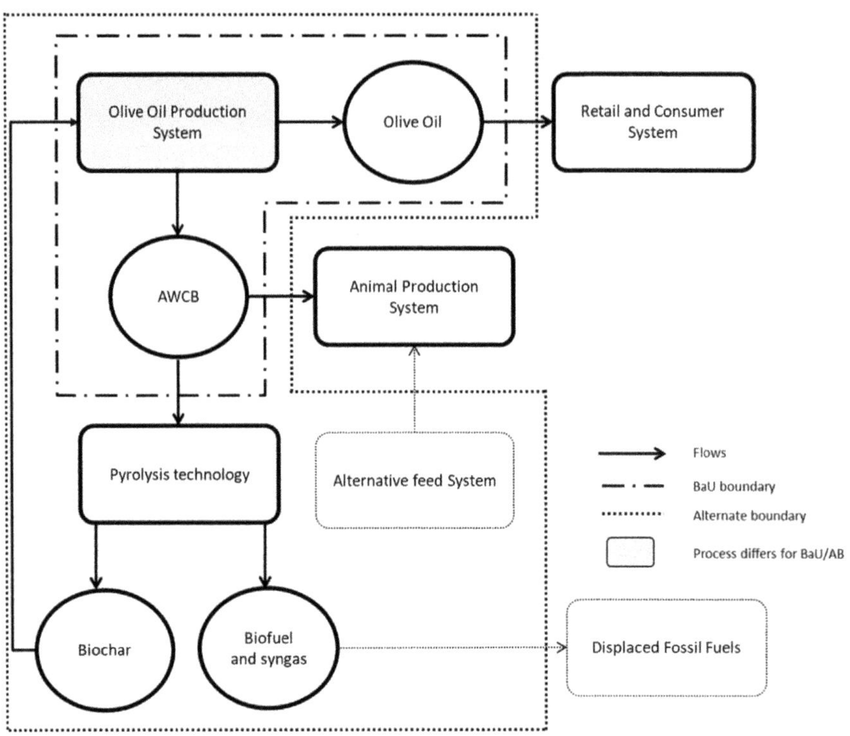

Figure 11.8 System diagram, with system boundary for olive pomace AWCB valorised *via* competing pathways.

Table 11.5 Life Cycle Inventory (LCI) data for carrying out AWCB valorisation *via* pyrolysis (adapted from Parascanu *et al.*)[42].

Agricultural Operations – Rice Production with Mineral Fertiliser[a]				
In	Machinery fuel consumption	Litres	Primary activity data – litres consumed	Database for emission factors
	Fertiliser (production and emissions)	kg	Primary activity data – amount and type	Database for emission factors
	Manure (production and emissions)	kg	Primary activity data – amount and type	Database for emission factors
	Pesticide (production and emissions)	kg	Primary activity data – amount and type	Database for emission factors
	Irrigation – fuel consumption	Litres	Primary activity data – litres consumed	Database for emission factors
	etc.			
Out	Olives	kg	Primary activity data	—
Food Processing – Oil extraction[a]				
In	Olives	kg	Primary activity data	—
	Energy consumption	kWh	Primary activity data – amount consumed	Database for emission factors
	Machinery fuel consumption	Litres	Primary activity data – litres consumed	Database for emission factors
	Other consumable materials	kg L^{-1}	Primary activity data	Database for emission factors
	etc.			
Out	Olive Oil	kg	Primary activity data	—
	Olive Pomace	kg	Primary activity data	—
	Wastewater	kg	Primary activity data	Database for emission factors
Transport of Product (*e.g.* Olive oil)[a]				
—	Type/distance	t km^{-1}	Primary activity data – distance/type	Database for emission factors
AWCB Processing/Valorisation – Pyrolysis[b]				
In	Olive pomace	kg	Primary activity data – amount required	Database for emission factors
	Energy consumption for pyrolysis	kWh	Primary activity data	Database for emission factors
	Other consumable materials	kg L^{-1}	Primary activity data	Database for emission factors
Out	Biochar	kg	Primary activity data	—
	Bio-oil	kg	Primary activity data	—
	Bio-syngas	m^3	Primary activity data	—
Agricultural Operations – Olive Production with Biochar[b]				
In	Machinery fuel consumption	Litres	Primary activity data – litres consumed	Database for emission factors
	Biochar (production and emissions)	kg	Primary activity data – amount and type	Database for emission factors
	Fertiliser (production and emissions)	kg	Primary activity data – amount and type	Database for emission factors
	Manure (production and emissions)	kg	Primary activity data – amount and type	Database for emission factors
	Pesticide (production and emissions)	kg	Primary activity data – amount and type	Database for emission factors
	Irrigation – fuel consumption	Litres	Primary activity data – litres consumed	Database for emission factors
	etc.			
Out	Olives	kg	Primary activity data	—
Alternative Feed System[b]				
—	Type based on nutritional composition	kg	Primary activity data – distance/type	Emission factor from previous study
Alternative Energy System[b]				
—	Type based on net calorific value	MJ	Primary activity data – type	Emission factor from studies

[a]Data for BaU.
[b]Data for new valorisation pathway.

the farm level, and therefore a functional unit of per kg of processed olive oil was used.

The results for such an analysis will differ depending on the AWCB value chain chosen, the valorisation technology, the location and distance to where the AWCB derived products are distributed. Figure 11.9a and b, are from a study by Parascanu *et al.*[42] on the life cycle assessment of olive pomace valorisation through pyrolysis and give an indication of what the results may look like. The two options should be presented, showing their total impact and the contribution for each stage of the valorisation system.

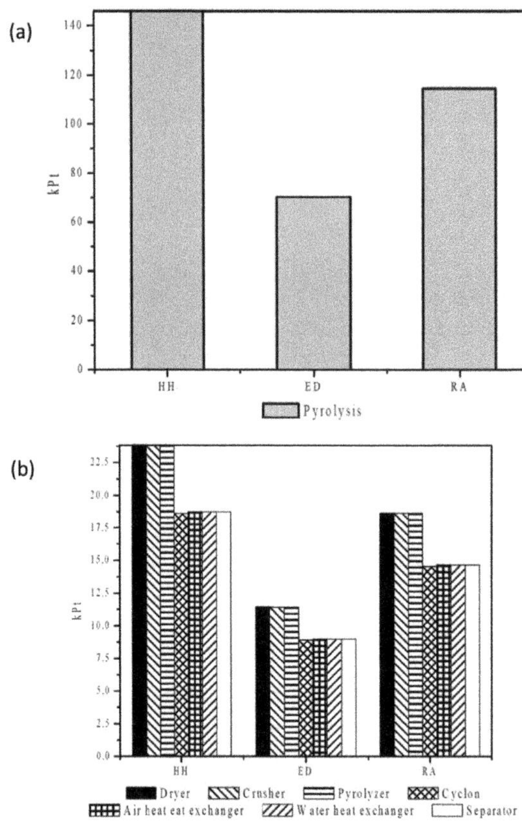

Figure 11.9 (**a**) Aggregated impacts (HH – Fossil depletion, ED – Damage to human health, RA – Damage to ecosystem diversity) according to the ReCiPe End-point methodology in Thousand eco-points (kPt) of pyrolysis system, calculated for 100 kg olive pomace. Reproduced from ref. 42 with permission from Elsevier, Copyright 2018. (**b**) Aggregated impacts in Thousand eco-points (kPt) according to the ReCiPe end-point methodology of each pyrolysis equipment (dryer, crusher, pyrolyzer, cyclon, air heat exchanger, water heat exchanger and separator), calculated for 100 kg olive pomace. Reproduced from ref. 42 with permission from Elsevier, Copyright 2018.

11.4.4 Crop Production (Potato) → Residue → Biocomposite

Unlike previous examples focusing on valorisation technology and system boundary in process-based LCA, this example explores the social-economic impacts of AWCB systems with a more complete system boundary. The potato-based starch production systems or potato processing systems generate a significant amount of potato pulp (AWCB) with a high fibre content. Currently, only a small part of potato pulp is valorised and used as cattle feed, while most of it is considered as an agricultural waste.[43] Potato pulp can be used as a bio-additive to replace the biomass currently used in biopolymer production,[44] thus reducing the biomass-associated impacts from the agricultural system. This proposed valorisation system could be categorised as an *extended-linear* system as it uses biomass from one system to replace the feedstock in another. The research has been carried out in AgroCycle project.[6] The alternative treatment of potato fibre in the AgroCycle system will not only change the social-economic impacts of the biopolymer production system, but also the impacts within the systems associated with biomass and biopolymer supply chains. To evaluate the feasibility of this valorisation approach, the comprehensive social-economic impacts must be investigated.

Figure 11.10 showed the system boundary for bioplastic production from a conventional and an extended-linear system. The conventional starch production system was also considered for allocating the upstream impacts of potato production in an AgroCycle extended linear system. The main objective of this LCA is evaluation of the social-economic effect of using potato fibre to replace biomass feedstock in biopolymer production. Therefore, the functional unit is one kg of biopolymer produced.

In order to capture the comprehensive social-economic impacts, the hybrid life cycle assessment approach was adopted.[45] Hybrid LCA combines

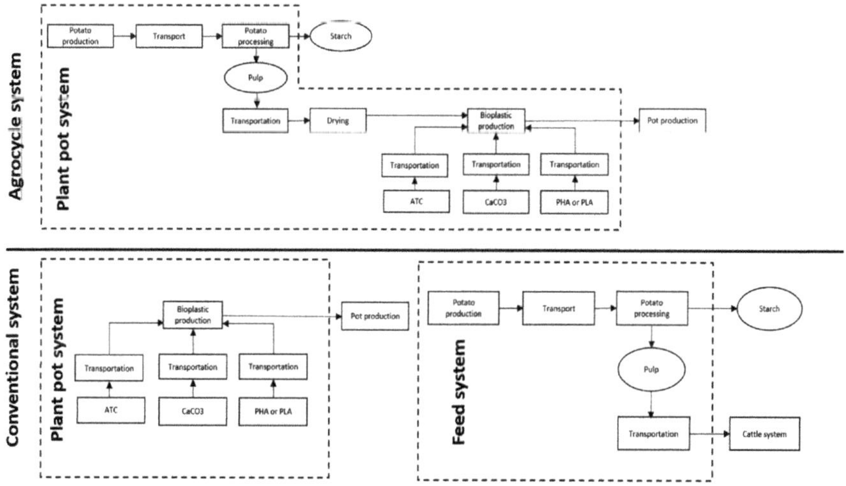

Figure 11.10 System diagram for conventional bioplastic and potato system, with extended linear AgroCycle system.

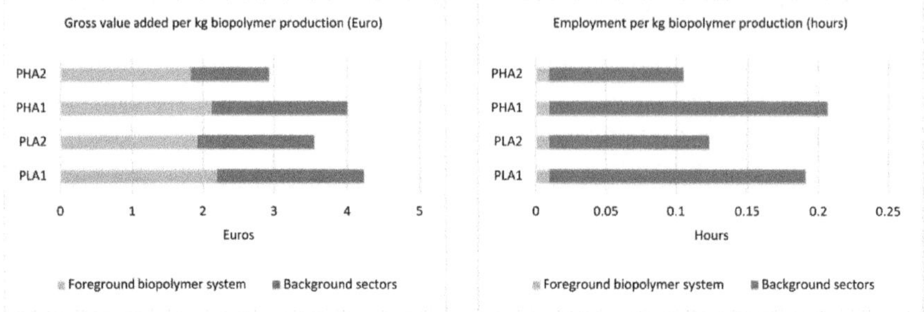

Figure 11.11 Gross value added and employment per kg biopolymer production.

national economic input-output (IO) data with process level engineering data. It can eliminate the issue of incomplete system boundary in process-based LCA, but also maintains the detailed engineering information of specific processes within the studied system.[46] To complete the LCI for a hybrid LCA, it requires the information of the amount of each physical input and the corresponding economic value, which serve as a bridge to link bottom-up engineering data with top-down IO data. The investigated social-economic impacts were employment and gross value added (GVA). There were four sets of experiment conducted. Bioplastic polylactic acid (PLA) and Polyhydroxyalkanoates (PHA) were both used in two groups. Only PLA1 and PHA1 used potato fibre in biopolymer production.

Figure 11.11 showed the GVA (euro) and employment (hours) per kg biopolymer production. If we only compare the GVA in foreground biopolymer system (system boundary of process based LCA), it seems potato fibre can increase the GVA in the PHA-based biopolymer and decrease GVA in the PLA-based system. However, combining the GVA in background sectors, the total GVAs were improved with potato fibre valorisation approach. This suggests that the extended linear approach for potato fibre through biopolymer production has positive effect on whole economic development. Similar to GVA, only focusing on employment in biopolymer production system will not reveal the total social impact of biopolymer production. Due to the use of agriculturally based biomass [polylactic acid (PLA), Polyhydroxyalkanoates (PHA)] was used in all groups, the main contributor of employment comes from background sectors, especially agriculture-related sectors in IO table. The employment of foreground system in four groups were almost the same. However, significant difference was noted between potato fibre added groups with non-added groups. The results also indicated the valorising potato fibre in biopolymer production also improved the social impact for the whole economy.

11.4.5 Crop Production (Tomato) → Residue → Biosolarization → Crop Production

This example is based on Oldfield *et al.*,[24] but differs in that it includes the upstream impact of the tomato pulp production, *i.e.* it does not follow the zero-burden assumption and uses a downstream functional unit, rather than per kg of pulp. This approach is more in line with the methodology (AgroCycle Protocol rules) presented in Section 11.4. This example is similar to the second example in its methodological approach. Again, we are analysing an AWCB that has already been valorised as a feed for livestock, as the BaU. However, this example differs in the alternative valorisation pathway, and we are now investigating a novel second-generation AWCB (tomato pulp) valorisation technology, biosolarization, a hydrothermal process of disinfesting soil of agricultural pests. Thus, displacing the need for chemical pesticide application. As per the second example, all functions of the system must be satisfied.

This example includes even more sub-systems (Figure 11.12) than the previous examples, which must be fulfilled in order to make the competing scenarios comparable. The valorisation system has potential to be called circular, but in its current approach it is more of an *extended-linear* system. Due to the tomato system exporting the pulp to be used as a feedstock for biosolarization in another field, *i.e.* for eggplant production. Therefore, any nutrient (NPK) that is exported will not be imported back to the tomato field, thus requiring nutrient top-up from other sources.

Biosolarization is a novel technology and sits at around TRL 3–4. There are limited data available for biosolarization, so this would have to be established from experiments. As this is low on the TRL, the LCA indicator included would be GWP in order to make sure that the technology has a positive effect

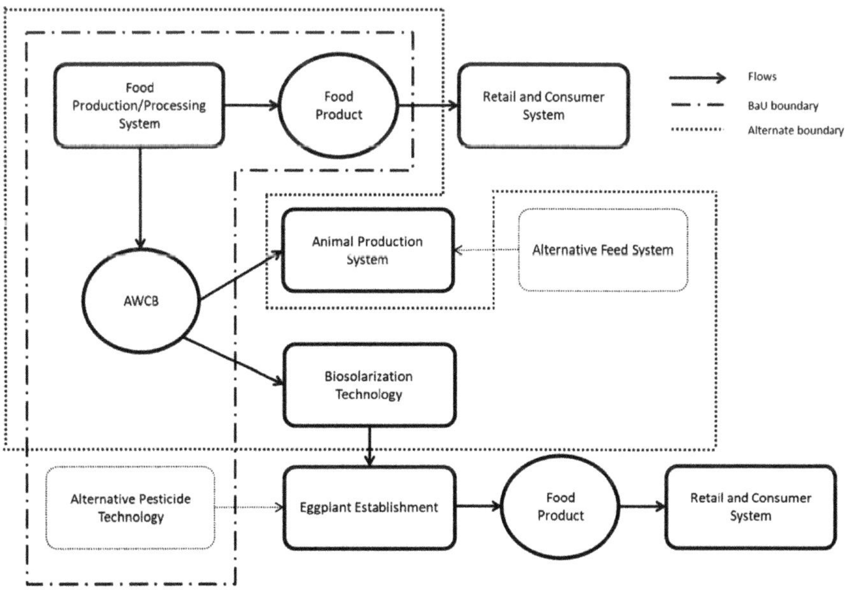

Figure 11.12 System diagram, with system boundary for AWCB valorised *via* competing pathways.

on climate change (*i.e.* it does not lead to an increase). It would also be important to include the indicator human-toxicity as this is the main harmful affect from pesticide, the BaU for biosolarization. Table 11.6 gives an indicative idea of the data requirements for carrying out such an analysis. The table is broken down into the different sub-systems and includes relevant processes.

A typical LCIA, which includes a sensitivity analysis and interpretation can be seen in Oldfield *et al.,* 2017. The results from such an analysis enable LCA practitioners to assist policy makers in making evidence-based decisions. This example found that a carbon saving could be realised by moving to a biosolarization system for dealing with pest control, but at the expense of increased water use through irrigation. This is an issue in the research area of California.

11.5 Discussion and Conclusions

There are three defining principles of circular economy[33] that can be used to assess our progress towards circularisation of food systems. These are, to design out waste in the system and the pollution it causes, to keep products and materials in use for as long as possible, and to create biological resources using restorative and regenerative practices for natural systems. It is possible to assess specific systems, or the contribution of technologies using a Venn diagram (Figure 11.13) with those options at the intersection of all three criteria offering real solutions that are circular. Of course, all of these must operate in the context of the limit to the planetary boundary, which can alternatively be considered the intersection of environmental, economic and social aspects of the system. Circular economy needs to be considered in the context of a strong sustainability model.[47] Taking the examples from this chapter, we can evaluate possible techno-economic solutions considering meeting both circularity and sustainability criteria.

11.5.1 Design Out Waste and Pollution

It is reasonable to state that, at present, progress on designing out waste and pollution is perhaps most advanced. A key driver has been the valorisation of AWCB to create new valuable resources, and the reduction of impacts such as eutrophication and climate change.[48] Taking the AgroCycle case studies as examples[48] we can see that the impact outcomes predicted using LCA, tend to be quite small marginal gains. It is important to note that few, if any, studies have attempted to assess all waste and all sources of pollution from a system, but the key indicators used to date have been typical life cycle impact methods such as climate change, eutrophication, acidification, ecotoxicity and biodiversity impacts.[49] Taking a life cycle thinking approach, we also have to ask whether the technologies proposed to reduce waste and pollution at a particular life cycle stage actually reduce impact over the whole system (*i.e.* avoid burden shifting). It is also important to ask what products are being

Table 11.6 Life Cycle Inventory (LCI) data for carrying out AWCB valorisation *via* biosolarization.

Agricultural Operations – Tomato Production[1/2]				
In	Machinery fuel consumption	Litres	Primary activity data – litres consumed	Database for emission factors
	Fertiliser (production and emissions)	kg	Primary activity data – amount and type	Database for emission factors
	Pesticide (production and emissions)	kg	Primary activity data – amount and type	Database for emission factors
	etc.			
Out	Tomato	kg	Primary activity data	—
Food Processing[1/2]				
In	Energy consumption	kWh	Primary activity data – amount consumed	Database for emission factors
	Other consumable materials	$kg\ L^{-1}$	Primary activity data – amount consumed	Database for emission factors
	Machinery	kg	Primary activity data – amount required	Database for emission factors
	etc.			
Out	Tomato paste	kg	Primary activity data	—
	Tomato pulp	kg	Primary activity data	—
Transport of Product (*e.g.* Tomato paste; feed)[1/2]				
—	Type/distance	$t\ km^{-1}$	Primary activity data – distance/type	Database for emission factors
Agricultural Operations – Egg Plant Production Soil Preparation *via* Pesticide Application[1]				
In	Machinery fuel consumption	Litres	Primary activity data – litres consumed	Database for emission factors
	Pesticide (production and emissions)	kg	Primary activity data – amount and type	Database for emission factors
	etc.			
Agricultural Operations – Egg Plant Production Soil Preparation *via* Biosolarization[1]				
In	Tomato pulp	kg	Primary activity data	—
	Biosolarization	—	Data from Experiment	—
	Machinery fuel consumption	Litres	Primary activity data – litres consumed	Database for emission factors
	Irrigation – fuel consumption	Litres	Primary activity data – litres consumed	Database for emission factors
	etc.			
Alternative Feed System[2]				
—	Type based on nutritional composition	kg	Primary activity data – distance/type	Emission factor from previous study

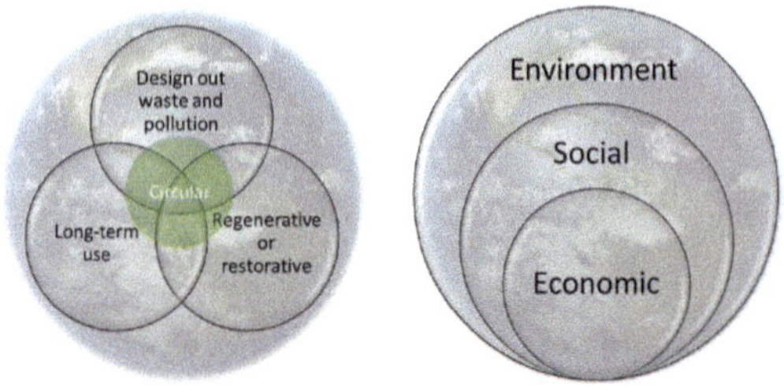

Figure 11.13 The ideal contribution of materials from a crop production system should operate at the intersection of minimum waste and pollution, maximum longevity and contributing to restoration or regeneration (left). This needs to occur within the limits of the planetary boundary, in the context of strong sustainability (right).

displaced in the market through valorisation.[50] It is equally important to ensure that displacement is focused on non-renewable resources with greater waste or pollution impact, otherwise the global benefit is negative.

11.5.2 Keep Products and Materials in Use for as Long as Possible

The design of products to have a long life has perhaps been of less significance for circular bioeconomy technologies that have emerged in the agri-food sector. A secondary, but just as important issue is how recoverable materials are for reuse. In the context of biological systems, this is a critical issue, because materials can become enmeshed in complex matrices that make recovery energy intensive. Cullen[30] proposed that the ratio of energy needed to recover materials to that needed for primary production is a good indicator of the effort needed to recover resources. In the case of AgroCycle examples, the energy ratio ranges from 1 : 0.41 (in:out) to 1 : 1.98 (in:out), indicating that some technologies will require additional energy to create value-added products, which perhaps helps to explain the small marginal gains reported for these systems. Extended linearisation might give the appearance of extending life but does not necessarily support circularisation. Evidence indicates that we cannot offset the impact of creating foods by merely making the created resources last longer and do more.[27]

11.5.3 Restorative and Regenerative Practices for Natural Systems

Reviewing studies that have looked at circularisation and valorisation technologies for the food system indicates that there has been very little, if any, attention on restorative or regenerative practices in the context of life cycle thinking. Such practices tend to be evaluated in a standalone manner, rather than in the context of food systems impacts. Indicators commonly used in LCA studies such as biodiversity impacts[51] could be interpreted in this context, but typically are not. Examples of current thinking on restorative indicators include total restored products, total restored materials (based on mass flow analysis) and supply chain regenerative flows (derived from mass flow analysis or life cycle assessment).[52] Taking the AgroCycle examples, 2 to 4, there is nothing in the protocol that considers on this aspect of circularity, as the entire focus has been on 'end-of-pipe' solutions to a 'waste' problem rather than a holistic, systems-based transition to circular bioeconomy.

To answer the question of whether we can really circularise the food system is quite difficult. In theory: (1) yes, we can design out waste and pollution, but to date we have focused more on eco-efficiency[32] rather than reducing impact to a fair-share; (2) yes, we can design a system that extends the life of bioresources and keeps the material in use for as long as possible;[33] and (3) yes, in theory we can design production systems to enable restorative and regenerative practices for natural systems. In practice: (1) we are making strides towards designing out waste and pollution from the system, but the rate of process is far from sufficient across all impacts. Taking national statistics for Ireland (an economy that is very reliant on the agri-food sector) as an example, we can see that little process has been made on absolute impact reduction (*e.g.* 53,54), even though eco-efficiency of products has greatly improved (*e.g.* 55); (2) there is some progress towards extending the life of bioresources and keeping materials in use for as long as possible;[27] and (3) we have made little meaningful progress on restorative and regenerative practices in the context of redesign of food systems to be circular, which is clear when we examine global reliance on relocation of water[56] and nutrient resources[57] and degradation of soil resources around the globe[58] to support unsustainable consumption. In this context, significant transformation of the system is still required to move to a sustainable footing.[1] These concepts of circularisation need to be enhanced with additional attributes relating to quantification of 'fair-share' of resources for a particular production/consumption activity and measures of acceptable absolute impact.

Acknowledgements

This project has received funding from the European Union's Horizon 2020 Research and Innovation programme under grant agreement No. 690142 (AgroCycle project), in collaboration with the People's Republic of China and the Hong Kong Government.

References

1. N. M. Holden, E. P. White, M. C. Lange and T. L. Oldfield, Review of the sustainability of food systems and transition using the Internet of Food, *NPJ Sci. Food*, 2018, **2**(1), 18.
2. T. Garnett, Food sustainability: problems, perspectives and solutions, *Proc. Nutr. Soc.*, 2013, **72**, 29–39.

3. http://ec.europa.eu/information_society/newsroom/image/document/2016-48/ward_-_circular_economy_applied_to_the_livestock_production_sector___brussels__2_40231.pdf (last accessed January 2019).
4. https://www.ucd.ie/t4cms/Is%20food%20waste%20waste.pdf (last accessed January 2019).
5. T. L. Oldfield, E. P. White and N. M. Holden, An environmental analysis of options for utilizing wasted food and food residue, *J. Environ. Manage.*, 2016, **183**, 826–835.
6. http://www.agrocycle.eu/ (last accessed January 2019).
7. http://agrocycle-dev.iris.cat/(last accessed January 2019).
8. http://multisite.iris.cat/agrocycle/files/2017/12/D1.3_FHolisticAnalysisAWCB.pdf (last accessed January 2019).
9. http://ec.europa.eu/environment (last accessed January 2019).
10. R. A. D. Arancon, C. S. K. Lin, K. M. Chan, T. H. Kwan and R. Luque, Advances on waste valorization: new horizons for a more sustainable society, *Energy Sci. Eng.*, 2013, **1**(2), 53–71.
11. C. S. K. Lin, L. A. Pfaltzgraff, L. Herrero-Davila, E. B. Mubofu, S. Abderrahim, J. H. Clark, A. A. Koutinas, N. Kopsahelis, K. Stamatelatou, F. Dickson, S. Thankappan, Z. Mohamed, R. Brocklesby and R. Luque, Food waste as a valuable resource for the production of chemicals, materials and fuels. Current situation and global perspective, *Energy Environ. Sci.*, 2013, **6**, 426–464.
12. M. Yan, J. Humphreys and N. M. Holden, An evaluation of life cycle assessment of European milk production, *J. Environ. Manage.*, 2011, **92**(3), 372–379.
13. J. K. Andersen, A. Boldrin, T. K. Christensen and C. Scheutz, Home composting as an alternative treatment option for organic household waste in Denmark: An environmental assessment using life cycle assessment-modelling, *Waste Manage.*, 2012, **32**, 31–40.
14. S. Belboom, J. Digneffe, R. Renzoni, A. Germain and A. Léonard, Comparing technologies for municipal solid waste management using life cycle assessment methodology: a Belgian case study, *Int. J. Life Cycle Assess.*, 2013, **18**, 1513–1523.
15. A. Bernstad and A. la Cour Jansen, A life cycle approach to the management of household food waste – A Swedish full-scale case study, *Waste Manage.*, 2011, **31**, 1879–1896.
16. S. S. Evangelisti, P. S. Lettieri, D. D. Borello and R. Clift, Life cycle assessment of energy from waste via anaerobic digestion: A UK case study, *Waste Manage.*, 2014, **34**, 226–237.
17. T. Fruergaard and T. Astrup, Optimal utilization of waste-to-energy in an LCA perspective, *Waste Manage.*, 2011, **31**(3), 572–582.
18. M. Grosso, C. Nava, R. Testori, L. Rigamonti and F. Viganò, The implementation of anaerobic digestion of food waste in a highly populated urban area: an LCA evaluation, *Waste Manage. Res.*, 2012, **30**, 78–87.

19. H. H. Khoo, T. Z. Lim and R. B. H. Tan, Food waste conversion options in Singapore: environmental impacts based on an LCA perspective, *Sci. Total Environ.*, 2010, **408**, 1367–1373.
20. M. H. Kim and J. W. Kim, Comparison through a LCA evaluation analysis of food waste disposal options from the perspective of global warming and resource recovery, *Sci. Total Environ.*, 2010, **408**, 3998–4006.
21. J. Martínez-Blanco, J. Colón, X. Gabarrell, X. Font, A. Sánchez, A. Artola and J. Rieradevall, The use of life cycle assessment for the comparison of biowaste composting at home and full scale, *Waste Manage.*, 2010, **30**, 983–994.
22. A. Saer, S. Lansing, N. Davitt and R. Graves, Life cycle assessment of a food waste composting system: environmental impact hotspots, *J. Cleaner Prod.*, 2013, **52**, 234–244.
23. http://www.cre.ie/web/wp-content/uploads/2010/12/Compost-Life-Cycle.pdf (last accessed January 2019).
24. T. L. Oldfield, Y. Achmon, K. M. Perano, R. M. Dahlquist-Willard, J. S. VanderGheynst, J. J. Stapleton, C. W. Simmons and N. M. Holden, A life cycle assessment of biosolarization as a valorization pathway for tomato pomace utilization in California, *J. Cleaner Prod.*, 2017, **141**, 146–156.
25. T. L. Oldfield, N. Sikirica, C. Mondini, G. Lopez, P. J. Kuikman and N. M. Holden, Biochar, compost and biochar-compost blend as options to recover nutrients and sequester carbon, *J. Environ. Manage.*, 2018, **218**, 465–476.
26. A. Contreras, E. Rosa, M. Pérez, H. Van Langenhove and J. Dewulf, Comparative Life Cycle Assessment of four alternatives for using by-products of cane sugar production, *J. Cleaner Prod.*, 2009, **17**(8), 772–779.
27. T. L. Oldfield, E. P. White and N. M. Holden, The implications of stakeholder perspective for LCA of wasted food and green waste, *J. Cleaner Prod.*, 2018, **170**, 1554–1564.
28. M. Pradel, L. Aissani, J. Villot, J. C. Baudez and V. Laforest, From waste to added value product: towards a paradigm shift in life cycle assessment applied to wastewater sludge - a review, *J. Cleaner Prod.*, 2016, **131**, 60–75.
29. J. Olofsson and P. Börjesson, Residual biomass as resource – Life-cycle environmental impact of wastes in circular resource systems, *J. Cleaner Prod.*, 2018, **196**, 997–1006.
30. C. M. Cullen, Circular Economy – Theoretical Benchmark or Perpetual Motion Machine? *Circ. Econ.*, 2017, **21**(3), 483–486.
31. D. Ness, Sustainable urban infrastructure in China: Towards a Factor 10 improvement in resource productivity through integrated infrastructure systems, *Int. J. Sustainable Dev. World Ecol.*, 2008, **15**, 288–301.
32. A. Bjørn and M. Z. Hauschild, Absolute versus Relative Environmental Sustainability. What can the Cradle-to-Cradle and Eco-efficiency Concepts Learn from Each Other? *J. Ind. Ecol.*, 2012, **17**, 321–332.

33. https://www.ellenmacarthurfoundation.org/assets/downloads/publications/Ellen-MacArthur-Foundation-Towards-the-Circular-Economy-vol.1.pdf (last accessed January 2019).
34. J. C. Mankins. Technology readiness levels. *White Paper*. 1995.
35. W. Chen and N. M. Holden, Tiered life cycle sustainability assessment applied to a grazing dairy farm, *J. Cleaner Prod.*, 2018, **172**, 1169–1179.
36. M. Ahiduzzama and A. K. M. Sadrul Islam, Energy Utilization and Environmental Aspects of Rice Processing Industries in Bangladesh, *Energies*, 2009, **2**, 134–149.
37. J. L. Nornberg, W. Stumpf Junior, J. Lopez and P. B. Costa, Value of rice bran as a fat source for Jersey cows in early lactation: apparent digestibility of nutrients, *Rev. Bras. Zootec.*, 2004, **33**, 2412–2421.
38. K. B. Cantrell, P. G. Hunt, M. Uchimiy, J. M. Novak and K. S. Ro, Impact of pyrolysis temperature and manure source on physicochemical characteristics of biochar, *Bioresour. Technol.*, 2012, **107**, 419–428.
39. D. Morris and J. Scurlock, Bioenergy, in *Renewable Energy: Power for a Sustainable Future*, Oxford University Press, Oxford, 2012.
40. A. Zabaniotou, D. Rovas, A. Libutti and A. Monteleone, Boosting circular economy and closing the loop in agriculture: Case study of a small-scale pyrolysis–biochar-based system integrated in an olive farm in symbiosis with an olive mill, *Environ. Dev.*, 2015, **14**, 22–36.
41. S. Capareda, *Introduction to Biomass Energy Conversions*, CRC Press Taylor & Francis Group, Florida, 2014.
42. M. M. Parascanu, M. Puig Gamero, P. Sánchez, L. Soreanub, J. L. Valverde and L. Sanchez-Silva, Life cycle assessment of olive pomace valorisation through pyrolysis, *Renewable Energy*, 2018, **122**, 589–601.
43. F. Mayer, Potato pulp: properties, physical modification and applications, *Polym. Degrad. Stab.*, 1998, **59**(1–3), 231–235.
44. N. Jabeen, I. Majid and G. A. Nayik, Bioplastics and food packaging: A review, *Cogent Food Agric.*, 2015, **1**(1), 1117749.
45. B. S. Koelbl, M. A. van den Broek, H. C. Wilting, M. W. J. L. Sanders, T. Bulavskaya, R. Wood, A. P. C. Faaij and D. P. van Vuuren, Socio-economic impacts of low-carbon power generation portfolios: Strategies with and without CCS for the Netherlands, *Appl. Energy*, 2016, **183**, 257–277.
46. S. Suh, Functions, commodities and environmental impacts in an ecological–economic model, *Ecol. Econ.*, 2004, **48**(4), 451–467.
47. J. Rockström, W. Steffen, K. Noone, A. Persson, F. S. L. Chapin, E. Lambin, T. M. Lenton, M. Scheffer, C. Folke, H. J. Schellnhuber, B. Nykvist, C. A. De Wit, T. Hughes, S. Van Der Leeuw, H. Rodhe, S. Sörlin, P. K. Snyder, R. Costanza, U. Svedin, M. Falkenmark, L. Karlberg, R. W. Corell, V. J. Fabry, J. Hansen, B. Walker, D. Liverman, K. Richardson, P. Crutzen and J. Foley, Planetary Boundaries: Exploring the Safe Operating Space for Humanity, *Ecol. Soc.*, 2009, **14**(2), 32.
48. www.agrocycle.eu/general-documents/#1513777244287-afff65fe-54ea (last accessed January 2019).

49. International Organization for Standardization, *Environmental Management – Life Cycle Assessment – Principles and Framework*, 2nd edn, ISO 14040:2006, 2006.
50. European Commission, *International Reference Life Cycle Data System (ILCD) Handbook – General Guide for Life Cycle Assessment – Detailed Guidance*, 1st edn, March 2010. EUR 24708 EN, Publications Office of the European Union, Luxembourg, 2010.
51. D. M. Souza, R. F. Teixeira and O. P. Ostermann, Assessing biodiversity loss due to land use with Life Cycle Assessment: are we there yet? *Global Change Biol.*, 2015, **21**, 32–47.
52. S. Pauliuk, Critical appraisal of the circular economy standard BS 8001:2017 and a dashboard of quantitative system indicators for its implementation in organizations, *Resources, Conservation & Recycling*, 2018, **129**, 81–92.
53. coe.epa.ie/ghg/data/inventories/2015/IE_2015_NIR2.pdf (last accessed January 2019).
54. www.epa.ie/pubs/reports/indicators/SoE_Report_2016.pdf (last accessed January 2019).
55. D. O'Brien, T. Hennessy, B. Moran and L. Shallo, Relating the carbon footprint of milk from Irish dairy farms to economic performance, *J. Dairy Sci.*, 2015, **98**, 7394–7407.
56. P. Goswami and S. N. Nishad, Virtual Water Trade and Time Scales for Loss of Water Sustainability: A Comparative Regional Analysis, *Sci. Rep.*, 2015, **5**(1), 9306.
57. J. Elser and E. Bennett, Phosphorus cycle: A broken biogeochemical cycle, *Nature*, 2015, **478**, 29–31.
58. H. Jianping, H. Yu, X. Guan, G. Wang and R. Guo, Accelerated Dryland Expansion under Climate change, *Nat. Clim. Change*, 2016, **6**, 166–171.

CHAPTER 12

Livestock Production

ALEXANDRE STRAPASSON,*[a,b] COLLINS ODUOR[c] AND BEATRIZ DOMENICONI[d]

[a] Imperial College London, Centre for Environmental Policy, Weeks Building, 16 Prince's Garden, London SW7 1NE, UK; [b] Harvard University, Belfer Center for Science and International Affairs, 79 JFK Street, Cambridge MA 02138, USA; [c] Department of Land Resource Management and Agricultural Technology (LARMAT), University of Nairobi, P.O. Box 29053-00625, Nairobi, Kenya; [d] Brazilian Roundtable on Sustainable Livestock (GTPS), Av. Paulista, 1754, Cj. 146, São Paulo, SP 01310-920, Brazil

*Email: alexandre.strapasson@imperial.ac.uk

12.1 Introduction

The circular economy (CE) concept, when applied to livestock production, provides a whole-systems perspective for the entire production chain of meat and meat by-products. Life Cycle Assessment (LCA) is a useful approach to understand the main inputs and outputs involved in the chain, but the CE concept goes beyond, by giving a view of the broad looping effects of livestock production systems. For example, the carbon, nutrient and water cycle involved in livestock production can be assessed under a broad systems analysis at the global scale, involving local, regional and cross-border impacts, which are interwoven worldwide. These complex interactions are the so-called telecoupling[†] effects.[1,2]

[†]Telecoupling is defined as socioeconomic and environmental interactions among coupled human and natural systems, at different scales over great distances.[3]

Life Cycle Assessment: A Metric for the Circular Economy
Edited by Aiduan Borrion, Mairi J. Black and Onesmus Mwabonje

Published by the Royal Society of Chemistry, www.rsc.org

Chemical elements and nutrients, such as carbon (C), water (H_2O), nitrogen (N), phosphorus (P), potassium (K), calcium (Ca), magnesium (Mg), sulphur (S) and others, are not destroyed in the livestock production system process. They are recycled or reallocated in different ways in natural systems. Sometimes elements are dispersed in ways that are hard to recover, *e.g.* when plant and animal nutrients are leached into water bodies and then into the ocean, or when organic residues are disposed of in landfills or open dumps. Sewage also contains nutrients that are rarely recovered as organic fertilizers and returned back to productive lands. The availably of fresh water is also a fundamental issue for the livestock sector.

Energy is also involved in every step of livestock production. The main source of energy used in the whole process is sunlight. The production system starts when solar radiation is fixed into the chemical structures of plants through the process of photosynthesis. Plants (mainly grasses and grain meals) are then eaten by herbivorous animals, which, in contrast to plants, cannot generate their own energy. Part of this energy intake is used by animals for maintaining their metabolism (*e.g.* respiration, movement, thermal balance, including transpiration), part is used for growing (*i.e.* cell growth, especially in young animals) and for maintaining body structures (*i.e.* meat, bones, organs and blood), and another part leaves their bodies, in the form of milk, eggs, urine and manure. Therefore, from the total feed intake, animals convert just part of this into edible meat. This conversion efficiency is called the Feed Conversion Ratio (FCR), as will be further explained in this chapter.

In addition to the energy associated with feeding animals, there are external energies associated with the livestock production chain, *e.g.* the energy used in tractors (usually diesel), in the transport of animal feed, meat, milk and living animals (*e.g.* road transport, trains, and maritime freight), in processing and conservation stores (slaughterhouses, refrigeration rooms), and in the retail market for the distribution of animal products to the consumers. The products are then consumed by the population, to maintain human growth and metabolism. Residual energy (*i.e.* chemical and thermal energy) from the original food intake is also released from the human body as waste. However, not all food that is produced is consumed. Approximately 19% of meat products and 24% of plant-based food, of all the food produced in primary energy terms, are discarded as residues in landfills and open dumps, burned, or converted into fertilizers (*e.g.* through organic composting).[4] Other than specifically composted waste or managed residues, little of these residues return to the pasturelands or croplands. Global data are scarce, but the United States alone for instance generated 36.43 million tons of food waste in 2012, 34.69 million tons of which went to landfill whilst 1.74 million tons were recovered or recycled, *i.e.* only 4.8% of the total organic waste.[5]

The production of animal products (meat, eggs and milk) involves several complex flows and environmental impacts, which can be reduced by implementing measures that can increase the total efficiency across the process, preferably as a closed-loop system. On the other hand, it is also possible to reduce these impacts by eating less meat and more plant-based food, including high protein crops such as pulses (*i.e.* leguminous plants, *e.g.* beans, peas and lentils). However, tackling dietary patterns involves addressing behavioural changes and cultural aspects, as well as the balance of the different nutritional values per type of food, *i.e.* proteins, carbohydrates, fats, sugars, vitamins and minerals. Food products, especially in urban areas, are becoming more industrialized globally over time, including the consumption of ultra-processed meat products with little or no whole food (*e.g.* sausages, salami, hamburgers, chicken nuggets and fish sticks), often increasing the total energy losses in the whole livestock system. This is before even considering the large amounts of wrapping materials (plastic films, bags, styrofoam, paper boxes *etc.*) usually involved in transportation and storage.

The environmental impacts of meat consumption are also different per type of meat. For example, ruminant animals (*e.g.* cows, sheep and goats) are less energy efficient than monogastric animals (*e.g.* pigs, chicken and other poultries) when converting feed into edible meat. Fish, on average, are also more efficient in converting feed into edible meat than ruminant animals, and production is not necessarily dependent on land resources (*e.g.* sea fishing). However, fish should also be assessed under an LCA approach, given that there are different inputs and outputs involved across its whole production chain, including marine biodiversity trade-offs in large commercial seafood operations, and the production of feedstocks used in commercial fish farming ponds (*e.g.* grain feed and fertilizers, such as animal manure, for plankton growth). Thus, there are different LCA methodologies[6,7] to assess fish production, depending on the aquaculture system involved.

One of the main environmental impacts of livestock production (excluding seafood) is on land use. As already mentioned, photosynthesis is the basis of the entire process and, hence, there is a need for land surfaces so that plant biomass can capture the sunlight. In addition, soil, water and climate conditions are essential factors. The land impact is not only related to the availability of pasturelands, but also for croplands used for producing grains as animal feed (*e.g.* soybean, maize, rapeseed, oats). Land use dynamics are also related to other types of productive lands, such as forestlands and lands for energy crops. Figure 12.1 summarizes these circular dynamics.

12.2 Fundamental Dynamics of Livestock Production

Food security and climate change are amongst the main challenges of our global society in this century. Both issues are interconnected: the capacity of food production is fundamentally dependent on climate conditions, whereas climate can be affected by the different techniques and plants used in the production of food, *e.g.* changes in evapotranspiration and albedo, as well as by the carbon cycles in soil and vegetation.

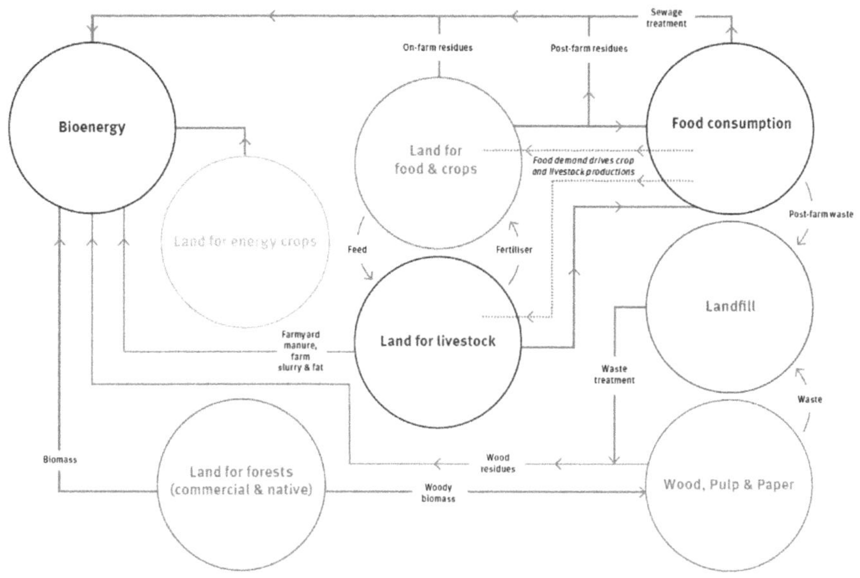

Figure 12.1 Livestock and land-use dynamics. Reproduced from ref. 8 with permission.

Ensuring food security is a challenge that faces many different aggravating factors, with different possible solutions. Therefore, it is important to have an integrated approach to increasing total food production, food quality and safety, guarantee of good (and fair) food distribution and the reduction of waste from the farm to transportation, storage, distribution and consumption levels. Likewise, the negative impacts of human actions on the climate can be reduced by actions in different segments, such as reduction of air and water pollution, adequate disposal and treatment of waste, reduction of land use change in forestlands and reduction of GHG emissions more broadly. The sense of urgency for implementing effective actions for climate change mitigation should also consider the need to maintain the living conditions of the future population globally, which may increase from 7.795 billion people in 2020 to approximately 9.735 billion in 2050 and 10.875 billion in 2100, according to the United Nations' World Population Prospects.[9]

The production of meat to meet future demand therefore poses a major challenge for land use and land use change, particularly because pasture represents the largest area of productive land in the world (it is larger than croplands, and commercial forestlands, which are used for producing wood, pulp and paper, and other forest materials). However, through substantial livestock yield increase, less pastureland would be needed for livestock production over time and more would be potentially available for other purposes such as the production of grains, forests or energy crops (*i.e.* designated as different land use types). Land areas that are no longer required for livestock production and are released to other activities are often called 'freed-up land' or 'surplus land'.[8,10]

There is a trend towards a gradual annual increase in livestock yields worldwide, particularly because of a significant yield gap in developing countries and the prevalence of extensive production systems, against intensive systems often used in developed nations. The FAO[11] estimates an annual livestock production growth of about 1.4% by 2030, and 0.9% from 2030 to 2050. However, such a growth rate is relative, because it depends on what is included in the estimates. In fact, several different issues can affect the overall livestock productivity besides pasture yields *e.g.* FCR, intensification of animals (in confined or semi-confined systems), and animal density on pasturelands (in the case of ruminants). Monogastric animals (*e.g.* chickens and pigs) do not require pasturelands to be reared,[‡] but do require a large enough area of land to produce animal feed, whereas ruminants (cattle, sheep, goats) are mostly reared on grasslands worldwide. Confined and semi-confined systems (including feedlots) are a growing alternative to free-range animals, but there are animal welfare implications that need attention. As a note of clarification, this book chapter was not aimed at discussing issues on animal ethics, but the authors recognize and highlight the importance of this perspective for discussing livestock systems and behavioural changes in dietary patterns.

The following sections briefly describe the three main issues related to livestock yields, namely: feed conversion ratio; animal intensification; and animal density on pasturelands. Other important issues directly related to these are the possibility of integrating different land uses in the same area, *e.g.* agro-livestock-forestry schemes, as well as rotational systems (*e.g.* crop-pasture) and multiple cropping, *i.e.* to produce more crops per year in the same area (*e.g.* summer crop and winter crop).[8]

12.2.1 Feed Conversion Ratio (FCR)

FCR represents the conversion efficiency that animals have for converting feed (*i.e.* inputs such as grains and grasses) into edible meat (carcass), milk and eggs. FCRs vary according to the type of animal, genetics, age, lifetime, region, feed quality and the animal husbandry system, *i.e.* the way farmers manage their livestock. According to Strapasson,[12] using several references,[13–16] the FCR in energy terms (rather than weight terms) of the main livestock groups produced worldwide are approximately, as a global average: 5% for cow and other bovines, sheep and goats; 24% for chicken and other poultry; 27% for pigs; 13% for chicken eggs; and 8% for milk products.

In fact, there is currently no comprehensive database available for FCR on the global scale per type of animal and these values are an approximation,

---

[‡]Except some type of herbivores with a monogastric stomach, *e.g.* equines, which can digest celluloses from grasses, thanks to symbiotic bacteria present in their adapted guts.

given the levels of uncertainty associated with this issue. It is noted that livestock systems in the Americas, Europe, Africa, Asia and Oceania have substantial differences in FCRs, and these may also vary on a local level, or even within a farm. Thornton[17] reports an example of a company in New Zealand that obtained an FCR for broiler production, in weight terms, as high as 1.55 (in a range of 1.40 to 1.60). The animals provided 1 kg of edible meat for each 1.55 kg of feed intake, equivalent to 65% efficiency, which even in energy terms is far superior to the world averages presented above. World FCR averages for cattle are substantially affected by the low livestock yields in India, which has the largest number of bovines in the world, thanks to religious observance. However, future expansion of beef production is expected to occur in several other countries, including Canada, USA, Australia, New Zealand, Argentina, Brazil, Paraguay and Uruguay, through greater use of livestock technology, especially animal breeding.

12.2.2 Intensification of Animals

The intensification of animal production means the proportion of animals raised in intensive livestock systems *versus* pasture (free-range). In terms of land use impacts, this has major implications in the production of ruminant animals. In some countries (*e.g.* Argentina, Brazil and Paraguay), grazing systems in large fenced areas are the mainstream, whereas in others (*e.g.* the USA, Canada and European countries), feedlot systems and semi-confined systems prevail. However, grazing systems are still the dominant practice worldwide, with only approximately 6% of cattle and 1% of sheep and goats reared in intensive systems, according to an estimate from FAO data.[11,13]

Although confined and semi-confined systems often present higher efficiencies, they also require substantial land use for producing feed, *i.e.* an external area (on the same farm or elsewhere) for cultivating animal feed products such as maize, soybean, oats, sorghum, barley and hay.[13] The international trade in animal feed can result in higher CO_2 emissions from transporting such products worldwide, *e.g.* by using road transport and long-distance freights. However, most emissions stem from fertilizer use, machinery use and land-use change. Confined systems are also the subject of controversial debates on animal welfare, and beef exporters using mostly grazing systems can take advantage of countries looking for 'natural beef'. On the other hand, feedlot systems usually maximize the production chain, *e.g.* some farmers specialize only in fattening animals acquired from other farmers before they go to slaughterhouses (abattoirs).[18]

12.2.3 Animal Density

Animal density represents the number of animals per unit of area, particularly ruminants in grazing systems (*e.g.* number of cows per hectare) and the concentration of animals depends on the carrying capacity of the pastureland. In several regions of Latin America and Africa, for example, the use of extensive systems for producing cattle, sheep and goats is quite common, leading to a low animal density, often much below the natural carrying capacity.[19] Thus, in some regions, it is not only possible to increase the number of animals on grasslands, but also to convert degraded pasturelands into efficient planted pastures in order to boost grassland performance.[20] Forages such as *Brachiaria spp.*, *Panicum maximum*, *Pennisetum spp.*, *Paspalum spp.* and *Medicago sativa* (alfafa or lucerne) can play a major role in increasing the animal density in several regions, particularly in tropical regions with significant availability of land resources.

Even in temperate countries such as the United Kingdom, grazed grass, when well managed, can be the cheapest alternative for livestock production, supplying about 90% of the energy requirements for cattle and sheep and 70% for dairy cows, as shown by Buckingham *et al.*[21] These authors also describe a grazing system of reference which uses a combination of high sugar grasses, white clover and chicory, and report the advantages of using rotational systems. Similarly, Balsom[22] shows that with modern grazing systems it is possible to significantly increase carrying capacity and improve soil conditions at the same time.

12.2.4 System Boundaries of Livestock Production

The use of LCA methodology for livestock production at the regional or national level must consider the boundaries involved in the assessed system. Looking at the entire production chain, the main stages are described as: livestock production; animal abatement (slaughterhouse); meat processing; distribution (including exports); storage; consumption; and waste disposal. In Europe, for example, the European Livestock and Meat Trades Union (UECBV[23]) suggests the system boundaries for a fresh meat production lifecycle (Figure 12.2), aiming at providing Footprint Category Rules for Red Meat (FCR RED MEAT) for individual companies in the red meat sector, so that they can compare their environmental performance for meat production within animal species *e.g.* beef *vs.* beef, pork *vs.* pork, lamb *vs.* lamb. The arrows in this figure represent potential transport steps. The flows of residues involved in the production system are not represented in the figure, which could partially close the loop in a CE approach. For example, animal manure can be used as a fertilizer for both crop and pasture cultivation, whilst bone flours and offal can be used as pet feed ingredients.

12.3 Case Study on Sustainable Beef Production in Brazil

This case study presents an overview of the lifecycle of beef production in Brazil. The lifecycle has some key physical inputs and outputs, as

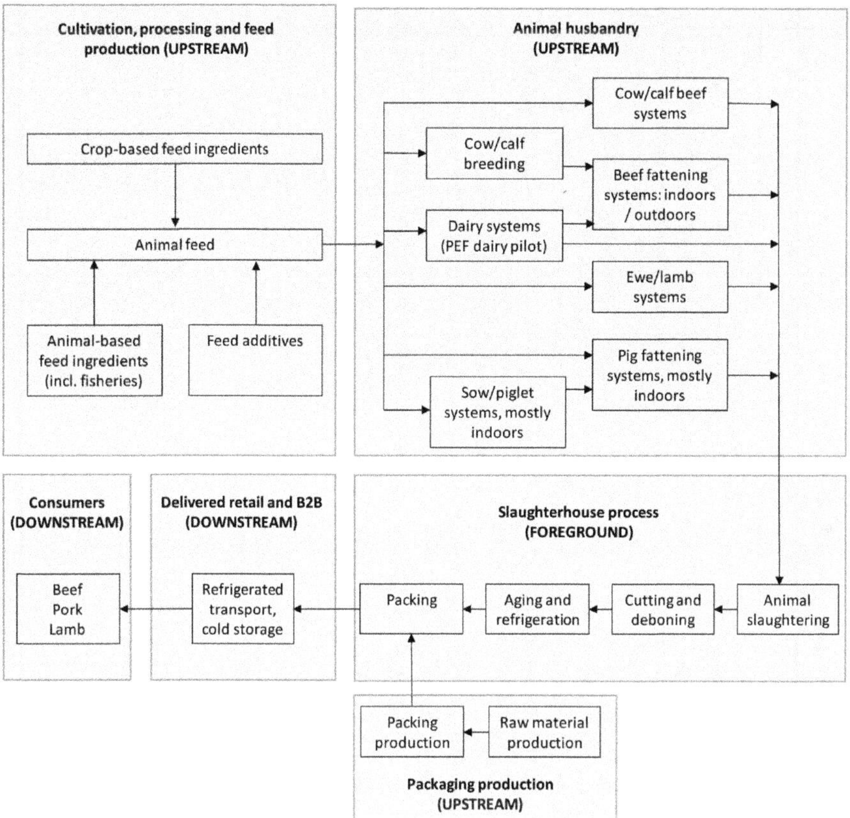

Figure 12.2 System boundaries for the life cycle of fresh meat.[23]

summarized in Table 12.1, which can be further detailed by focusing on each stage and assessing all the main variables involved in greater detail. This can also be sketched as a large flow diagram, representing the main sets of dynamic relationships in the whole system. Moreover, several variables are interdependent; for example, carbon and nutrient dynamics from livestock production may be integrated with crop production within the same farm, through agro-livestock schemes or agro-livestock-forestry schemes, among many other possible interactions, further including several cross-boundary effects, such as potential impacts on logistics and rural infrastructure.

Therefore, LCA for livestock production must have a clear systems boundary in order to offer a consistent assessment. The boundary can be, for example, the carbon dynamics of pasturelands at farm level or, more broadly, the main environmental impacts (*e.g.* carbon and water footprint, land use change) from the beef production chain at country level. Once the boundary has been established, it is necessary to identify the main flows involved in the production system and to obtain reliable data for the main inputs and outputs, linked to the main assessed stages (*e.g.* pasture system, animal birth, growth,

Table 12.1 Summary of the main physical inputs and outputs of the livestock production chain.

Stages	Main inputs	Main outputs
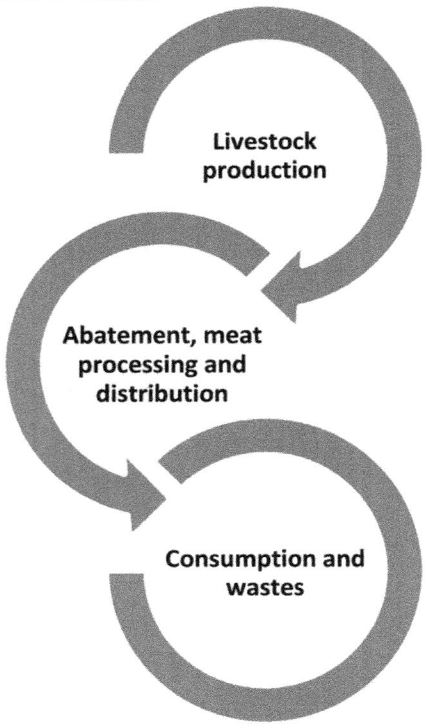	• Land resources, either pastureland or cropland used to produce animal feed. • Human labour and machinery. • Plant fertilizers and agrochemicals. • Animal vaccines and other sanitary medicines. • Water for plant and animal production, washing and industrialization. • Energy inputs, *e.g.* fuel used in tractors, heating, meat processing, transport and refrigeration. • Sunlight and CO_2 used in the photosynthesis process for plant production consumed by the animals. • Feed supplements.	• Meat, milk and other animal products (*e.g.* leather). • Agricultural and livestock residues (*e.g.* manure), which can be either totally or partially recycled in the production chain, including as a bioenergy source. • Greenhouse gas emissions, including CO_2 from fuel use and potential land use change; CH_4 from enteric fermentation and manure; and N_2O from nitrogen fertilizers. • Potential carbon sequestration in soils. • Food waste and sewage. The wastes may be treated, recycled or allocated in landfills. Some residues may end up in open dumps.

fattening, reproduction, abatement, meat processing, distribution *etc.*). Another important issue is to have a clear understanding of the functional units to be evaluated, *i.e.* what does the LCA look to assess? It may be carbon emissions per kg of meat produced, the water balance for fattening animals, the nutrient cycle on pasturelands, among many other possible relationships that may meet the objectives of the required LCA study.

At the national scale, the LCA approach must also consider the context of the assessed country and its historical changes and trends. Brazil, for example, moved from being a net beef-importing nation about 30 years ago, to the largest beef exporter in the world.[24] The livestock sector plays an important role in the Brazilian economy, accounting for approximately 8.7% of the Brazil's total Gross Domestic Product (GDP) in 2018, including all types of meat (beef, pork, chicken *etc.*).[25] About 20% of its beef production has been exported in different forms to several nations: 80% as raw meat to countries such as China, Hong Kong, Egypt and Chile; 12% as processed meat (industrialized), particularly to the European Union and the United States; and 8% in the other forms, especially to Hong Kong.[25]

A LCA approach for livestock production can look not only at physical variables, but also at the economic dynamics involved in the production cycles, which may vary according to the level of technology applied. The costs can be split into capital costs (*e.g.* land value, machinery, installations and equipment) and operational costs (*e.g.* labour, fuel, maintenance, veterinary and administrative costs, *etc.*). As an example, Table 12.2 shows the main

Table 12.2 Summary of LCA for main operation costs of livestock production in Brazil, per technology level.[25a]

Item	Extractive	Low tech	Medium tech	Medium-high tech	High tech	Intensive
	kg of livestock produced per hectare					
	15–45	45–90	90–180	180–270	270–390	390–570
	Operational costs in US$ per kg					
Nutrition	0.27	0.31	0.39	0.63	0.69	0.79
Sanitary programme	0.06	0.06	0.07	0.07	0.07	0.06
Fertilizers and soil pH correction	0.00	0.08	0.34	0.36	0.43	0.41
Agrochemicals	0.00	0.11	0.06	0.05	0.02	0.01
Fuels and lubricants	0.23	0.14	0.13	0.12	0.11	0.10
Animal breeding	0.00	0.02	0.07	0.13	0.14	0.12
Employees	0.39	0.34	0.25	0.19	0.16	0.12
Maintenance	0.32	0.15	0.11	0.10	0.09	0.07
Administrative costs	0.04	0.03	0.03	0.02	0.02	0.01
Electricity	0.01	0.01	0.01	0.01	0.01	0.01
Depreciation	0.89	0.50	0.36	0.21	0.16	0.11
Total operational costs	**2.20**	**1.76**	**1.81**	**1.87**	**1.90**	**1.83**

[a]Values estimated for 2018 base year and originally available in Brazilian Reals. Exchange rate: US$ 1.00 = R$ 3.90.

operational variables involved in the whole livestock production system (animal nursery, growing, fattening and abatement, as well as reproduction) according to different technology levels used in Brazil. The data show that the more technology is used, the lower is the marginal cost for livestock production, *i.e.* cost per kg unit, which serves as the functional unit for this LCA example. Analogously, another important LCA functional unit for livestock production is GHG emissions (*e.g.* gCO_2eq) per meat unit (*e.g.* kg), rather than per area unit (*e.g.* hectare), given that it is possible to obtain significant productivity gains (production per area) by using agronomical assistance and new technologies, including improvements in animal breeding and feed quality, and agro-livestock-forestry integration.[26,27]

A broad question might be: is Brazilian beef sustainable? Brazil is a country where scale is a constant factor for both food production and forest conservation, as it includes biomes such as the Amazon and Cerrado (a savannah-like biome). The challenge is to ensure that efficient crop and livestock production occur alongside the provision of environmental services and conservation of native ecosystems.

Livestock farming has shown important advances in production efficiency. Between 1990 and 2017, productivity increased by 150% in Brazil, while pastureland area reduced significantly. Since 1990, livestock farming has released 30 million hectares that were either used for the production of other crops or have regenerated native vegetation. Best management practices for pasture, linked to integrated production systems technologies can play an important role in improving the carbon balance of cattle production. Brazil has reduced the production cycle of animals in the last few decades and, hence, presented a negative growth rate of methane emissions from enteric fermentation per unit of beef produced, although the total enteric methane emissions increased in the same period.[28] Another important characteristic of Brazilian cattle ranching is that their production models are predominantly dependent on tropical pastures, a type of vegetation that plays an important role in the carbon balance between atmosphere and soil, as well as the physical, chemical and biological quality of the soil itself. Well-managed pasturelands in these areas usually have very high contents of soil carbon.

These positive values for livestock in Brazil represent national averages, which means that although most farmers are not causing deforestation, there are negative cases in some areas, especially in extensive pasture systems on the border of forest areas, usually in places with poor infrastructure and control. Most livestock farms in the country are technology-intensive and competitive, meeting sustainability standards at the international level. However, deforestation is still a major issue and involves several other variables, *e.g.* the development of road infrastructures, although livestock production from critical areas represents a small percentage of total production in the country. The country is formally committed to eliminate illegal deforestation in the coming years, but problems clearly persist. Part of the recent soybean expansion, for example, occurred in Cerrado areas in the

North and Northeast States in the so-called MATOPIBA Region.[29,§] Brazil is the largest exporter of soybean globally, which is one of the main animal feedstocks. China is Brazil's largest soybean importer. Like Brazil, Argentina has been facing similar issues and challenges. Therefore, sustainability standards and life cycle assessment tools, with a circular economy view, must be part of the routine of livestock farmers and policies.

At the global level, Brazil undertook a bold commitment to reduce GHG emissions during COP 21 in Paris. Most of the Brazilian GHG emissions come from deforestation and land use change, followed by the energy sector. The major forest biome in the country is the Amazon, which had its deforestation rate significantly drop from 2004 to 2013, from about 27 000 km^2 per year to 6000 km^2 per year, respectively,[30] whilst also increasing renewable energy sources nationwide in the same period. However, the Amazon deforestation rate has increased in the past few years and so has the land conversion of the Cerrado biome. Since 2019, after the presidential election, the federal government has reported controversial views on climate change, with several high-level officials neglecting scientific evidence, although the country has kept its commitment to the Paris Agreement, due to internal and external pressures.[31]

In 2012, Brazil published a new Forest Code (Federal Law no. 12651), a bill that enforces and regulates the maintenance of native vegetation within rural properties. Forests in Brazil are not only protected in national parks or public areas, but also within private properties in which the owners are responsible for maintaining between 20% and 80% of the property area, depending on the biome and zoning system applied. Brazilian environmental law is one of the strictest environmental legislation globally, but the country still struggles to halt illegal deforestation nationwide, especially in its Northern region. It is worth noting that deforestation has occurred for several reasons in Brazil, such as: illegal logging; land tenure issues; mining; corruption; infrastructure development (*e.g.* new roads, hydropower plants, settlements); as well as agriculture and livestock expansion, although net pastureland area has been decreasing in the country over the last several decades. It is relevant to mention that deforestation is not necessarily illegal; it is evaluated on a case-by-case basis. When it is illegal, it is often connected to local corruption and lack of efficient law enforcement, punishment, surveillance system, police intelligence, funding and political will. At the same time, deforestation is also related to the opportunity costs associated with the land use alternatives. Hence, the market and the economy also play an important role in deforestation, in addition to an often-inefficient institutional system more broadly.[32]

As in many countries, social pressure on environmental issues is gaining momentum. Consumers are becoming more aware of the consequences of

§MATOPIBA is a geographical region in Brazil, comprising the State of Tocantins (TO) and parts of the States of Maranhão (MA), Piauí (PI), and Bahia (BA). The term MATOPIBA is an acronym for the abbreviation of these four states.

their choices. New initiatives on sustainable land use in Brazil have arisen at the private sector level, universities and NGOs, so that livestock can develop in a more sustainable way. The federal government also launched the ABC Plan (a low carbon emission agriculture programme) about 10 years ago, and business commitments have been undertaken, with the establishment and monitoring of sustainability criteria within supply chains.

Like any large-scale production, there is a heterogeneity in all the criteria that one wants to establish. Brazil certainly has a large potential to contribute to global food security, as well as in climate change mitigation. The challenge is to promote sustainable land use with agricultural technology, income distribution, and biodiversity conservation. At the local level, the use of LCA to assess and therefore address environmental impacts has been proven useful. Dick *et al.*[33] carried out a LCA for grassland systems in Southern Brazil, comparing an extensive system against an improved system. These authors obtained variation of results for GHG emissions, whilst also identifying key areas to be addressed according to regional singularities in order to reduce environmental impacts. A similar analysis was recently performed for beef production in the United States.[34]

12.4 Case Study on Carbon Cycles in Kenya's Pasturelands

This section provides a discussion on the fundamental issues required for preparing a LCA, focused on the carbon (C) dynamics of pasturelands, taking a rural area in Kenya as the example. The systems boundary was delimited to farms assessed in Kenya's West Pokot County, under fenced and non-fenced pasturelands systems. The functional units included were C emissions per area of pasture production, and GHG emissions associated with the dairy sector. Primary data were collected on site and laboratory analysis of soil samples carried out at the University of Nairobi.[35]

Pasturelands in the Kenyan context refer to rangelands, which are generally arid and semi-arid lands (ASALs), with vegetation comprising of natural/native grass species and scattered trees. The ASALs host over 60% of the county's livestock population, employing approximately 90% of the population of that region. Pastoralism¶ and extensive production systems dominate in these areas that comprise over 80% of the country's land area, with farmers owning considerably large herds that graze in communally owned grazing lands. According to the 2009 census report, the population of cattle and goats accounts for 26.8% and 42.5% of the total livestock population (excluding domestic birds and donkeys) in the country, respectively,

¶Pastoralism is the practice of herding livestock involving pastoral mobility on rangelands, either through nomadism or transhumant pastoralism, often carried out by traditional communities in developing nations.[38]

Table 12.3 Classification of agro-climatic zones in the pasturelands of Kenya.[36]

ACZ	Description of climate	% of total area	Annual rainfall (mm)	Rainfall/Evaporative Potential (%)
IV	Semi-arid	8.5	500–1000	25–39
V	Arid	52.9	300–600	15–24
VI	Very arid	19.8	200–400	<15

followed by sheep and camels. The livestock production system is also characterized with low adoption of improved inputs such as concentrated feeds, medicines and artificial insemination. Increased incidents of drought, livestock diseases (*e.g.* trypanosomiasis) and poor animal nutrition are major constraints of production and livelihoods in the ASALs.

Pasture production systems in Kenya fall under three agro-climatic zones (ACZ),[36] as shown in Table 12.3. The variation in climatic conditions in the ACZ reflects the differences in altitude, soil characteristics, and temperature and rainfall reliability. ACZ IV, also referred to as the semi-arid production system, has a medium potential for plant growth and livestock production and covers parts of Northern and Eastern Kenya and the coastal strip to the west. The lowest potential for biomass and livestock production is in the arid and very-arid production systems that are characterized by high rainfall variability and high prevalence of livestock diseases.

Effective C cycling in agricultural systems, including pasturelands, depends on plant diversity, soil macro- and microorganisms. Most of the agricultural production systems in Kenya mainly depend on rainfall, which is typically unreliable. As most livestock production systems in the country are characterized by low external inputs of feed and substrates, the majority of livestock depend on natural pastures that occur in rangelands/ASALs as a combination of grass, woody species and crop residues. Other sources of animal feed include exotic planted woody fodder species, such as *Leucaena leucocephala* and *Gliricidia sepium*,[37] forage crops, such as Napier grass (*Pennisetum purpureum*), Lupins (*Lupinus albus*) and other marketed fodders.

12.4.1 Pathways of Carbon (C) Losses in the Pasturelands of Kenya

This section focuses on the C dynamics associated with pasturelands in Kenya, as an actual reference for LCA studies related to this type of system boundary and production scheme. It briefly shows the main data sources required for the LCA and the results obtained for the proposed functional unit (C emissions per pasture area) used by Oduor.[35] The methodology employed in this case study is similar to an Eco-LCA,[39] which is an ecologically based lifecycle assessment aimed at the interpretation and aggregation of environmental services. This case study is, more specifically, focused on the soil C dynamics on pasturelands, according to different production systems, excluding GHG emissions related to enteric fermentation of ruminant animals. Hence, the system boundary is the pastureland itself, *i.e.* the assessment does not involve other parts of the production chain, such as animal abatement, meat processing, distribution, storage, consumption, and waste disposal, as previously described.

Most pastures in Kenya are natural grasses and forages, and there is no consolidated fertilizer manufacturing industry in the country. Hence, the contribution of CO_2 and N_2O emission from fertilizers' manufacturing and from the on-farm use of artificial fertilizers (*i.e.* non-organic) is relatively low. However, livestock-related land use changes, such as the expansion of grazing lands into forests, may emit much of the organic C in most grazing lands.[40–42] Illegal logging and increased demand for fuelwood, charcoal and timber, as well as the conversion of forests into grazelands and fodder production areas have also contributed to the significant (6.5%) reduction of forest cover in Kenya during the period between 1990 and 2010.[43,44] These changes release large quantities of C from the soil and vegetation into the atmosphere, because forest usually holds more C compared to croplands or pastures, in total terms, *i.e.* soil C and above-ground vegetation C combined. However, data on the net reduction in C stocks due to the conversion of forest to pasture are lacking for Kenya, probably because the estimation of C fluxes due to forest conversion is the most complex of the emissions inventory components.[45]

Soils remain the largest pool of C in the terrestrial carbon cycle, holding two times more C than in living vegetation. Spatial estimates of Kenya's soil organic C content show that the top 30 cm of soil stores approximately 2.4 Gt of C,[46] with the highest and lowest soil C stocks being the humid-to-subhumid and ASAL areas of the country, respectively. Thus, a small change in C stored in soil can have a considerable influence on the C balance.[47] Anthropogenic management activities accelerate the loss of C stored in soil. Ritchie[48] reported that continuous grazing in the Northern rangelands has significantly reduced soil organic C in the past 30 years. Overgrazing is reported to be the major cause of soil organic C losses in in the rangelands of Kenya.[49–51]

Nevertheless, proper management practices, such as reduced grazing intensity, can reclaim the soil organic C at a capacity of 0.3–0.5 $t\,C\,ha^{-1}$ per year across a range of soil types.[48] Maize stover is commonly used as animal feed both in the high-potential and semi-arid parts of the country. Supposing an annual loss of 100 $kg\,CO_2\,ha^{-1}$ per year is derived from crop residues,[52] the nearly 1.6 million hectares of land cultivated with maize in the Kenya's Rift Valley Region adds 160 million kg of CO_2 emission annually to the livestock balance. Also, soil amendment practices, such as the addition of cattle manure in the maize cropping systems, contribute to GHG emissions. Much of the CH_4 emissions from cropping fields originates from animals, because the manure applied in soil is a significant source of atmospheric methane.[53] However, the average C content of soils in the tropics, including Kenya, is relatively low compared with the temperate region,[54] implying lower carbon emissions, although this depends, case by case, on the type of soil, vegetation and climate. Conversely, the continuous expansion of grazing areas into uncultivated areas and cropping land[55] can escalate CO_2 emission.

Land degradation due to poor livestock management in natural grasslands usually reduces vegetation cover, causing changes in C and nutrient cycling of the ecosystem. In Kenya, a large proportion of the ASALs are highly degraded and another 2% are completely desertified.[56] A recent study in Kenya[57] showed that a small change in above-ground biomass caused a reduction in soil organic carbon storage over time in the rangelands, enhancing the decomposition of organic matter. It is difficult to approximate the contribution of livestock in these losses, but it is certainly high, given that rangelands occupy over 80% of Kenya's land area and that land degradation and desertification is considerably higher under pasture than other land uses.[58] Because water availability is the main limiting factor of production in rangelands, the majority of the restoration initiatives in Kenya comprise of land and water conservation technologies that reduce erosion, increase water infiltration and improve seed germination. Rangeland reseeding is practiced in some areas.[59] Establishment of pasture enclosures by fencing-off grazing areas is also gaining cognizance due to the potential to enhance biomass production and restore degraded soils.[51,60] Other initiatives that design artificial obstacles to reduce run-off have also generated good results.

12.4.2 Direct Carbon Emissions from Livestock

Compared to the wildlife population, the average number of livestock animals in natural grasslands has significantly increased during the period between 1977 and 2016.[61] What is the contribution of these animals to emissions of GHGs? Enteric fermentation is the major contributor of agricultural GHG emissions in Kenya, accounting for about 56% of the emissions (3.53 MtCO_2e) between 1990 and 2013, followed by manure left on pasture (37%).[62] A study in Kajiado County revealed that enteric fermentation produced a total of 220 tons of CH_4 with a global anthropogenic effect of 5,510 tons CO_2e, due to its higher global warming potential (GWP). The type of animal, region, feed and nutrition have a great influence on ruminal fermentation. Methane produced from manure depends on season, animal age and the quality of the manure, which greatly depends on the quality of feed. Ndung'u *et al.*[63] showed that cattle in the lower highlands emitted more methane than cattle in the upper midlands, with enteric CH_4 showing a positive relationship with feed intake. Availability of feed is the main limiting factor for livestock owners in Kenya, especially during the dry seasons. This results in high rates of methane emissions during the wet season, when feed is abundant, and low emissions during the dry season, when feed intake decreases.[63]

12.4.3 Carbon Emissions from Processing Livestock Products: Reflections on the Dairy Sector

It is estimated that there are about 4.3 million dairy cattle producing 3.4 billion litres of milk in the Kenya, with the sector accounting for approximately 12.3 million tonnes CO_2e.[62] The emissions are expected to increase as the demand for processed milk is likely to increase, with the 5% projection in annual growth increasing the demand for milk to 12 billion litters by 2030.[64] The mean emission intensity for milk produced in Kenya is 3.8 kg CO_2e per kg of fat and protein corrected milk (FPCM) with the lowest (4.1 kg CO_2e per kg FPCM) and highest (7.1 kg CO_2e per kg FPCM) emission expected in intensive and extensive systems, respectively. Nonetheless, research shows that improving animal husbandry, health, feed quality and supplemental feeding can help reduce emission intensity and increase dairy productivity.[63,65]

Transport-related CO_2 emissions for milk are also considered to be very high.[45] Besides the delivery of milk from the farm-gate to the processors, and from the processors to retailers and consumers, poor packaging that under-exploits the available transport capacity may significantly increase milk-related CO_2 emissions. The Government of Kenya has, therefore, considered improving the efficiency of heavy-duty vehicles, the use of biodiesel, and the road networks in the rural areas to be among the abatement options to reduce CO_2 emissions in the transport sector.[66]

12.5 Case Study on Global Livestock Production: a Whole-system Approach

In order to assess the LCA systems boundaries for livestock production and the choices made in a lifecycle inventory, this section describes a practical modelling approach aimed at understanding the complex dynamics of food production systems at global level. These dynamics include simulations of livestock production and corresponding land use, and resulting GHG implications from the present to 2050, in the context of the Paris Agreement of the United Nations Framework Convention on Climate Change (UNFCCC), and the UN Sustainable Development Goals (SDGs). It is worth noting that LCA for industrial products, such as electronics, is relatively different to biological products. For instance, a LCA for manufactured goods usually involves the following stages: resources; material processing; product manufacturing; distribution; use; and end of life. For biological 'goods' such as meat, resources are linked to the livestock production stage rather than manufacturing; a meat processing stage (slaughterhouses) rather than material processing; distribution involving the additional factor of refrigeration; and instead of end of life, there is a food consumption or disposal, resulting in further consideration of the management of the respective organic waste.

The model described here as a LCA case study for livestock production is called the Global Calculator‖ (GC). The calculator is a whole-system model that includes all sectors of the world economy (land use, food, industry, energy, building, transport) and demography. The approach employed in the land use, bioenergy and food module of the GC, applies a mathematical model for balancing the necessary expansion in the production of food crops, livestock, biofuels and other bio-based products, with resources conservation. It allows users to simulate many trajectories of land use change and its associated GHG emissions, according to different demands for land-dependent products and services by 2050. Users can then develop their preferred pathways to 2050 by varying the weight of a selected set of parameters ('levers') according to their GHG mitigation objectives ('levels' 1 to 4, with several intermediate levels, and increasing levels of ambition). These include:

- Food calories consumed per person per day;
- Meat consumed per person per day, which is split into meat calories and meat types;
- Crop yields;
- Livestock yields, which include changes in feed conversion ratio, the share of feedlot systems, and animal density in pasture systems;
- Bioenergy yields and types, *i.e.* solid biomass and biofuels (biogas was modelled as a fixed estimate);
- Surplus land for forest and/or energy crops;
- Wastes and residues; and
- Land use efficiency (or land use integration), *i.e.* multi-cropping effects and integrated farming schemes, such as agro-forestry and agro-livestock systems.

The GC model** also considers several additional variables for the calculations, including the use of fertilizers, agricultural losses, GHG emission factors, feed conversion ratios, the proportion of animals raised in intensive production systems (feedlots), animal density (*i.e.* concentration of animals in grazing systems) and limiting factors for land distribution. The accuracy of each trajectory is limited by the availability of, and uncertainty associated with, data for global scale estimates, and the restricted number of input parameters in the calculator, given the high complexity and uncertainty of all these levers. The model draws on several data sources, primarily FAO, IEA and IPCC statistics, and representative international references on land use modelling, for the purpose of providing not only a robust and credible methodology, but also a simple and user-friendly calculator for the lay user.

‖The Global Calculator webtool is available at: http://tool.globalcalculator.org

**Further information about the GC land use model, including its methodological approach and additional references, is available in Strapasson *et al.*[10]

The GC is presented as a webtool and was built on a database generated by a multiparadigm language (Ruby), using a C language programme based on a comprehensive matrix in MS Excel format. The model has several input parameters and variables, which are used for estimating future land use distributions, as well as the associated CO_2, N_2O and CH_4 emissions. Land use change is determined by a hierarchy of land use types. Priority is given to food production (croplands and pasturelands), and the remaining land area is allocated to forestation, natural regeneration and/or energy crops. Thus, with the GC it is possible to simulate a large number of trajectories for food, bioenergy and forest land by 2050. The land use dynamics presented by the GC includes both direct and indirect Land Use Change (dLUC and iLUC), given that it is based on a global balance (closed system) of several land use allocations. Thus, it provides a circular economy perspective, with allowances for several uncertainties involved.

12.5.1 The Global Calculator's Simulations on Livestock and Dietary Changes by 2050

Two 2050 scenarios are presented, which have been simulated in the GC (version no. 23): a low meat-consumption pathway and a high meat-consumption pathway. Both scenarios assume that all sectors will follow the same moderate climate change mitigation pathway, varying only the *per capita* quantity of meat and type of meat consumed by 2050. These scenarios were based on Chatham House's simulations[††] for carbon mitigation through dietary changes in terms of meat consumption, and the *per capita* calories consumed were FAO global projections for 2050, for both scenarios (level 2 in the model, 2330 kcal per person per day), while the current caloric intake is approximately 2180 kcal per person per day. The current world meat consumption is about 187 kcal per person per day and, based on historical trends, it is likely to continue increasing in the coming years, if significant behavioural changes are not adopted. All the caloric values presented exclude food wastes, *i.e.* the calorific values represent only the food intake (eaten by humans). The current share of meat types is also considered at approximately 22% consumption of beef, lamb and goat (*i.e.* meat from ruminant animals), and 78% pork, chicken and other poultry (*i.e.* meat from monogastrics). Fish consumption was assumed to follow equal global trends to both scenarios. The differences of the two simulated scenarios are described:

- High-meat scenario: global meat consumption by 2050 would reach similar levels to the current European meat consumption level

[††]The Chatham House's simulations for climate mitigation are available in the list of 'example pathways' described on the GC webtool. The simulations were prepared by its own staff, as an independent exercise (rather than a forecast), based on internal analysis and research paper.[67]

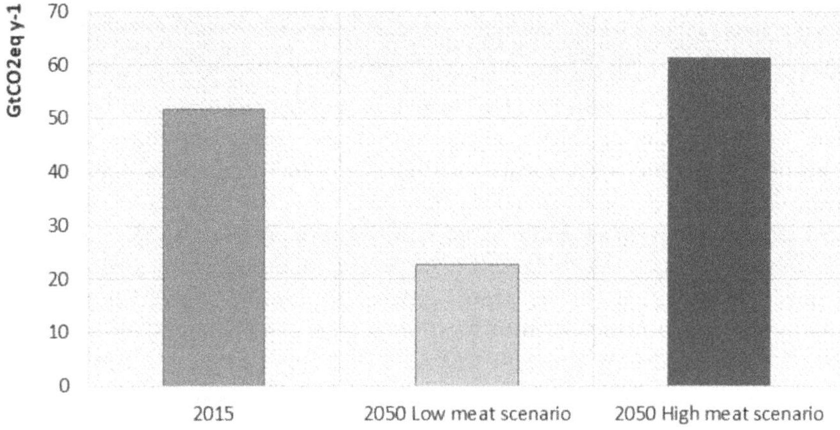

Figure 12.3 Scenarios of GHG emissions using the Global Calculator.

(level 1 in the model, 281 kcal per person per day). The share of meat from ruminant animals from the total *per capita* meat consumption would increase from the current 22% to 25% by 2050, similar to the share presently observed in countries such as Canada.

- Low-meat scenario: global meat consumption would drop to a level which is assumed, by the World Health Organization (WHO[68]), to be relatively safer for human health, *i.e.* about 90 grams of meat per person per day (level 3 in the model, approximately 152 kcal per person per day). The current world meat consumption is about 187 kcal per person per day and, based on historical trends, it is likely to keep increasing in the coming years, if significant behavioural changes are not adopted. Regarding the share of meat from ruminant animals, the current 22% share would decrease to 10% by 2050, similar to the share currently observed in China.

Based on the assumptions described for both scenarios, it is possible to identify that changes in dietary patterns can play a major role in climate change mitigation (Figure 12.3). Dietary choices can either exacerbate or ameliorate the situation. In the high-meat scenario, for example, current GHG emissions increased from approximately 51.7 $GtCO_2eq$ per year in 2015 to 61.2 $GtCO_2eq$ per year in 2050 (+18%), whereas for the low-meat scenario, emissions drop from 51.7 $GtCO_2eq$ per year in 2015 to 22.7 $GtCO_2eq$ per year in 2050 (−56%). These simulations are selected scenarios that could be re-calibrated, but which show the order of magnitude of GHG impacts that behavioural changes could achieve.

An alternative simulation using the GC assumes all sectors would keep a Business-as-Usual (BaU) trend, assuming current policies towards a moderate climate change mitigation by 2050, similar to the International Energy

Agency's 4 °C scenario[‡‡] available on the GC webtool. A sensitivity analysis is then run, by changing only the quantity of *per capita* meat consumption by 2050 from level 2 (220 kcal per person per day), which is analogous to the FAO global trend, to level 3 (152 kcal per person per day), which is the WHO recommendation already mentioned. In this case, the change in consumption would reduce the estimated global emissions (all sectors combined) from a 53.6 $GtCO_2eq$ per year in 2050 (level 2, BaU) to 36.6 $GtCO_2eq$ per year in 2050 (level 3, WHO recommendation).

If, instead of changing the *quantity* of meat consumed, we change the *type* of meat from level 2 (BaU), which represents the FAO projection of increasing the consumption of ruminant meat from the current 22% to 25% by 2050, the expected BaU emissions of 53.6 $GtCO_2eq$ per year in 2050 would change as follows: 67.4 $GtCO_2eq$ per year (level 1, 28% meat from ruminants, as Canada currently), 47.5 $GtCO_2eq$ per year (level 3, 22% meat from ruminants, as the current global level) and 27.8 $GtCO_2eq$ per year (level 4, 10% meat from ruminants, as China currently) in 2050, with changes of +26%, −11% and −48%, respectively, against the BaU case.

Finally, changes in livestock yields have been simulated, *i.e.* the number of animals per hectare on pasturelands. Current global levels are approximately 0.60 cows per hectare, and 3.07 sheep and goats (combined) per hectare. BaU in the GC suggests that this density is likely to increase by 50%, reaching about 0.95 cows per hectare and 4.60 sheep and goats (combined) per hectare (level 3). If, instead, it is increased it to 80%, reaching 1.14 cows per hectare and 5.53 sheep and goats per hectare, global GHG emissions could drop from an expected 53.6 $GtCO_2eq$ per year in 2050 (under BaU) to 40.2 $GtCO_2eq$ per year in 2050 (−25%), including consequential carbon dynamics from land use change worldwide.

Both behavioural changes in diet and livestock productivity can, therefore, substantially affect climate change, positively or negatively. These strong impacts are not only due to enteric emissions of methane (CH_4), which release significant amounts of CH_4 in the atmosphere, but also (and mainly) due to the associated land use change impacts from these dynamics, and other agricultural emissions (*e.g.* fertilizers, energy use). For example, if a higher meat intake is not associated with higher crop and livestock yields, then we would need more land to meet the projected food market in order to ensure food security, resulting in more land use change and net deforestation, releasing substantial amounts of GHG into the atmosphere, due to changes in above-ground vegetation and soil carbon. Instead, if we increase crop and livestock productivity, whilst preferably also reducing meat

[‡‡]The IEA simulations for climate mitigation are available in the list of 'example pathways' described on the GC. There are three different IEA scenarios by 2050, IEA2DS (high mitigation effort), IEA4DS (moderate mitigation effort) and IEA6DS (little or no mitigation effort), which reflect an approximate representation of the scenarios and projections proposed by IEA,[69] aligned with an expected global mean surface temperature change by 2 °C, 4 °C and 6 °C, respectively, by 2100.

consumption, land would become available for other uses (including bioenergy, and afforestation/reforestation), resulting in a higher carbon mitigation effect.

12.6 Conclusions

Livestock production involves several complex interactions of natural resources and international trade flows, with positive and negative impacts on both food security and climate change mitigation, whereas the LCA approach gives the possibility of assessing these dynamics under a systems perspective towards a circular economy. Some of the key variables associated with this sector are the dynamics of plant nutrients, energy, water and carbon, with several inputs and output involved across the entire livestock production chain, which may substantially vary according to the type of animal (*e.g.* cattle, sheep, goats, pig, chicken, fish) and production system.

For preparing a robust LCA, it is fundamental to have clear objectives, system delimitation (boundaries) and identification of the main inputs, outputs and flows involved, as well as reliable data and consistent functional units to meet the project's objectives. Thus, it is possible to obtain results that are comparable over time in order to reduce environmental impacts systemically, whilst also promoting climate change mitigation and sustainable rural development.

Acknowledgements

The authors thank the International Support Network for African Development (ISNAD Africa) for the support provided in part of this research, and the valuable comments made by Dr Mairi Black (University of Surrey) and Dr Onesmus Mwabonje (Imperial College London) in the preparation of this book chapter.

References

1. J. Liu, V. Hull and M. Batistella, *et al.*, Framing Sustainability in a Telecoupled World, *Ecol. Soc.*, 2013, **18**(2), 26.
2. J. Woods, R. Chaturvedi, A. Strapasson, J. Millington and S. Petersen, *Assessing the climate impacts of Chinese dietary choices using a telecoupled global food trade and local land use framework*, presented in part at Global Land Project 3rd Open Science Meeting, publisher Global Land Programme, Beijing, 2016.
3. UKRI – United Kingdom Research and Innovation, *Food Security and Land Use: The Telecoupling Challenge*, research project led by the Centre for Environmental Policy at Imperial College London, funded by NERC, ref. no. NE/M021548/1, Principal Investigator J. Woods, London, available online https://gtr.ukri.org/projects?ref=NE%2FM021548%2F1 (accessed Dec 2019).

4. B. Lipinski, C. Hanson, J. Lomax, L. Kitinoja, R. Waite and T. Searchinger, *Reducing Food Loss and Waste*, Instalment 2 of 'Creating a Sustainable Food Future', WRI and UNEP Working paper, June 2013.
5. Quartz, *These maps show how the world composts*, article produced by Quartz Marketing team on behalf of BASF, 3 June 2014, available online https://qz.com/216261/these-maps-show-how-the-world-composts/ (accessed Dec 2019).
6. P. J. G. Henriksoon, J. B. Guinée, R. Kleijn and G. R. Snoo, *Int. J. Life Cycle Assess.*, 2012, **17**, 304–313.
7. F. Ziegler, S. Hornborg, B. S. Green, O. R. Eigaard, A. K. Farmery, L. Hammar, K. Hartmann, S. Molander, R. W. R. Parker, E. S. Hognes, I. Vázquez-Rowe and A. D. M. Smith, *Fish Fish.*, 2019, **17**, 1073–1093.
8. A. Strapasson, J. Woods and K. Mbuk, *Land Use Futures in Europe: How Changes in Diet, Agricultural Practices and Forestlands Can Help Reduce Greenhouse Gas Emissions*, Imperial College Grantham Institute, London, Briefing Paper no. 17, 2016.
9. UN – United Nations, *World Population Prospects 2019*, UN Department of Economic and Social Affairs: Population Dynamics, available online at: https://population.un.org/wpp/ (accessed Dec 2019).
10. A. Strapasson, J. Woods, H. Chum, N. Kalas, N. Shah and F. Rosillo-Calle, *Glob. Change Biol. Bioenergy*, 2017, **9**, 1721–1735.
11. FAO, *World Agriculture Towards 2030/2050: The 2012 Revision*, ESA Working Paper No. 12-03, ed. N. Alexandratos and J. Bruinsma, Global Perspective Studies Team at the Agricultural Development Economics Division of the UN Food and Agriculture Organization, Rome, 2012.
12. A. B. Strapasson, *The Limits of Bioenergy: A Complex Systems Approach to Land Use Dynamics and Constraints*, PhD thesis, Imperial College London, UK, 2014, 233p.
13. FAO, *Livestock's Long Shadow: Environmental Issues and Options*, Technical Report, LEAD & FAO, Rome, 2006.
14. J. N. Galloway, M. Burke, G. E. Bradford, R. Naylor, W. Falcon, A. K. Chapagain, J. C. Gaskell, E. McCullough, H. A. Mooney, K. L. L. Oleson, H. Steinfeld, T. Wassenaar and V. Smil, International Trade in Meat: The Tip of the Pork Chop, *Ambiol*, 2007, **36**(8), 622–629.
15. P. Best, Poultry performance improves over past decades, in *WATT Executive Guide to World Poultry Trends*, 2011.
16. S. Wirsenius, *Human Use of Land and Organic Materials: Modeling the Turnover of Biomass in the Global Production System*, PhD thesis, Chalmers University of Technology and Goteborg Universit, Goteborg, Sweden, 2000.
17. G. Thornton, New Zealand's Tegel Poultry achieves world's best feed conversion, in *WATT Poultry USA*, Watt Global Media, e-Newsletter, published on 6 Sep 2012, updated on 21 May , 2013.
18. R. Price, Integrated beef supply chain drives efficiency, *Farmers Weekly*, 2013.

19. J. N. Landers, *Tropical Crop-livestock Systems in Conservation Agriculture: The Brazilian Experience*, Integrated Crop Management, UN FAO, vol. 5, 2007.
20. L. R. A. Rodrigues, D. G. Quadros and A. K. B. Ramos, Recuperação de Pastagens Degradadas, in: *Annals of 'Simpósio Pecuária 2000 – Perspectivas para o III Milênio'*, FZEA, Pirassununga, Brazil, 2000.
21. S. Buckingham, H. McCalman and H. Powell, Grazing Systems, *Farming Connect*, Fact sheet published by the IBERS Grassland Development Centre, United Kingdom, 2013.
22. A. Balsom, Grazing on long grass boosts soil and performance, *Farmers Weekly*, 17 Oct 2013, United Kingdom, 2013.
23. UECBV - European Livestock and Meat Trades Union, *Footprint Category Rules Red Meat: Version 1.0*, Report, Technical Secretariat for the Red Meat Pilot. Brussels, Belgium, 2019.
24. A. B. Strapasson, A. Ramalho Filho, D. Ferreira, J. N. S. Vieira and L. C. M. A. Job, Agro-ecological Zoning and Biofuels: the Brazilian experience and the potential application in Africa, in *Bioenergy and Sustainable Development: The Role of Sugar Cane in Africa*, ed. F. X. Johnson and V. Seebaluck, simultaneously published in the USA and Canada by Routledge (Taylor and Francis Group), 2012, pp. 48–65.
25. ABIEC – Brazilian Association of Beef Exporting Industries, *Perfil da Pecuária no Brasil*, Report, ABIEC, Brazilian Beef, ApexBrasil, Sao Paulo, Brazil, 49p, 2019.
26. E. D. Assad, L. A. M. Cordeiro and R. L. Marchão *et al.*, Potencial de mitigação da emissão de gases de efeito estufa por meio da adoção da estratégia de integração lavoura-pecuária-floresta, in *Integração lavoura-pecuária-floresta: o produtor pergunta, a Embrapa responde*, ed. R. L. Cordeiro *et al.*, Embrapa, Brasilia, Brazil, 2015.
27. A. Bernt, L. A. Romero-Solórzano and L. S. Sakamoto, *Proceedings of the Symposium of Ruminant Nutrition*, UNESP, Botucatu, SP, Brazil, 2013.
28. D. D. Millen, R. D. L. Pacheco, P. M. Meyer, P. H. M. Rodrigues and M. B. Arrigoni, Current outlook and future perspective of beef production in Brazil, *Anim. Front.*, 2011, 46–52.
29. A. Strapasson, J. Falcão, T. Rossberg, G. Buss and J. Woods, Land Use Change and the European Biofuels Policy: The expansion of oilseed feedstocks on lands with high carbon stocks, *Oilseeds Fats, Crops Lipids*, 2019, **26**, 39.
30. INPE – Instituto Nacional de Pesquisas Espaciais, *Monitoramento do Desmatamento da Floresta Amazônica Brasileira por Satélite*, PRODES, INPE, Brazil, available online, http://www.obt.inpe.br/OBT/assuntos/programas/amazonia/prodes (accessed Dec 2019).
31. M. Planelles, El Acuerdo de París resiste (de momento) a Trump, *El País* (newspaper), Spain, 8 June 2019.
32. A. Strapasson, Das relações entre o humano e a natureza num país em transe, *Revista Política Democrática Online*, Fundação Astrojildo Pereira, Brazil, ISSN 1518-7446, 2019.

33. M. Dick, M. A. Silva and H. Dewes, Life cycle assessment of beef cattle production in two typical grassland systems of southern Brazil, *J. Cleaner Prod.*, 2015, **96**, 426–434.
34. S. Asem-Hiablie, T. Battagliese, K. R. Stackhouse-Lawson and C. A. Rotz, *Int J. Life Cycle Assess.*, 2019, **24**, 441–455.
35. C. Oduor, *Managing Soil Organic Carbon and Greenhouse Gas Emissions through the Establishment of Pasture Enclosures in West Pokot County*, Master thesis, University of Nairobi, Kenya, 2018.
36. R. Jaetzold and H. Schmidt, *Farm management handbook of Kenya Vol II/ B- Natural Conditions and Farm Management Information-(Rift Valley and Central Provinces)*, Report, Ministry of Agriculture, Kenya, in Cooperation with the German Agricultural Team (GAT) of the German Agency for Technical Cooperation (GTZ), 1983, pp. 181–202.
37. R. Roothaert and S. Franzel, Farmers' preferences and use of local fodder trees and shrubs in Kenya, *Agroforestry systems*, 2001, **52**, 239–252.
38. S. Dong, Overview - Pastoralism in the World, in *Building Resilience of Human-Natural Systems of Pastoralism in the Developing World: Interdisciplinary Perspectives*, ed. S. Dong *et al.*, Springer, ch. 1, 2016, pp. 1–37.
39. S. Singh and B. R. Bakshi, Eco-LCA: A tool for quantifying the role of ecological resources in LCA, in *Proceedings of the 2019 IEEE International Conference on Environment and Electrical Engineering*, Phoenix, AZ, USA, DOI: 10.1109/ISSST.2009.5156770, 18–20 May, 2009.
40. R. F. Follett, J. Kimble, C. Rice, D. Reed, S. Mooney and R. Lal, Economic and Societal Benefits of Soil Carbon Management: Cropland and Grazing Land Systems, in *Soil Carbon Management: Economic, Environmental, and Societal Benefits*, CRC Press, Taylor and Francis Group, 2007, pp. 99–128.
41. E. Milne and S. Williams, *Grazing Lands, Livestock and Climate Resilient Mitigation in Sub-Saharan Africa: State of the Science*, Report for the USAID, project contract no. AIDOAAL1000001, 2015, available online: http://www.vivo.colostate.edu/lccrsp/reports/GrazingLandsLivestock ClimateMitigation_Paper1_Final6Aug2015editedv4a.pdf (accessed 16 June 2019).
42. C. Garcia, P. Nannipieri and T. Hernandez, The Future of Soil Carbon, in *The Future of Soil Carbon Its Conservation and Formation*, Elsevier, 2018, pp. 239–267.
43. FAO, *Global Forest Resource Assessment Country*, Kenya, Report, Rome, 2010, p. 9.
44. Government of Kenya, Forestry, in *Kenya's Climate Change Action Plan: Mitigation*, ch. 4, 2012.
45. H. Steinfeld, P. Gerber, T. Wassenaar, V. Castel, M. Rosales, M. Rosales and C. de Haan, *Livestock's long shadow: environmental issues and options*, Report, UNFAO, 2006.
46. B. Minasny, B. P. Malone, A. B. McBratney, D. A. Angers, D. Arrouays, A. Chambers, V. Chaplot, Z.-S. Chen, K. Cheng and B. S. Das, Soil carbon 4 per mille, *Geoderma*, 2017, **292**, 59–86.

47. C. W. Rice, Subcommittee on production and price competitiveness hearing on carbon cycle research and agriculture's role in reducing climate change, 1999, document available online, https://soilcarboncenter.k-state.edu/docs/Soil%20C%20testimony.pdf (accessed Dec 2019).
48. M. Ritchie, *Soil Carbon Dynamics in Northern Kenya Rangelands*, Report, USAID, 2014.
49. A. Verdoodt, S. M. Mureithi and E. Van, Ranst, Impacts of management and enclosure age on recovery of the herbaceous rangeland vegetation in semi-arid Kenya, *J. Arid Environ.*, 2010, **74**, 1066–1073.
50. S. M. Mureithi, A. Verdoodt, C. K. Gachene, J. T. Njoka, V. O. Wasonga, S. De Neve, E. Meyerhoff and E. Van Ranst, *J. Arid Land*, 2014, **6**, 561–570.
51. C. Oduor, N. Karanja, R. Onwonga, S. Mureithi, D. Pelster and G. Nyberg, *BMC Ecol.*, 2018, **18**, 45.
52. J. L. Sauvé, T. W. Goddard and K. R. Cannon, A preliminary assessment of carbon dioxide emissions from agricultural soils, in *Annual Alberta Soil Science Workshop*, Alberta, 2000, pp. 22–24.
53. A. Chan and T. Parkin, *J. Environ. Qual.*, 2001, **30**, 1896–1903.
54. J. T. Houghton, Y. Ding, D. J. Griggs, M. Noguer, P. J. van der Linden, X. Dai, K. Maskell and C. Johnson, *Climate Change 2001: The Scientific Basis*, The Press Syndicate of the University of Cambridge, 2001.
55. R. Reid, S. Serneels, M. Nyabenge and J. Hanson, The changing face of pastoral systems in grass-dominated ecosystems of Eastern Africa, in *Grasslands of the World*, ed. J. M. Suttie, S. G. Reynolds and C. Battello, FAO Plant Production and Protection Series 34, Rome, 2005, pp. 19–76.
56. G. A. Keya, Alternative policies and models for arid and semi-arid lands in Kenya, Scandinavian Institute of African Studies, in *When the Grass is Gone: Development Intervention in African Arid Lands*, Uppsala, 1991, pp. 90–114.
57. T. G. Vagen, L. A. Winowiecki, C. Neely, S. Chesterman and M. Bourne, Spatial assessments of soil organic carbon for stakeholder decision-making–a case study from Kenya, *Soil*, 2018, **4**, 259–266.
58. M. Tolba, *Status of Desertification and Implementation of the United Nations Plan of Action to Combat Desertification*, Document no. UNEP/GCSS 3, 15, Nairobi: UN Environment Programme, 1991.
59. D. Kinyua, L. E. McGeoch, N. Georgiadis and T. P. Young, Short-term and long-term effects of soil ripping, seeding, and fertilization on the restoration of a tropical rangeland, *Restor. Ecol.*, 2010, **18**, 226–233.
60. G. Nyberg, P. Knutsson, M. Ostwald, I. Öborn, E. Wredle, D. J. Otieno, S. Mureithi, P. Mwangi, M. Y. Said and M. Jirström, Enclosures in West Pokot, Kenya: Transforming land, livestock and livelihoods in drylands, *Pastoralism*, 2015, **5**, 25.
61. J. O. Ogutu, H. P. Piepho, M. Y. Said, G. O. Ojwang, L. W. Njino, S. C. Kifugo and P. W. Wargute, Extreme wildlife declines and concurrent increase in livestock numbers in Kenya: What are the causes? *PLoS One*, 2016, **11**, e0163249.

62. FAO, *Options for low emission development in the Kenya dairy sector - reducing enteric methane for food security and livelihoods*, Report, UN Food and Agriculture Organization and the New Zealand Agricultural Greenhouse Gas Research Centre, Rome, 2017.
63. P. W. Ndung'u, B. O. Bebe, J. O. Ondiek, K. Butterbach-Bahl, L. Merbold and J. P. Goopy, *Anim. Prod. Sci.*, 2018, DOI: 10.1071/AN17809.
64. Government of Kenya, *National Dairy Development Policy*, Report, Ministry of Agriculture, L.A.F., State Department of Livestock (Ed.), Nairobi, Kenya, 2013.
65. E. Bryan, C. Ringler, B. Okoba, J. Koo, M. Herrero and S. Silvestri, Can agriculture support climate change adaptation, greenhouse gas mitigation and rural livelihoods? Insights from Kenya, *Clim. Change*, 2013, **118**, 151–165.
66. Government of Kenya, Transportation, in *Kenya's Climate Change Action Plan: Mitigation*, ch. 7 [Government of Kenya], Nairobi, 2012.
67. R. Bailey, A. Froggatt and L. Wellesley, *Livestock – Climate Change's Forgotten Sector: Global Public Opinion on Meat and Dairy Consumption*, Research Paper of the Chatham House's Energy, Environment and Resources, London, 2014.
68. WHO – World Health Organization, in *Protecting health from climate change: World Heath Day 2008*, Annex 1: Reducing your carbon footprint can be good for your health – a list of mitigating actions, UN WHO paper, Geneva, 2008.
69. IEA – International Energy Agency, *Scenarios and Projections*, IEA, OECD, Paris, 2014.

CHAPTER 13

Case Study – Bioenergy Life-cycle Analysis and Implications on Bioenergy-driven Circular Economy

HAO CAI* AND MICHAEL WANG

Systems Assessment Group, Energy Systems Division, Argonne National Laboratory, 9700 South Cass Avenue, 60439, Lemont, USA
*Email: hcai@anl.gov

13.1 Background

Circular Economy (CE) has evolved into various concepts based on a common philosophy envisioning an economy that is entirely circular. CE promotes the development of an optimal system that utilizes limited resources for delivering products and services with minimum generation of wastes and minimum environmental impacts primarily through material reducing, reuse, and recycling. The CE thinking takes a system-level approach that helps spot opportunities for 'closing the loops' of a given product or service system, giving rise to different strategies of redesigning the system that promotes system designs where the waste of one system becomes the resource of another.[1] In a natural resource-constrained world, CE gains its popularity as a potential solution to tackling issues such as resource availability and sustainable development by promoting alternative practices and business models to minimize the use of virgin materials. As a result, the degree of success of a CE strategy is often measured by the degree of recycling and reuse of process

Life Cycle Assessment: A Metric for the Circular Economy
Edited by Aiduan Borrion, Mairi J. Black and Onesmus Mwabonje

Published by the Royal Society of Chemistry, www.rsc.org

materials of a product or service system. Metrics such as the Material Circularity Indicator (MCI) have been developed to quantify the restoration of material flows, taking into account a combination of product characteristics including use of virgin materials, use of re-used materials and recycled content in the product, the efficiency of recycling, the mass of unrecoverable waste that goes to landfill or incineration, and a utility factor that accounts for the duration and intensity of the product's use.[2] With a focus on maximizing material circularity through strategies to realize loop-closing opportunities, quality and environmental issues of the product or service systems may emerge. For example, using recycled materials could impact the lifetime of a product due to quality loss or introducing and accumulating contamination from recycled streams, and reuse of waste streams from an existing system may add to impacts through the additional acquiring, transporting, and purifying processes required in a separate system. Under these circumstances, questions about whether the CE strategies improve the system-level environmental performance in relation to specific CE strategy that may offer the most environmental benefits need to be answered.

Life-cycle analysis (LCA), which assesses the environmental impacts of products or services from cradle to grave, is a useful technique for evaluating options for CE solutions to ensure a positive balance of efforts and benefits in both new product designs and increased recycling. LCA possesses the capabilities to track material and energy flow along the supply chain of a product system and to identify opportunities for improvement of resource utilization efficiencies, waste reduction, redesign of supply chain processes and operations through recycling and reuse of process materials and energy. Combining both the robustness of the LCA methodology and the aspirational principles of CE could enable identifying and choosing the best strategy of designing products and societal service systems to achieve optimal resource efficiency and minimal environmental impacts.

Biofuels have been pursued as an alternative energy sources to fossil-derived fuels primarily in the transportation sector in the United States, China, Brazil, European Union, among other countries. Biofuels with various fuel properties and thus varied applications can be produced from a wide range of biomass feedstock types *via* biochemical, thermochemical or biological conversion technologies. Nevertheless, one commonality of biofuels is that the carbon in the fuel comes from carbon in the biomass feedstock that is sequestered during the plant growth by photosynthesis. Once the biofuel is combusted, the embedded carbon is converted to CO_2 emissions and released back into the atmosphere. This closed loop biogenic carbon cycle between uptake by plant growth, conversion to biofuels, and release to the air from biofuel combustion, as shown in Figure 13.1, makes biofuels commonly considered carbon neutral and thus a promising solution to reducing CO_2 emissions from using fossil fuels. Because of that, tapping carbon from biomass presents a promising option to recycle carbon in the atmosphere, reuse the carbon as part of an energy source, and reduce fossil carbon emissions.

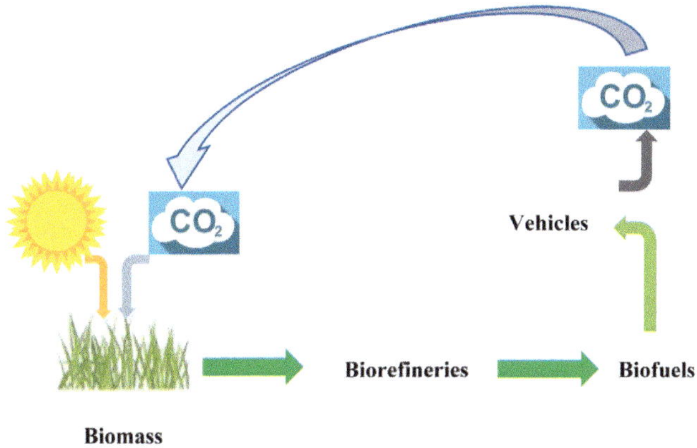

Figure 13.1 Biogenic carbon cycle associated with biofuel production for transportation use.

Promoting a biofuel-driven low-carbon economy presents an opportunity to promote a circular economy for carbon. The net carbon impacts of this circular carbon economy, however, need careful evaluation because biofuels of different types may exhibit unique carbon cycles and implications depending on the source of the feedstock carbon, conversion technologies, carbon end of life, energy and emission burdens of all the activities involved in the circular carbon economy, as well as indirect effects such as land use change (LUC) induced carbon emissions or sequestration. LCA has been a useful method to address these key issues and holistically evaluate the net carbon effects and implications of various biofuel technologies. Here, we present a range of biofuel pathways as case studies to discuss the key issues that impact the synergies and trade-offs of the carbon implications of biofuels from a carbon recycle, reuse, and reduce point of view.

13.1.1 Case Study 1: Corn Ethanol

Corn ethanol is the single largest biofuel alternative by volume to petroleum gasoline in the United States. Through use of atmospheric CO_2 and releasing it back to the air from corn production, ethanol production, and ethanol combustion, the corn ethanol industry can be regarded as a type of CE.[35,36] The original goals of the U.S. ethanol program were to reduce the nation's dependence on imported oil and to reduce air pollutant emissions from motor vehicles, such as volatile organic compounds and carbon monoxide, and later included the reduction of greenhouse gas (GHG) emissions. Numerous LCA studies on the energy balance and GHG emission impacts of corn ethanol showed results in both ways as a result of applying different system boundaries, data sources that represent different technology improvement status, and the methodology and assumptions adopted to address the controversial LUC effects.[3–6] ANL researchers have conducted LCA of corn

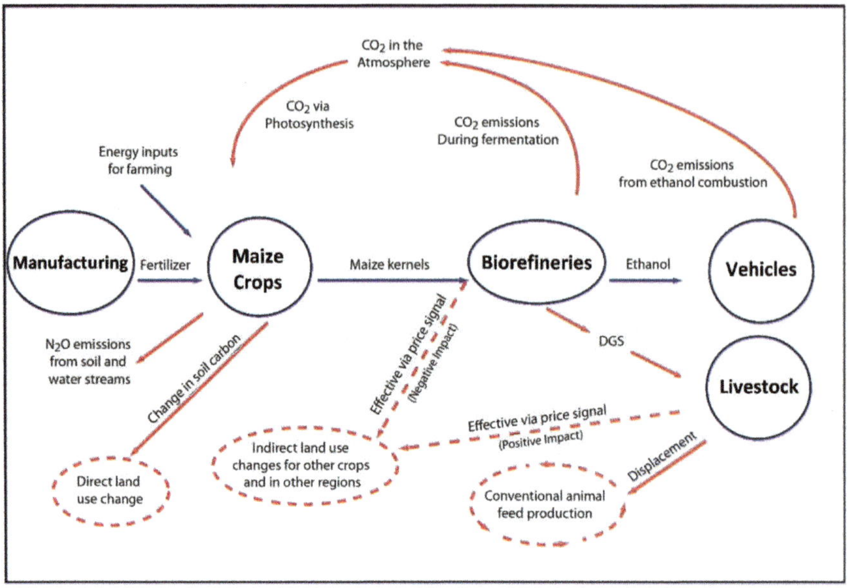

Figure 13.2 System boundary of life-cycle analysis of corn ethanol carbon intensity considering its biogenic carbon cycle, external energy and chemical inputs, co-product displacement effects, and land use change-induced emissions.

ethanol and have addressed a few emerging issues in a series of publications.[3,7–11] These issues include LCA system boundary, farming energy use and fertilizer inputs, energy use at corn ethanol plants, co-product and its displacement effects, and LUC emission impacts. Figure 13.2 shows the key supply chain processes and the carbon and energy flows among the supply chain of corn ethanol production and consumption as a transportation fuel. It shows that extra energy and chemical inputs are required to make ethanol from corn besides the biogenic carbon cycle that happens during corn farming and growth, fermentation of corn kernels into ethanol, and ethanol combustion. These extra energy and chemical inputs could produce GHG emissions (including, not only CO_2 emissions, but also more potent N_2O emissions) directly or indirectly from energy production, chemical manufacturing, soil amendment activities, and transportation of these energy and chemicals to cornfields or corn ethanol plants for use. Besides, below-ground soil carbon might decrease as a result of direct LUC due to change of pastureland, grassland, or marginal land to corn fields, as well as of indirect LUC due to additional land conversion to make up food, feed, or fiber production that was displaced by corn ethanol production. These complex dynamics and changes of soil carbon go beyond the biogenic carbon cycle for crop growth. Therefore, the corn ethanol pathway is not a simple, closed-loop biogenic carbon cycle that does not introduce extra carbon emissions or sequestration. To evaluate whether a given mega-joules (MJ) of corn ethanol offers any emission reduction benefits compared to petroleum-derived gasoline, an LCA takes into account both direct emissions from process energy use over the course of each stage of its life-cycle, *e.g.* natural gas or coal as sources of process heat for fermentation at the corn ethanol plant, and indirect emissions from manufacturing and transportation of process energy and chemicals, *e.g.* synthetic nitrogen fertilizers, of the corn ethanol pathway.

One key issue that greatly impacts the energy and carbon flows of the corn ethanol pathway is technology improvement of corn farming and ethanol production. Over the past several decades, corn yield in the U.S. has been consistently and greatly improved by the positive impact of improved crop genetics and crop production technologies.[12] These agricultural technological advancements contribute to a concurrent decline of fertilizer application rates for corn farming, as shown in Figure 13.3.

Meanwhile, the corn ethanol industry in the U.S. has matured, in that it continuously increases its ethanol production yield per bushel of corn processed, as shown in Figure 13.4, and reduces its energy consumption per unit of ethanol output, as shown in Figure 13.5.

These technological improvements have collectively contributed to a lower carbon intensity of corn ethanol overtime, as shown in Figure 13.6. Most recently, life-cycle GHG emissions of corn ethanol have decreased to 52 g CO_2e/MJ, compared with 94 g CO_2e/MJ for petroleum gasoline.

In a corn ethanol-based CE that takes advantage of the biogenic carbon cycle of corn as a biomass feedstock, LCA studies show that recycling and

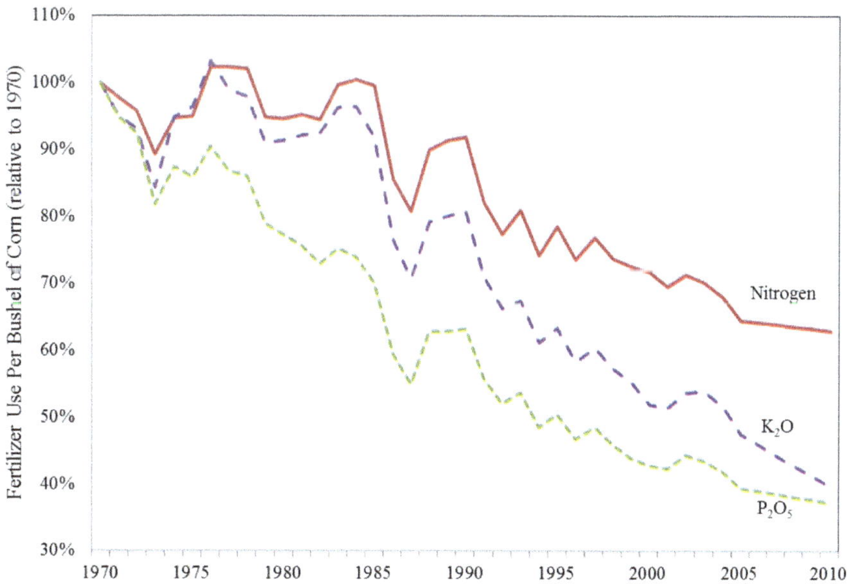

Figure 13.3 Trends of application rates of key fertilizers per bushel of corn for corn production in the U.S.
Reproduced from ref. 37 with permission from Elsevier, Copyright 2011.

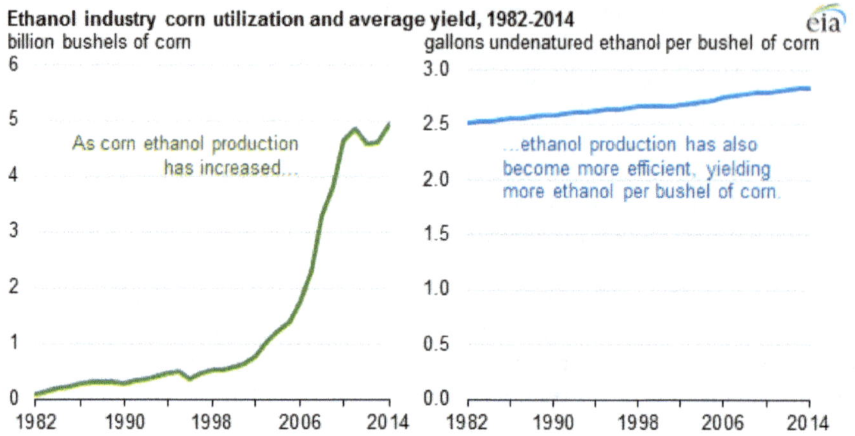

Figure 13.4 Trends of corn ethanol production and average ethanol yield per bushel of corn in the U.S.[13]

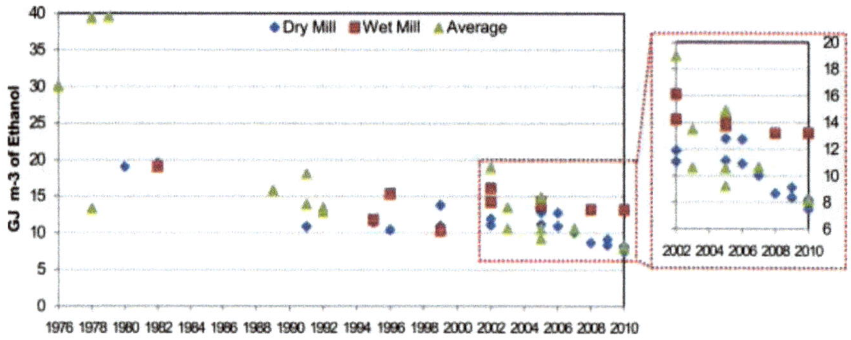

Figure 13.5 Trends of energy use at dry mill and wet mill ethanol plants in the U.S.[14] Reproduced from ref. 37 with permission from Elsevier, Copyright 2011.

reuse of biogenic carbon to displace fossil carbon offers corn ethanol a significant biogenic carbon credit, which is the most important factor that corn ethanol is less carbon-intensive than petroleum gasoline, as shown in Figure 13.6. Meanwhile, these studies reveal that utilizing biogenic carbon does not come as energy- or emission-burden free. In fact, a significant amount of external energy and chemical inputs is needed to sustain the biogenic carbon cycle for corn ethanol production, and indirect effects such as LUC emissions may occur. As a result, the carbon intensity of corn ethanol is driven by additional carbon flows embedded in the process energy and chemicals, as well as LUC effects that are caused by the biogenic carbon cycle. The carbon intensity of corn ethanol-based CE is greatly impacted by technological improvement in corn farming, energy efficiencies and fermentation technologies at corn ethanol plants, and use of co-products as

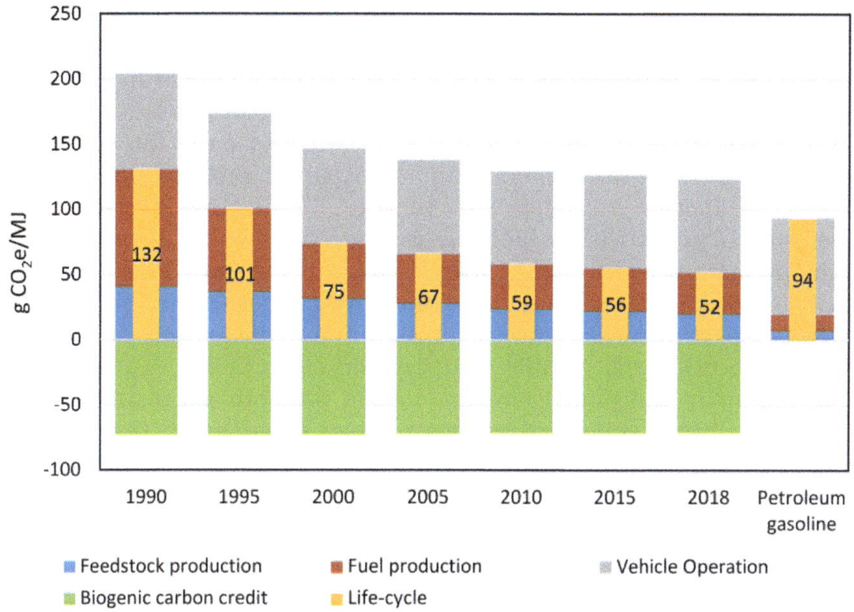

Figure 13.6 Trends of life-cycle carbon intensities of U.S. corn ethanol. Results are generated with the Argonne National Laboratory's Greenhouse gases, Regulated Emissions, and Energy use in Transportation (GREET®) model[15] that incorporates the historical trends of technological advancement of corn farming and corn ethanol industries.

animal feeds to displace other types of animal feeds that could be energy- and emission-intensive to produce. Greatly improved corn yield per acre of land not only reduces the farming energy and fertilizer requirement to produce a bushel of corn but also a MJ of corn ethanol. It shows that reducing the external energy and emission burdens of any given input stream of the corn ethanol pathway from technology improvement contributes to reduction in carbon intensity of the finished fuel overtime. This means that a CE should consider, not only minimizing the quantity of, but also the carbon intensities of the virgin feedstock, energy, and material inputs to achieve minimum carbon impacts. As the corn ethanol industry keeps advancing, regular LCA that evaluates the impacts of technological advancement on the carbon impacts of a corn ethanol-based CE, which are driven by external energy and emission burdens and affected by indirect LUC emissions as a result of sustaining the biogenic carbon cycle to produce corn ethanol, is warranted.

13.1.2 Case Study 2: Sugarcane Ethanol

Brazilian sugarcane has been used as feedstock for production of ethanol on a large scale in Brazil for over three decades, where most of the sugarcane mills produce sugar, ethanol, and electricity.[13] Presently, the common production model is ethanol and sugar produced in an integrated process to meet the market demands for sugar and ethanol. Nevertheless, dedicated ethanol distilleries have been built recently, to meet the expected higher demand of ethanol in the future, which is partly driven by the Brazilian government's ethanol mandate of an ethanol blending level at 27% in the fuel used by light-duty vehicles.[14] Like corn, sugarcane is a renewable biomass with a biogenic carbon cycle. Unlike corn, it is the sugar in the sugarcane juice, instead of starch in corn kernels that is converted to ethanol *via* fermentation. A typical sugarcane mill receives and prepares the cane and extracts juice from the cane, which are common unit operations to the ethanol distillery and sugar factory. The extracted juice is sent to the juice treatment system for both sugar and ethanol production that is subject to its own product specificities. The extracted juice is blended with molasses, a concentrated residual solution obtained after sugar crystallization in the sugar factory, and is fermented using yeast. The fermentation product containing ethanol is sent for distillation and dehydration, and anhydrous ethanol is made. Process steam and electricity needed for the fermentation is produced *via* a combined heat and power (CHP) unit at the mill using sugarcane bagasse as the fuel. On average, surplus electricity is available and is sold to the grid, displacing the conventional electricity generation in Brazil, which is dominated by low-carbon hydropower.

One unique issue of the Brazilian sugarcane ethanol industry is that part of the sugarcane straw during the sugarcane harvest was historically burnt, releasing most of the embedded biogenic carbon back into the atmosphere as CO_2 emissions and part of the carbon as black carbon that is known as a strong climate forcer.[15] The black carbon emissions from open field burning of sugarcane straw present a biogenic carbon leakage in the sugarcane biogenic carbon cycle because it cannot be used during sugarcane growth *via* photosynthesis, and its climate impacts need to be assessed. It has been recently quantified with respect to the impact of this biogenic carbon leakage in the form of black carbon emissions on the carbon intensity of sugarcane ethanol.[16] The results showed that this source of carbon emissions could add to as much as 57% of the life-cycle GHG emissions of sugarcane ethanol (85% by volume)-driven vehicles when the 20-year Global Warming Potential of black carbon was taken into account. This highlights the importance of not only tracking any additional carbon flows coming into the biogenic carbon cycle, but also tracking any leakage of biogenic carbon and assessing its fate and impacts to paint a complete picture of the carbon impacts of a biofuel system and the CE based on it. Meanwhile, open burning of sugarcane fields are phasing out in Brazil with mechanical harvest of sugarcane. The straws left at the sugarcane field may become available for alternative uses, such as being cellulosic feedstock for ethanol production. This option is not yet practiced on a large scale in Brazilian ethanol facilities due to high recovery and conversion costs and questions about short- and long-term soil implications.[17] However, the impacts of any practices to manage and potential utilization of this source of carbon may positively impact the carbon effects of the existing Brazilian sugarcane ethanol industry. Such impacts require new LCA with an appropriately defined system boundary to fully quantify.

Another key issue of the sugarcane ethanol pathway is to feed bagasse to CHP for process heat and electricity generation. Although the CO_2 emissions from the combustion process are considered carbon-neutral because the carbon embedded in the bagasse is part of the sugarcane biogenic carbon cycle, the resulting surplus electricity that displaces grid electricity from the Brazilian electricity generation mix has its carbon impacts. From a CE's perspective, reuse of the bagasse as an intermediate material stream to aid the production of the designed product of the system eliminates the need for external sources of heat and electricity, which improves the resource utilization and energy efficiencies of the system. Meanwhile, the full picture of carbon impacts of reusing the bagasse stream becomes clear only when the avoided generation of the same amount of surplus electricity with the existing electricity generation mix and the resulting avoided emissions are taken into account. Such shifts of emission burdens in a CE system can be fully captured and quantified by LCA that adopts an appropriate system expansion approach.

Finally, the Brazilian sugarcane ethanol industry has a history of recycling nutrients from nitrogen-rich vinasse and filter cake, two sugar mill byproducts, for soil amendments at sugarcane fields. Such practices reduce the need of synthetic nitrogen fertilizer application for sugarcane farming, and avoid the emission burdens of manufacturing nitrogen fertilizers and the resulting N_2O emissions when they are applied to the soil. On the other hand, transportation of the liquid vinasse in open channel could produce methane, another potent greenhouse gas. Such intended benefits and somewhat unexpected negative emission impacts of the practices are captured in an LCA.[3] The LCA highlights the importance of not only focusing on reducing, recycling, and reusing the waste stream whenever possible, but also taking into account any direct and indirect consequences of such practices in order to fully understand the carbon impacts of a CE system featuring ethanol production from sugarcane.

Addressing the key issues mentioned above, we found that additional energy and chemical inputs for sugarcane biomass production are the key emission drivers to the life-cycle GHG emissions of sugarcane ethanol, as shown in Figure 13.7.[3] Unlike fossil carbon in petroleum fuels that takes millions of years to form and is therefore considered to be non-renewable carbon, the biogenic carbon credit resulting from rapid biogenic carbon recycling of sugarcane ethanol offers a significant carbon reduction benefit, as shown in Figure 13.7. This demonstrates the carbon benefits of the sugarcane ethanol industry, compared with the fossil energy sector that does not recycle fossil carbon in current practices.

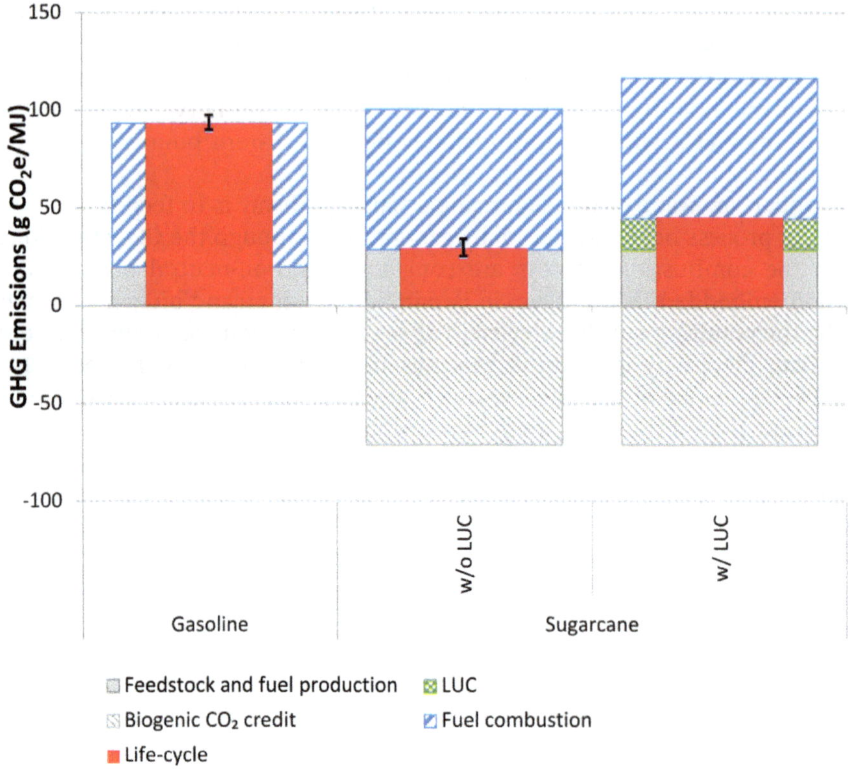

Figure 13.7 Life-cycle GHG emissions of Brazilian sugarcane ethanol that benefits from carbon neutral biogenic carbon cycle, in comparison to petroleum gasoline.

13.1.3 Case Study 3: Corn Stover Ethanol

As part of the corn crop that participates in the corn biogenic carbon cycle, corn stover that consists of the leaves, stalks, and cobs of corn is about the same weight as the corn grain on a dry matter basis. Corn stover can be used as ensilaged or non-ensilaged animal feed, as a residue left in the field to provide organic matter for soil amendment, or as feedstock for bioenergy production including bioethanol. Studies found that one-third to one-half of corn stover in cornfields in the U.S. could be sustainably removed without causing erosion or deteriorating soil quality.[18–20] When corn stover is used as a feedstock for bioethanol production, it is considered an agricultural crop residue and the embedded carbon is considered burden-free because the energy and chemical inputs during corn farming is intended for corn production. However, when corn stover is removed, nutrients therein such as N, P and K are removed, too, and require supplement with synthetic fertilizers to maintain the soil fertility. Nevertheless, corn stover provides a biogenic carbon source with less additional energy and chemical burdens than other dedicated cellulosic biomass feedstocks do, such as miscanthus, switchgrass, willow, or poplar, which all require a significantly larger amount of farming energy and fertilizers to grow.[21,22] At the cellulosic ethanol plant, corn stover goes through processes including dilute-acid pretreatment, enzymatic saccharification, and fermentation to produce ethanol. About 30% of the biogenic carbon in corn stover is eventually converted to ethanol, with about 52% of the carbon converted to CO_2 from combustion of the non-fermentable lignin stream and other intermediates, and about 15% of the carbon converted become vent CO_2 during the fermentation process.[23] The conversion processes consume chemicals, enzymes, and yeast that could be energy- and emission-intensive to make. Combusting of the lignin portion of the feedstock produces sufficient steam and electricity to meet the process demand. Like the sugarcane ethanol plant, surplus electricity is available and sold to the grid to displace the grid electricity. Once the ethanol is used as a transportation fuel, the biogenic carbon therein is converted to CO_2 and released into the atmosphere, ready to be utilized *via* photosynthesis by additional corn crop for growth.

Our LCA found that corn stover ethanol has a low life-cycle GHG emission intensity of about 5 g CO_2e/MJ when LUC effects are not considered, as shown in Figure 13.8. This is partly owing to the by-product nature of corn stover that only carries small emission burdens associated with production of supplementary fertilizers, and partly owing to the energy self-sufficient process design that converts the lignin to process energy, resulting in

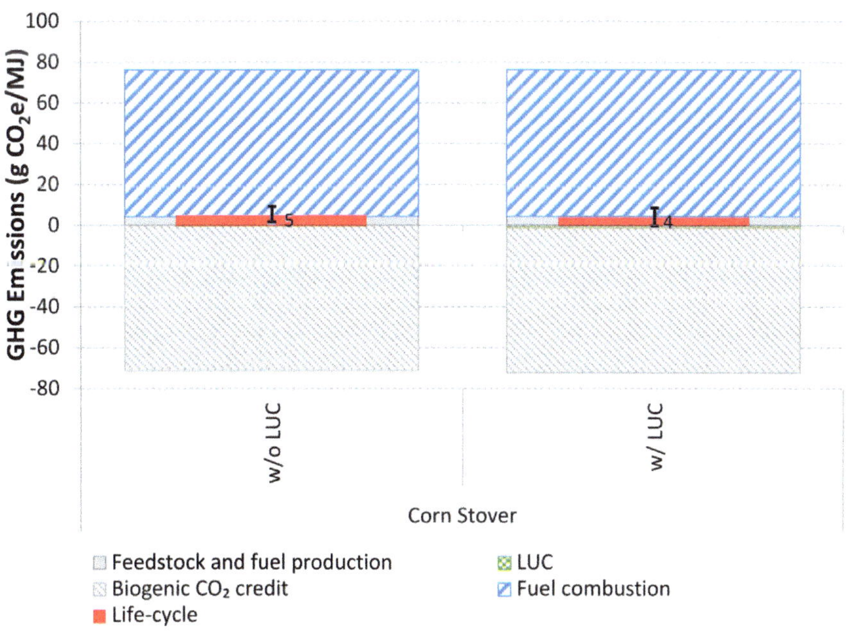

Figure 13.8 Life-cycle GHG emissions of corn stover ethanol that benefits from carbon neutral biogenic carbon cycle.

additional emission displacement credit from the surplus electricity. Like other bioethanol production pathways, the biogenic carbon credit resulting from rapid biogenic carbon recycle offers corn stover ethanol a significant carbon reduction benefit.

The corn stover ethanol technology shows that agricultural crop residues provide a low emission-burden source of biogenic carbon, and tapping part of the biogenic carbon for process heat and power generation purpose further minimizes the need for potentially emission-intensive external energy sources. From a carbon recycling point of view, leveraging the biogenic carbon source available from the crop biogenic carbon cycle to meet the demand for both carbon and energy sources provides a low carbon CE prototype with minimum requirement of external sources of energy and materials that carry their own carbon burdens.

Meanwhile, technologies that aim to convert the non-fermentable lignin stream from cellulosic feedstocks including corn stover to value-added bioproducts are being developed to enhance the economics of the bioenergy production. One example is to convert the lignin to adipic acid (AA), as shown in Figure 13.9.[24]

With such an alternative design of utilizing the lignin, a significant amount of external grid electricity and chemicals are required to produce the renewable diesel blend stock (RDB), the fuel product, and AA. As a result, life-cycle emissions of RDB is greatly impacted and depends on how the emission burdens of additional energy and chemicals inputs are allocated between the fuel and non-fuel products. Meanwhile, the lignin-derived AA shows a significant carbon reduction benefit compared to the conventionally fossil-derived AA, as shown in Figure 13.10, which is mostly driven by a significant biogenic carbon credit it has. This biogenic carbon credit of lignin-derived AA is not affected by the end-of-life of the embedded biogenic carbon, which could be complete

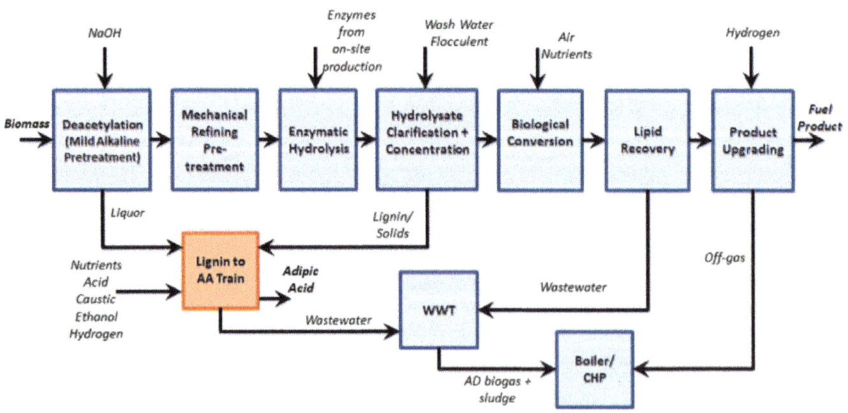

Figure 13.9 Integrated biorefinery process flow diagram of co-production of lignin-derived adipic acid with renewable diesel blendstock *via* biochemical conversion.

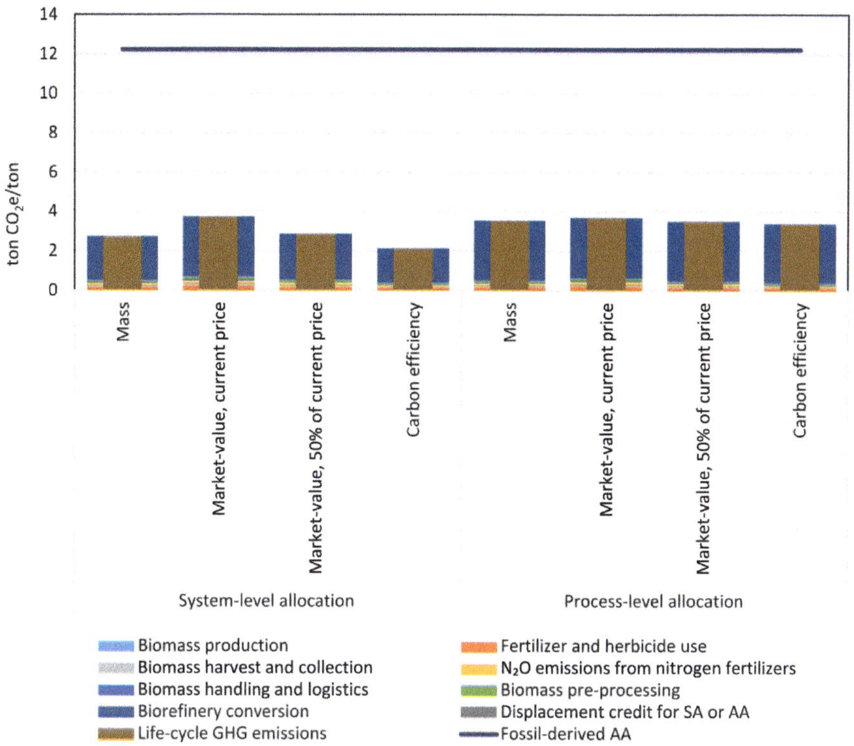

Figure 13.10 Life-cycle GHG emissions of co-produced lignin-derived AA, in comparison to those of fossil-derived counterpart, using system-level and process-level allocation methods.

decomposition as biogenic CO_2 emissions, or remaining a biogenic carbon sequester without decomposition, or somewhere in between. Regardless of the end-of-life, lignin-derived AA always offers a biogenic carbon credit compared to its fossil-derived counterpart.

In this example of biorefinery redesign that features using biomass to produce biofuels with a significant amount of co-produced bioproduct, it illustrates the need to re-evaluate the carbon impact of converting part of the biogenic carbon into value-added products on both the bioproduct and the originally designed biofuel product. A system-level analysis that takes into account, not only new product systems that take advantage of the biogenic carbon cycle to the maximum, but also any conventional product systems that are subject to replacement, is warranted to fully account for the carbon impacts of the biogenic carbon-based product systems.

13.1.4 Case Study 4: Soybean Biodiesel

Besides bioethanol, biodiesel has gained significant popularity worldwide over the past two decades as one of the most commercially produced and consumed biofuels.[25] Among various biomass feedstocks for biodiesel production, soybean is the most used feedstock, accounting for approximately 55% of all reported feedstock inputs.[26] Like ethanol production from corn, sugarcane, corn stover, among other feedstocks, soybean biodiesel production requires use of external energy and chemicals. Thus, LCA that considers not only the biogenic carbon cycle, but also external energy and chemical inputs and their carried emission burdens is needed to quantify the carbon intensity of soybean biodiesel relative to conventional petroleum diesel. Given the technology advancement over time, particularly the energy-efficiency improvement of biodiesel due to increases in soybean yield and biodiesel conversion efficiency,[27] as well as emerging issues such as indirect land use change (ILUC) effects, LCA of soybean biodiesel are updated periodically.[25] A recent LCA[25] that considers recent improvement of soybean farming (*i.e.* decreased fertilizer and chemical inputs, decreased farming energy use, and increased soybean yields), soy oil recovery (*i.e.* reduced electricity consumption), and biodiesel production *via* transesterification (*i.e.* improved biodiesel yield and chemical usage efficiency), as well as unintended ILUC effects, found that external energy and chemical inputs resulted in a carbon intensity of about 22 g CO_2e/MJ for soybean biodiesel, which is about 79% reduction in GHG emissions compared with petroleum diesel. The biogenic carbon credit for soybean biodiesel is the biggest difference between its life-cycle GHG emissions and those of petroleum diesel, as shown in Figure 13.11. When the ILUC effects are considered, which could be significant depending on the modeling methods and data, up to 32 g CO_2e/MJ ILUC emissions could be induced beyond the soybean biogenic carbon cycle, as shown in Figure 13.11.

One key issue associated with the soybean biodiesel LCA is handling of a significant amount of soy meal as a by-product from the soy oil extraction process. The soy meal is not fed to the transesterification process to produce biodiesel, but rather used as an animal feed. However, soy meal is a valuable co-product from the soy oil extraction step and it stems from soybeans as the feedstock. Therefore, it is reasonable to allocate the farming and transportation energy and emission burdens of soybean between soy oil and soy meal, and take into account the shared burden in any subsequent operations involving each intermediate product. Because of this, the energy and emission impacts of recycling and reusing any intermediate stream that may carry upstream burdens for other purposes on the finished product need to be carefully evaluated and quantified. As shown in previous studies, assigning an appropriate energy and emission burden to any intermediate or finished product in a multiple co-product system may be challenging, and the allocation methods used to do so can greatly impact the results.[10,24,28] Pros and cons of possible allocation methods[10,24] should be considered to help make the most reasonable decision, and the results could help decision-making on whether or not tapping any intermediate material stream of an existing product system for new products would help deliver any energy or emission benefits.

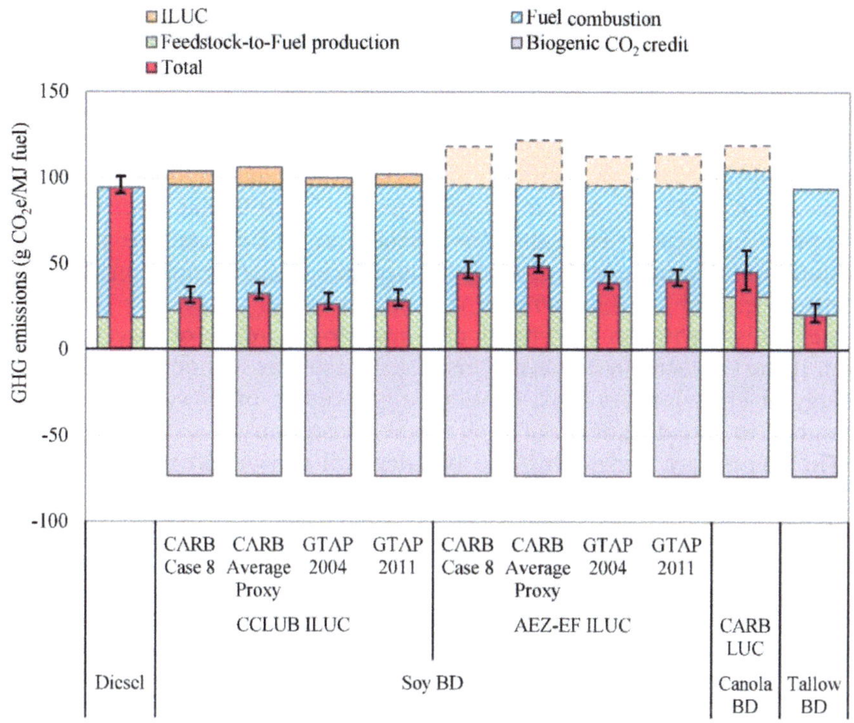

Figure 13.11 Life-cycle GHG emissions of soybean biodiesel, in comparison with petroleum diesel and biodiesel from canola and tallow. ILUC emissions in dotted yellow bars represent those modelled by Agro-ecological Zone Emission Factors (AEZ-EF).
Reproduced from ref. 25 with permission from Elsevier, Copyright 2017.

13.1.5 Case Study 5: Algae Renewable Diesel

A great diversity of organisms from microscopic cyanobacteria to giant kelp referred to as algae converts sunlight into carbon and energy sources using photosynthesis like many other biomass feedstocks. Unlike them, algae's potential as a renewable fuel source lies in increasing algal biomass productivity per acre of land, which, according to some researchers, could be 10 or even 100 times more productive than traditional bioenergy feedstocks.[29] Once harvested, algae can be converted into fuels *via* biochemical or thermochemical conversion technologies. One key to realizing algae's potential as a biomass feedstock is to the source of high concentration of CO_2 that is needed to support algae cultivation demands. Advanced carbon capture and transport of flue gas with high CO_2 concentration from point sources, such as power plants, corn ethanol plants, among others, are being considered as viable options by the U.S. Department of Energy. Meanwhile, saline algae species open up broader opportunities for siting the algae ponds along the coastal areas and cause no stress to limited fresh water resources.

Numerous LCA have been conducted to assess the carbon intensity of renewable diesel or biodiesel produced from algae.[30–33] The boundary for the LCA typically includes all energy and chemicals associated with algae biomass growth and dewatering, nutrient supply, algae conversion, production of any co-products, fuel transportation and distribution, and combustion of the biofuel in vehicles. A recent harmonization study by the U.S. Department of Energy's national laboratories[34] found that electricity consumption for algae growth and dewatering and for CO_2 capture and transportation to the algae farm is one major source of GHG emissions for both thermochemical and biochemical conversion routes to produce renewable diesel. This highlights the energy and emission burdens of recycling CO_2, even if the CO_2 stream *per se* is considered waste gas without any upstream energy or emission burdens, cannot be ignored in other systems that are designed to take advantage of such waste gas streams.

The co-product design with the biochemical conversion route has a significant impact on the carbon intensity of the fuel product. For example, a significant amount of polyurethane co-product can be produced from upgrading unsaturated fatty acids isolated from the extracted lipid fraction of algae. This co-produced biochemical, which possesses a significant biogenic carbon credit compared with its fossil-derived counterpart, provides a significant displacement emission credit. When this credit is normalized to the fuel product in order to capture the carbon emission reduction benefits of both the fuel and non-fuel biochemical products in the existing fuel-focused carbon regulation such as Renewable Fuel Standard and Low Carbon Fuel Standard, the carbon intensity of the fuel is greatly altered, as shown in Figure 13.12.[24]

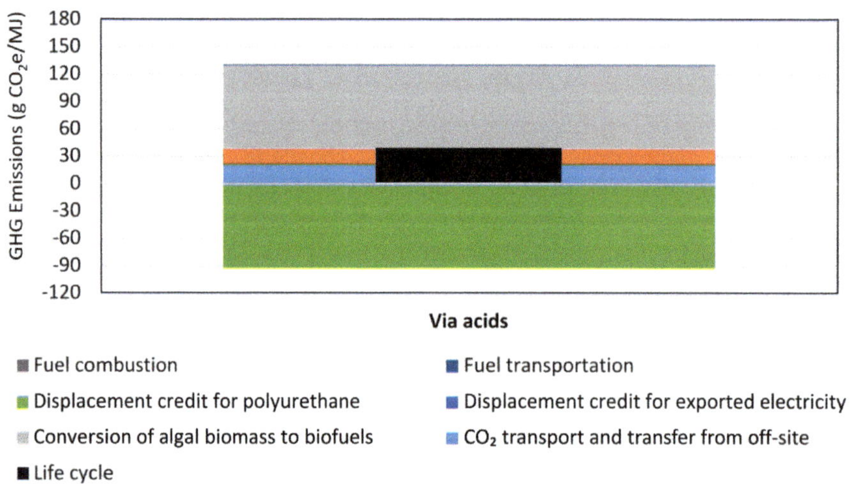

Figure 13.12 Life-cycle GHG emissions of renewable diesel produced from combined processed algae that features a significant co-production of renewable carbon-derived polyurethane.[38]

Algae cultivation presents a more land-efficient option of utilizing biogenic carbon cycle to produce biomass for bioenergy production purposes. Given that algae could be potentially productive and has small footprint of terrestrial land, other unintended indirect impacts, such as ILUC effects that could be significant for other biomass feedstocks, might be much smaller for algae. Such benefits of recycling CO_2 sources to produce algae-based fuels and chemicals, as well as the other emission impacts mentioned, can be assessed holistically through LCA. Meanwhile, the algae biofuel case highlights the need to consider land availability and land use efficiency of producing biomass through biogenic carbon cycles for biofuel production. Improving resource efficiency contributes to building a CE oriented system that aims to address resource scarcity issues, and minimizing its potential environmental impacts. LCA can help identify opportunities and bottlenecks to move a waste resource-based biofuel system towards a CE with minimum environmental impacts.

13.2 Discussion and Conclusion

Case studies of a wide range of biofuel production pathways show a commonality of utilizing biogenic carbon cycle to produce biomass and eventually to biofuels. Therefore, these biofuel production systems can all be regarded as a CE system of their own, where biogenic carbon is being recycled and reused to produce a targeted biofuel. However, each biomass type and its biofuel production pathway have their key issue to address that greatly impacts the net GHG effects of the finished biofuels, and the resulting CE models.

With LCA, potential burden shifting, or the transfer of an environmental impact from one life-cycle stage or one product stream to another, can be identified, and its impacts on specific products or the entire system can be quantified. External energy and chemical requirement, surplus electricity, co-produced bioproducts, ILUC effects, and recycling waste CO_2 stream as shown in the case studies represent key burden shifts associated with the biofuel production pathways discussed. Quantifying the carbon impacts of such burden shifting and identifying opportunities to mitigate it through LCA is key to ensuring a low-carbon profile of the biofuel, as well as its derivative CE.

Biofuel LCA shows that tapping part of the biogenic carbon that is not best suited for biofuel production to produce process energy or alternative chemicals may be beneficial for reducing the carbon intensity of the biofuel product. With an expanded energy or product system, the goal of the system is not only to reach circularity, but also to reduce environmental impacts based on LCA. With a reduced demand for energy or chemicals that might be carbon intensive in conventional production ways, targeting the displacement credit presents an alternative way that expands the system boundary of a CE model that is designed to be regenerative within the system to minimize environmental impacts. In such an alternative system, it is important to consider use of a displacement coefficient that takes into account the quantity and the market availability when evaluating the displacement benefits. To make sure a co-product design provides better environmental benefits, system-level LCA, rather than product-specific LCA, which takes into account both the main product and the co-product would be a more effective approach.

Biofuels that focus on biogenic carbon cycles and those divert part of the carbon in the biogenic carbon cycle to produce new products highlight trade-offs of environmental performance between circularity and displacement of the biogenic carbon. Circularity oriented biofuel pathways often require external energy and chemical sources to sustain the biogenic carbon cycle, while those that also make new biogenic carbon-derived products to displace fossil-derived products offers a biogenic carbon credit, which reduces the environmental impacts of the biofuel pathway. In both cases, LCA can be a useful tool to quantify the environmental impacts of different systems emphasizing either circularity or displacement or both, and provide insights on mitigating the environmental impacts of the system.

Acknowledgements

This work was supported by the Bioenergy Technologies Office (BETO) of the Office of Energy Efficiency and Renewable Energy of the U.S. Department of Energy. The effort at Argonne National Laboratory was funded under contract DE-AC02-06CH11357.

References

1. Why circular economy business models need LCA (part 1), PRé Sustainability, https://www.pre-sustainability.com/news/why-circular-economy-business-models-need-lca, (accessed January 4, 2019).
2. Circularity Indicators, https://www.ellenmacarthurfoundation.org/resources/apply/circularity-indicators, (accessed January 23, 2019).
3. M. Wang, J. Han, J. B. Dunn, H. Cai and A. Elgowainy, *Environ. Res. Lett.*, 2012, **7**, 045905.
4. T. Searchinger, R. Heimlich, R. A. Houghton, F. Dong, A. Elobeid, J. Fabiosa, S. Tokgoz, D. Hayes and T.-H. Yu, *Science*, 2008, **319**, 1238–1240.
5. S. Tokgoz and D. Laborde, *Curr. Sustainable/Renewable Energy Rep.*, 2014, **1**, 104–110.
6. Y. Yang, J. Bae, J. Kim and S. Suh, *Environ. Sci. Technol.*, 2012, **46**, 3671–3678.
7. Z. Qin, J. B. Dunn, H. Kwon, S. Mueller and M. M. Wander, *GCB Bioenergy*, 2016, **8**, 1136–1149.
8. J. B. Dunn, S. Mueller, H. Kwon and M. Q. Wang, *Biotechnol. Biofuels*, 2013, **6**, 51.
9. M. Wang, M. Wu and H. Huo, *Environ. Res. Lett.*, 2007, **2**, 024001.
10. M. Wang, H. Huo and S. Arora, *Energy Policy*, 2011, **39**, 5726–5736.

11. S. Arora, M. Wu, M. Wang and E. Systems, *Update of Distillers Grains Displacement Ratios for Corn Ethanol Life-cycle Analysis*, Argonne National Lab. (ANL), Argonne, IL (United States), 2011.
12. R. Nielsen, *Historical Corn Grain Yields for the U.S.*, 2017.
13. M. O. de Souza Dias, R. Maciel Filho, P. E. Mantelatto, O. Cavalett, C. E. V. Rossell, A. Bonomi and M. R. L. V. Leal, *Environ. Dev.*, 2015, **15**, 35–51.
14. *USDA Foreign Agricultural Service*, Biofuels Annual, Brazil, https://www.fas.usda.gov/data/brazil-biofuels-annual-3, (accessed October 19, 2018).
15. S. J. Smith and A. Mizrahi, *Proc. Natl. Acad. Sci. U. S. A.*, 2013, **110**, 14202–14206.
16. H. Cai and M. Q. Wang, *Environ. Sci. Technol.*, 2014, **48**, 12445–12453.
17. T. de F. Cardoso, O. Cavalett, M. F. Chagas, E. R. de Morais, J. L. N. Carvalho, H. C. J. Franco, M. V. Galdos, F. V. Scarpare, O. A. Braunbeck, L. A. B. Cortez and A. Bonomi, *Sci. Agric.*, 2013, **70**, 353–360.
18. J. Sheehan, A. Aden, K. Paustian, K. Killian, J. Brenner, M. Walsh and R. Nelson, *J. Ind. Ecol.*, 2003, **7**, 117–146.
19. S. J. Birrell, D. L. Karlen and A. Wirt, *BioEnergy Res.*, 2014, **7**, 509–516.
20. D. J. Muth, K. M. Bryden and R. G. Nelson, *Appl. Energy*, 2013, **102**, 403–417.
21. Z. Wang, J. Dunn, J. Han and M. Wang, *Material and Energy Flows in the Production of Cellulosic Feedstocks for Biofuels for GREET1_2013*, https://greet.es.anl.gov/publication-feedstocks-13, (accessed January 24, 2019).
22. C. Canter, Z. Qin, H. Cai, J. Dunn and M. Wang, *Update to Herbaceous and Short Rotation Woody Crops in GREET® Based on the 2016 Billion Ton Study*, https://greet.es.anl.gov/publication-feedstocks-13, (accessed January 24, 2019).
23. D. Humbird, R. Davis, L. Tao, C. Kinchin, D. Hsu, A. Aden, P. Schoen, J. Lukas, B. Olthof, M. Worley, D. Sexton and D. Dudgeon, *Process Design and Economics for Biochemical Conversion of Lignocellulosic Biomass to Ethanol: Dilute-Acid Pretreatment and Enzymatic Hydrolysis of Corn Stover*, National Renewable Energy Lab. (NREL), Golden, CO (United States), 2011.
24. H. Cai, J. Han, M. Wang, R. Davis, M. Biddy and E. Tan, *Biofuels, Bioprod. Biorefin.*, 2018, **12**, 815–833.
25. R. Chen, Z. Qin, J. Han, M. Wang, F. Taheripour, W. Tyner, D. O'Connor and J. Duffield, *Bioresour. Technol.*, 2018, **251**, 249–258.
26. D.-S. Kim, M. Hanifzadeh and A. Kumar, *Environ. Prog. Sustainable Energy*, 2018, **37**, 7–19.
27. A. Pradhan, D. S. Shrestha, J. Van Gerpen, A. McAloon, W. Yee, M. Haas and J. A. Duffield, *Trans. ASABE*, 2012, **55**, 2257–2264.
28. H. Huo, M. Wang, C. Bloyd and V. Putsche, *Environ. Sci. Technol.*, 2009, **43**, 750–756.
29. Algal Biofuels, Department of Energy, https://www.energy.gov/eere/bioenergy/algal-biofuels, (accessed January 19, 2019).
30. H. Passell, H. Dhaliwal, M. Reno, B. Wu, A. B. Amotz, E. Ivry, M. Gay, T. Czartoski, L. Laurin and N. Ayer, *J. Environ. Manage.*, 2013, **129**, 103–111.

31. A. F. Clarens, E. P. Resurreccion, M. A. White and L. M. Colosi, *Environ. Sci. Technol.*, 2010, **44**, 1813–1819.
32. P. K. Campbell, T. Beer and D. Batten, *Bioresour. Technol.*, 2011, **102**, 50–56.
33. L. Lardon, A. Helias, B. Sialve, J.-P. Steyer and O. Bernard, *Life-cycle Assessment of Biodiesel Production from Microalgae*, ACS Publications, 2009.
34. R. Davis, A. Coleman, M. Wigmosta, J. Markham, C. Kinchin, Y. Zhu, S. Jones, J. Han, C. Canter and Q. Li, *2017 Algae Harmonization Study: Evaluating the Potential for Future Algal Biofuel Costs, Sustainability, and Resource Assessment from Harmonized Modeling*, 2018.
35. Ethanol Alternative Fuels Data Centre, *U.S. Department of Energy*, https://afdc.energy.gov/fuels/ethanol.html, (accessed February 21, 2020).
36. Annual Energy Outlook, 2019, Transportation Sector Energy Use by Fuel Type Within a Mode, U.S. Energy Information Administration, https://www.eia.gov/outlooks/aeo/data/browser/#/?id=46-AEO2019&cases=ref2019&sourcekey=0, (accessed February 21, 2020).
37. M. Wang, J. Han, Z. Haq, W. Tyner, M. Wu and A. Elgowainy, *Biomass Bioenergy*, 2011, **35**(5), 1885–1896.
38. H. Cai, T. Benavides, U. Lee, M. Wang, E. Tan, R. Davis, A. Dutta, M. Biddy, J. Clippinger, N. Grundl, L. Tao, D. Hartley, M. Roni, D. Thompson, L. Snowden-Swan, Y. Zhu and S. Jones, *Supply chain sustainability analysis of renewable hydrocarbon fuels via indirect liquefaction, ex situ catalytic fast pyrolysis, hydrothermal liquefaction, and biochemical conversion: update of the 2018 state-of-technology cases and design cases*, Argonne Technical Report: ANL/ESD-18/13, 2018, https://greet.es.anl.gov/publication-supply_renewable_hc.

CHAPTER 14

Resource Impacts of Fully Renewable Energy Systems: The Case of Metals

V. MOREAU

Ecole Polytechnique Fédérale de Lausanne, Laboratory of Environmental and Urban Economics, ENAC-IA-LEURE, Station 16, 1015 Lausanne, Switzerland
Email: vincent.moreau@epfl.ch

14.1 Introduction

In the face of climate change, the combustion of fossil fuels should end by 2050 to stand a chance of limiting global warming to 2 degrees.[1] This has multiple implications, one of which is the decarbonization of the energy system. Low-carbon energy sources, including solar, wind, hydropower, geothermal energy and biomass, are plentiful. However, renewable energy sources must be converted before they can supply energy services such as heating and cooling, and transportation, as well as electricity. The technologies necessary in this conversion process have to be manufactured and maintained, and require a flow and stock of mineral resources, in particular metals. Both the current fossil energy system, as well as the future renewable energy system rely on the extraction of resources, such that specialty metals might very well become the oil and gas of the future.[2–4] Nevertheless, the transition towards a renewable energy system has several advantages over the current burning of fossil fuels. First, renewable energy sources not only have lower carbon emission per unit of output, but they can also play an

Life Cycle Assessment: A Metric for the Circular Economy
Edited by Aiduan Borrion, Mairi J. Black and Onesmus Mwabonje

Published by the Royal Society of Chemistry, www.rsc.org

important role in capturing existing carbon in the atmosphere for synthetic fuels and storage purposes.[5] Second, while combustion changes the nature of carbon, metals necessary to manufacture renewable energy (RE) technologies conserve their properties for future uses. Through reuse and recycling, a renewable energy system can be made circular, contrary to the current linear system based on fossil fuel combustion.

An important characteristic of renewable energy systems from the perspective of mineral resource availability, is the use of specialty metals in both supply and demand side technologies. In addition to conversion technologies, the intermittency of solar and wind energy requires some form of storage, which in turn needs metals, alloys and catalysts, depending on the technology. On the demand side, hydrogen and battery electric vehicles, for example, also require specialty metals. Mineral reserves and resources are therefore coveted by industrial activities for manufacturing a range of products that either produce, store or consume energy. In this chapter, only the supply side of energy is accounted for, including storage, in terms of metal resources and the impacts of a fully renewable energy system.

From a circular economy perspective, such a renewable energy system requires an initial stock of anthropogenic materials large enough for reuse, remanufacturing or recycling strategies to be implemented.[6–8] The question, therefore, is can a renewable energy system be built given supply constraints on materials and especially metals? The reserves of metal are defined as what can be extracted economically with current technology and available energy. In terms of impacts, a commonly used measure of resource depletion, is the amount of ore known (proven, probable or potential) in the Earth's crust which become available as technology and prices evolve. The term "resource" is ambiguous here, as it describes more than natural resources, their geological concentration, such that ultimate reserves is often used instead in environmental impact assessment.[9] Moreover, these definitions mean that future reserves are found in current resources, but changes in the energy cost of extraction, for example, could also mean that current reserves become future resources.[10] The transition towards a fully renewable energy system places additional pressure on the main mineral resources, as well as on specialty metals.[11–14] Much research already exists on the metal stocks and flows associated with new technologies, including renewable energy ones.[15–19] However, none addresses the issue of metal scarcity from a comprehensive perspective; that is, by evaluating the resource impacts of the demand from base and specialty metals for multiple RE technologies. Among the key attempts, Kleijn *et al.*[20] and Vidal *et al.*[21] assess the material and mineral implications of switching from fossil fuels to low-carbon technologies. While their results show that low-carbon sources of electricity require more metals and minerals, they did not evaluate the impacts or compare them with current reserves and resources.

Finding a common indicator of resource depletion across metals and among stakeholders in the supply chain has been all but straightforward. The supply of metals depends on specific geological, physical and industrial characteristics, such that one metric of scarcity or impact does not fit all metals. Thus, the impacts are often evaluated individually and per application, such as cadmium (Cd), tellurium (Te), indium (In), gallium (Ga) or selenium (Se) for thin film solar cells, such as cadmium telluride (CdTe) and copper indium gallium selenide (CIGS).[22–25] In addition, the supply of these metals depends to a large extent on the extraction of parent metals such as copper, zinc, tin or aluminum, confounding the potential impacts.[26–28] Impact assessment suffers from multiple challenges, resistance from vested interests, confusion among analysts, lack of knowledge and data unavailability.[9] In life cycle impact assessment (LCIA), abiotic depletion potential (ADP) has served as a relatively consensual approach to evaluate resource depletion, that is, the rate of extraction divided by reserves or resources.[29] Although the issue of resource depletion in LCIA remains highly debated, other existing methods are based on the energy cost of extraction, substitution technology or exergy, to name a few.[30,31] More recent impact assessment indicators attempt to capture the criticality of certain resources, especially in the context of renewable energy technologies.[32] Depletion horizons, or the ratio of reserves or resources over future extraction, expressed as depletion year, were also calculated across base and specialty metals and showed that the supply of cadmium, cobalt (Co), lithium (Li) and nickel (Ni) are potentially limiting factors in the deployment of renewable energy systems by 2050.[33]

Criticality assessment of metals and minerals for renewable energy systems are increasingly common, for both supply and demand side technologies.[34] Specialty metals and rare earths are often the most critical elements in deploying large-scale, low-carbon technologies.[14,35–37] Yet, basic and precious metals are also affected and criticality assessment is not directly part of LCIA as it encompasses more than environmental impacts. Nevertheless, results on critical metals are relevant, such as Kavlak *et al.*,[38] who quantified the metal requirements for large-scale deployment of photovoltaics (PVs) and showed that the demand for indium (In), selenium (Se) and tellurium (Te) might constrain the thin film PV industry as early as 2030. Grandell and Thorenz[39] reached a similar conclusion for silver (Ag). Elshkaki[40] concluded, through dynamic material flow analysis, that resources of platinum (Pt) would be depleted before the end of this century.

Among the more comprehensive evaluation of metal use for RE technologies, Harmsen *et al.*[41] evaluated the impact of copper (Cu) scarcity for long-term global renewable energy scenarios (2050). Elshkaki and Graedel[15] quantified the availability of several metals used for wind turbines, PV panels, concentrated solar power (CSP), hydropower, geothermal, biomass, coal, oil, gas and nuclear power under policy- and market-based scenarios up to 2050. They concluded that the manufacturing and replacement of wind turbines may not face major resource constraints, whereas the production of PV panels is potentially more problematic. Tellurium, in particular, might become critical in terms of resource availability and production capacity for CdTe panels. Yet, this study fell short of accounting for the metal constraint of a fully renewable energy system and excludes an important subset of metals: the platinum group metals (PGMs), which might prove critical.

In this chapter, we build on the works of Vidal *et al.*[13,21] and Habib *et al.*,[42] who estimated the demand for the main metals and rare earths, respectively, in several renewable energy scenarios. We expand the analysis by quantifying the impacts of 29 metals necessary for large-scale deployment and maintenance of renewable energy technologies. Well-established scenarios of energy supply and demand up to 2050 are used, since they rely on 100% renewable energy sources, or with small shares of non-renewables. We also account for battery technologies necessary for short-term storage of intermittent sources, namely PV and wind. Finally, we simulate different combinations of renewable energy and battery storage technologies, to estimate the impact of technology mixes on metal resources. The impacts are evaluated with several life cycle impact assessment methods to stress the potential depletion of metal resources from fully renewable energy systems.[43] The current resource impacts of metal extraction can also be compared with the additional demand from renewable energy technologies to identify where bottlenecks might be most acute. We assume that the supply chain of metals are equally global as that of oil and gas, with regional disparities in resource supply and potentially new cartels.[4,44]

This chapter extends previous work on the depletion potential of metals for renewable energy systems by evaluating the impacts from a life cycle perspective.[33] In Section 14.2, we detail the methodological approach. Section 14.3 presents the results, in particular, the impacts of specific metals, and energy and technology scenarios. Section 14.4 discusses the results and we conclude in Section 14.5.

14.2 Materials and Methods

Reserves of metals are economic measures of availability and are regularly re-evaluated in the interest of continued extraction. Resources, or ultimate reserves, as they are often defined in impact assessment, are a more absolute measure of metal availability in the Earth's crust and are, therefore, more suitable for evaluating impacts across metals. Overall, the current demand for metals from renewable energy technologies can be considered marginal compared with incumbent activities. We consider that the future demand from the deployment of renewable energy technologies is essentially additional. Moreover, renewable energy technology industries will increasingly compete, at a comparative disadvantage, for the same specialty metals as electronics, aerospace and other economic activities.

The main renewable energy technologies, as well as battery energy storage, to compensate for intermittent renewables, are taken into account. Although battery electric vehicles can also store electricity temporarily, in their first or second life, only stationary battery storage is considered. As explained below, we model different technology mixes for photovoltaics, wind, biomass and storage to assess substitution opportunities for the renewable energy industry. More specifically, only intraday electricity storage requirements are taken into account and covered by five different battery chemistries.

The methodological approach can be divided into 4 steps. First, 5 scenarios of global energy supply or demand for 2050 are selected and broken down by energy sources. These scenarios account for changes in the demand for energy as a result of population and economic growth, as well as energy efficiency, and are subject to the potential of renewable sources. Second, sub-scenarios of renewable energy technologies are modeled for solar, wind and biomass, as well as for batteries. Third, we quantify the amount of each metal needed to manufacture and maintain renewable energy technologies in each of the 5 main and sub-scenarios. The life cycle inventory (LCI) of each technology and metal (in kg kWh^{-1}) is scaled according to each scenario to obtain the metal requirements. We ignore potential resource productivity gains, as the rate of changes required in the energy system to meet emissions targets by 2050 will likely outpace productivity. In the fourth and last step, the resource impacts are evaluated with established life cycle impact assessment methods based on the LCIs. The underlying assumption is that renewable energy systems are deployed linearly by 2050. Scenario combinations can easily be compared, including the impact of storage and recycling. Although recycling has little impact as demand grows and the energy system is implemented, we nonetheless included several recycling scenarios. Indeed, PV technologies, for example, have already witnessed their first generation and end of life.

14.2.1 Energy Scenarios

Two well-established scenarios estimate global energy demand based exclusively on renewables: an IPCC scenario and a scenario that results from the work of Ecofys and WWF.[45,46] Three additional scenarios are also included for their high share of renewables: the International Energy Agency (IEA) High REN and "2 degree" scenarios, as well as the International Renewable Energy Agency (IRENA) "REMAP" scenario.[47,48] All scenarios have the same time horizon of 2050, but differ in their estimates of global energy demand. They also differ in the share of each renewable (and non-renewable) energy sources, such that together they provide a measure of uncertainty in the demand for energy and thus for metals required to manufacture and maintain renewable energy technologies (see Figure 14.1).

14.2.2 Renewable Energy Technology Scenarios

The scenarios in Figure 14.1 estimate global energy demand in 2050, together with the supply per energy source. Forecasts per technology for each of the resources such as the mix of PV technologies are challenging to estimate, not only because technology improves continuously, but more importantly because a range of technologies normally co-exist, in particular for solar, wind, biomass or batteries. Given the metals required to manufacture each

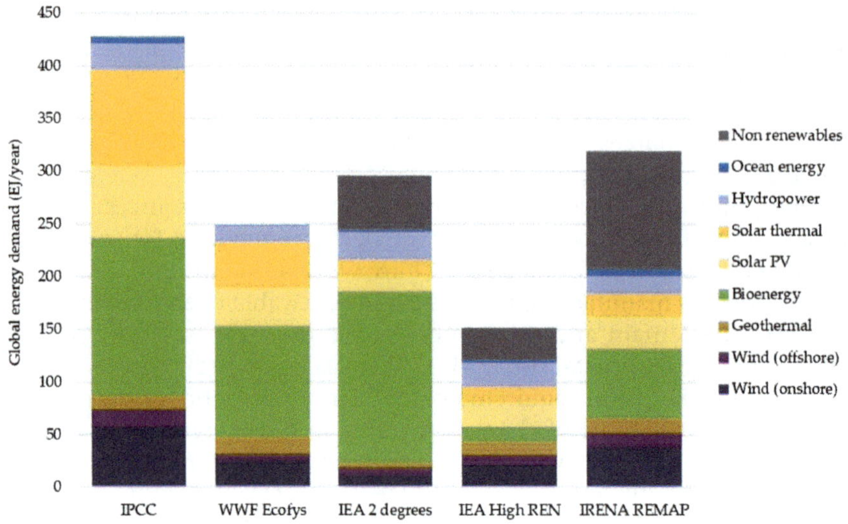

Figure 14.1 Five scenarios of global energy supply/demand in 2050 with high shares of renewables.[33]

technology, the technology mixes require different quantities and often also qualities of metals. Thus, sub-scenarios of renewable energy technologies for solar PV, biofuels and battery energy storage were needed, as explained below. All scenarios already accounted for the potential of onshore and offshore wind.

14.2.2.1 Solar PV Scenarios

The current market for photovoltaic (PV) panels is dominated by mono- and multi-crystalline silicon solar cells, for which life cycle inventory (LCI) data exist.[49] IRENA forecasts that market shares of second-generation thin film solar cells (CdTe, CIGS) will slowly increase to make up approximately 10% of the market in 2030.[50] The PV technology mix included here consists of close to 90% crystalline silicon, with multi-layered panels accounting for two thirds. The remainder is split approximately equally between CdTe and CIGS technologies, for which data also came from the ecoinvent LCI database.[51] In these inventories, the lifetime of PV panels is approximately 30 years. Thus, we concentrate on commercial PV technologies, and do not consider future technologies such as Perovskites, given the uncertainty regarding their deployment.

14.2.2.2 Biofuels

Second-generation biofuels are considered only, with the corresponding LCI data being relatively well documented. Although the impact of biomass would have to be regionalized, we only consider the metal requirements, especially catalysts, for biorefinery processes and no regional disparities. The use of biogas for transportation remains marginal, and a mix of biodiesel and ethanol dominates the market.[52] In Figure 14.1, energy from biomass, including biofuels and the generation of heat and electricity are aggregated into a single category.

14.2.2.3 Storage Technologies

Intermittent renewable energy sources, namely solar and wind, require some forms of storage to balance supply and demand. The need for storage technologies increases non-linearly with the penetration of intermittent renewables. Many storage technologies, including batteries, also require specialty metals.[37] In Europe, the electricity surplus from renewable energy sources (approximately 40 TWh per year) is either curtailed, sold to neighboring countries or stored in pumped hydro and storage (PHS) plants.[53] Storage clearly depends on the regional potential and installed capacity of solar and wind, and competes with alternatives such as reinforcement of the electric grid, standing power reserves and demand response. Therefore, in order to estimate the storage requirements regionally and globally, we derived the daily solar and wind profiles for a full year, as well as the loads of five regions (EU, US, India, Brazil and South Africa), representative of the 5 continents.[54,55] India represents Asia Pacific on its own for the lack of public data to model such profiles for China. Storage requirements were estimated by the daily surplus of renewable electricity in each region. The regionalized versions of the scenarios in Figure 14.1 were then used to scale up storage requirements. The global estimates are shown in Table 14.1. For comparison purposes, the estimates by IRENA, the Energy Watch Group and the World Bank are also included in brackets.

The estimates in Table 14.1 account for intraday electricity storage only, no seasonal or heat storage was considered. Nevertheless, they can be covered by several technologies, centralized and decentralized.[57] We assumed that 50% of the estimated storage in Table 14.1 would be covered by decentralized batteries and simulated three different mixes of five battery chemistries, as shown in Table 14.2, which include one non-lithium technology, for which LCI data are available.

NCA batteries[37,58] are based on lithium nickel cobalt aluminum technology and have lower energy densities than NCM622 technology, which substitute manganese and higher nickel content for aluminum.[59] NCM811 batteries are currently at the R&D stage and add some silicon at the anode

Table 14.1 Global storage requirements per scenario in TWh.

IPCC	WWF Ecofys	IEA 2 degrees	IEA High REN	IRENA REMAP	Energy Watch Group	World Bank
14 ± 5	7.5 ± 3	3.9 ± 2	4.8 ± 2	(12.4)[48]	(48)[56]	(6.6)[3]

Table 14.2 Battery chemistry scenarios in 2050.

Battery chemistry	MIX 1	MIX 2	MIX 3
NCA	20%	25%	20%
NCM622	0%	5%	0%
NCM811	30%	45%	80%
Zebra (NaNiCl)	50%	10%	0%
LFP	0%	15%	0%

for yet greater energy density. Zebra are sodium metal hydride batteries designed to operate at high temperatures and for long periods of discharge, and exclude Li.[51] Finally, LFP batteries are based on Li iron phosphate technology for high cycling life and safety at the expense of energy density.[59]

14.2.3 Metal Inventories

The metals embodied in each renewable energy technology were quantified based on a life cycle perspective. In LCI, products (goods and services) are modeled with respect to their functions and measured by the functional unit. In the case of RE technologies, the functional unit is 1 kWh, such that LCI data sets provide the life cycle metal requirements per kWh. Data sets were sourced from the ecoinvent LCI database version 3, which inventories 29 metals found in RE technologies.[51] The data sets are so called "cradle to gate", meaning they account not only for the metals in the technology itself, but also those used for its manufacturing, installation, end of life, as well as its installation and connection to the closest point of distribution (*e.g.* connection to the grid for wind turbines and PV panels) over the lifetime of the technology. Thus, metal intensities in kg per functional unit and technology were used to scale up to global renewable energy systems according to the energy and technology scenarios above. For technologies that are not available in the ecoinvent database (*e.g.* emerging battery chemistries), their metal inventories were based on literature references.[37,59–61]

Recycling was modeled as open loop, displacing a fraction of the primary metal supply. We tested the sensitivity of recycling rates, using four different scenarios, shown in Table 14.3: (1) the current recycling rates for each metal remain unchanged to 2050 (conservative estimate);[62] (2) a moderate and uniform 5% increase in recycling rates for each metal, between now and 2050; and (3) an additional 50% increase in recycling rates; (4) specialty metals are recycled at the current rates of their parent metals, which are assumed to remain unchanged to 2050, according to the wheel of metals by product.[63,64] Table 14.3 summarizes the recycling rates per metal.

14.2.4 Resource Impacts

The impacts of resource depletion are characterized in several ways in LCIA. A common definition of scarcity hardly exists, but essentially it means steady

Table 14.3 Recycling scenarios.

Metal	Current	+5%	+50%	Parent metal
Aluminum (Al)	50%	53%	76%	50%
Bromine (Br)	1%	6%	53%	50%
Cadmium (Cd)	10%	15%	57%	50%
Chromium (Cr)	50%	53%	76%	50%
Cobalt (Co)	50%	53%	76%	94%
Copper (Cu)	50%	53%	76%	50%
Gallium (Ga)	1%	6%	53%	50%
Gold (Au)	50%	53%	76%	50%
Indium (In)	1%	6%	53%	50%
Iron (Fe)	50%	53%	76%	50%
Lanthanum (La)	1%	6%	53%	50%
Lead (Pb)	50%	53%	76%	50%
Lithium (Li)	1%	6%	53%	81%
Magnesium (Mg)	25%	29%	64%	25%
Manganese (Mn)	50%	53%	76%	50%
Molybdenum (Mo)	25%	29%	64%	25%
Neodymium (Nd)	1%	6%	53%	25%
Nickel (Ni)	50%	53%	76%	50%
Palladium (Pd)	50%	53%	76%	50%
Platinum (Pt)	50%	53%	76%	50%
Rhenium (Re)	50%	53%	76%	50%
Rhodium (Rh)	50%	53%	76%	50%
Silver (Ag)	50%	53%	76%	50%
Tantalum (Ta)	1%	6%	53%	50%
Tellurium (Te)	1%	6%	53%	50%
Tin (SN)	1%	6%	53%	50%
Titanium (TiO2)	50%	53%	76%	50%
Zinc (Zn)	50%	53%	76%	50%
Zirconium (Zr)	1%	6%	53%	25%

decrease in global average grade of ores extracted over time and increase in prices of extracted metals, which cannot be compensated by improving and upgrading mining technologies.[65] Aside from the problem of depletion and scarcity, other impact assessment methods evaluate the potential damages from resource extraction in order to cover the full cause-and-effect chain. The improvement in mining technologies, for example, compensates for declining ore grades, but the energy cost of extraction keeps on growing, with potential impacts on renewables themselves.[66] Following the recommendations of the reference ILCD handbook, two impact assessment methods were chosen here, CML at the resource depletion level and ReCiPe at the damage level.[67]

The abiotic depletion potential (ADP) of a metal resource is the ratio of extraction rate over the square of the ultimate reserves. The reference substance in the CML assessment method is Antimony (Sb) used in deriving the ADP as in eqn (14.1).

$$\mathrm{ADP} = \frac{\text{rate of extraction}}{\text{ultimate reserves}^2} \times \frac{\text{ultimate reserves of Sb}^2}{\text{rate of extraction of Sb}} \quad (14.1)$$

The ultimate reserves are squared to capture the fact that extracting a given metal fraction from a small deposit has a larger impact than from a large deposit. A greater impact based on ADP can mean one of four cases: (1) a greater use of resources; (2) resources which will be depleted sooner than others; (3) resources with low ultimate reserves; and (4) combinations thereof. Data quality is the main limitation of these methods, as ultimate reserves and extraction rates tend to have high uncertainties. CML characterization factors are updated, as are the reserves and resources data available from the US Geological Survey (USGS).[68] The USGS defines *resources* as the "concentration of naturally occurring solid, liquid, or gaseous material in or on the Earth's crust in such form and amount that economic extraction of a commodity from the concentration is currently or potentially feasible". Where USGS estimates are aggregated as for rare earths and PGMs, literature references also exist.[69]

ReCiPe also quantifies abiotic depletion potentials using copper as the reference substance. This method also goes one stage further by quantifying the potential damages in monetary terms, that is, the additional cost for society to extract a given quantity of resources. One of two key features of ReCiPe is to integrate value choices in impact assessment through an egalitarian, a hierarchical and an individualistic perspective.[70] For example, the individualist takes reserves in order to calculate depletion potentials, whereas the hierarchist and the egalitarian consider ultimate resources. A second feature of ReCiPe is to consider the extraction of ores, which typically contain several metals, as in parent and child metals. In terms of copper equivalents or at the damage level, a higher ReCiPe score means that the product (good or service) has more impact over its life cycle on metal depletion or on the marginal cost of extraction.

Both methods are applied in this chapter to evaluate the depletion of metal resources due to the deployment of large-scale renewable energy technologies. The resulting scores are compared, per scenario but also per metal, and together provide a measure of uncertainty. A linear rate of deployment between now and 2050 means a steep growth in the implementation of renewables, and we assume that productivity gains are essentially added value past 2050, given the long lead time and scaling of near-future technologies.

14.3 Results

The results show that the additional pressure on metal resources varies per scenario and might set an upper limit to both metal extraction and the implementation of renewable energy capacity. In this section, the results are discussed first per scenario and a more detailed analysis at the metal level is presented second.

14.3.1 Resource Impacts of Renewable Energy Scenarios

The resource impacts of the extraction of metals required for deploying renewable energy systems are evaluated with the characterization factors of

ReCiPe and CML. At the resource level, once the impact scores are in copper or antimony equivalents, they can be added across metals to yield a score per scenario. Given four recycling scenarios, four levels of storage and three mixes of battery chemistries, displaying all combinations is somewhat cumbersome and not necessary, as shown here. In Figure 14.2a and b, we emphasize the impact of storage with the base case that does not include any forms of battery energy storage, and one of the extreme cases where all battery chemistries are represented (Mix 2 in Table 14.2) at a high level of storage. The scores of four different methods are shown in relative terms, three perspectives for ReCiPe, I for individualist, H for hierarchist and E for egalitarian, as well as CML scores. Comparison between impact assessment methods at the resource level is only meaningful in relative terms, since the absolute scores for ReCiPe and CML differ by about three orders of magnitude.

The highest impact, with and without storage, clearly lies with the IPCC scenario, which is fully renewable, as well as assuming a high energy demand in 2050, almost 430 EJ per year. Without storage, the WWF Ecofys and IEA 2-degree scenarios have the same impact with ReCiPe, but the latter has almost half the impact of the former, according to CML. Outside the IPCC scenario, the other four are within a similar range across all assessment methods. In absolute terms, there is little difference in impact between the value choices in ReCiPe, even if the individualist perspective fares slightly better. Although the lowest impact scenarios are the IEAs, only one of them has the lowest energy demand forecast for 2050. One of the least metal-intensive renewable energy technologies is biomass conversion, which covers almost half of the energy demand in the IEA's 2-degree scenario. In other words, the resource impacts of energy scenarios are related more to the choice of technology and necessary resources than their total energy supply.

Now the ranking among scenarios changes slightly when intraday battery energy storage is factored in. Suddenly, the IRENA REMAP scenario has more impact that all others, except for IPCC. This illustrates the potential impacts of batteries on resource depletion, as IRENA estimated relatively high storage requirements. Yet, no significant differences can be observed in the CML assessment, which means that ReCiPe places additional impact on metals that are commonly associated with batteries. Indeed, the characterization factors for lithium vary significantly. The lowest impact scenario with battery energy storage becomes the IEA 2 degrees, as evaluated with CML. Again, the share of biomass in its energy mix contributes to the reduction in the depletion potential of metal resources and products of biorefineries are typically liquids that do not need batteries for storage.

Impacts at the damage level, from the extraction of metal resources to deploy a fully renewable energy system can be evaluated with ReCiPe's characterization factors in monetary terms (USD2013). In Figure 14.3, the results are given in absolute terms (Figure 14.3a) and relative terms (Figure 14.3b) for the base case without storage.

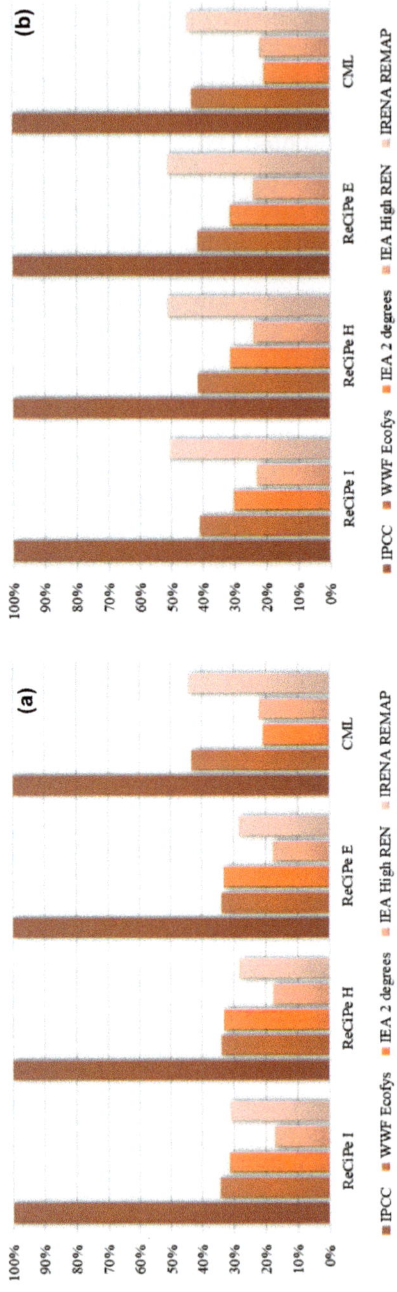

Figure 14.2 Resource depletion impacts in relative terms for all scenarios without storage (a) and with a high level of storage (b).

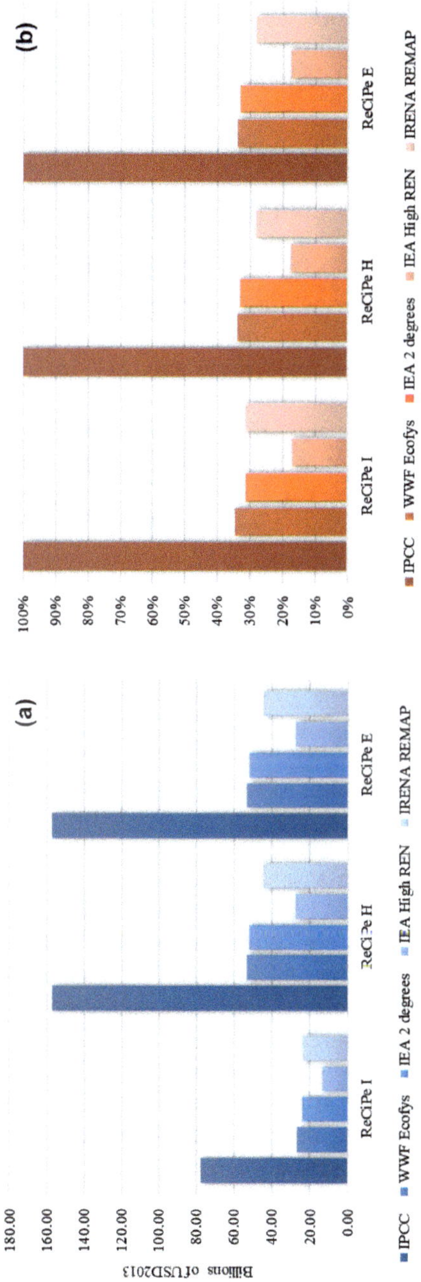

Figure 14.3 Damage assessment for all scenarios using ReCiPe, in absolute (a) and relative (b) terms.

The individualist perspective on damage assessment is limited compared to that of the other two perspectives, with about half the costs as shown in Figure 14.3a. But in relative terms (Figure 14.3b), the damages change very little across value choices. Moreover, the results per damage follow a similar trend as in Figure 14.2, with the IPCC scenario having a much larger impact than all others and storage also increasing the damage costs proportionally.

14.3.2 Disaggregated Impact Scores

To explain the contributions from the 29 metals inventoried in the assessment of renewable energy technologies, we can disaggregate the impact scores of the different scenario. ReCiPe and CML differ in that the first uses Cu as the reference substance, while the latter uses Sb. The most remarkable result shown in Figure 14.4 is that impact scores per metal vary little across scenarios for a given impact assessment method, in spite of the different technology mixes. For the sake of clarity, the scale in Figure 14.4 is log transformed and the impacts according to ReCiPe Hierarchist (Figure 14.4a) and CML (Figure 14.4b) are shown. The four metals with the largest scores, as evaluated by ReCiPe are Co, Cu, Fe and Ni, with Al and Ag not far behind. In the case of CML, the top four scores are for Cd, Cu, Au and Ag, with Pb and Te close behind. This confirms existing research that Cd, Co, Cu and precious metals Au and Ag are among those that could potentially limit the large-scale deployment of renewable energy technologies.

Where results are missing or incomplete, characterization factors are missing, such as Br for ReCiPe and the rare-earth elements (lanthanum, neodymium, rhodium) for CML. This disaggregation of the results at the metal level shows the importance of technology mixes, and that substitution of metals or technologies might reduce overall impacts. When adding a high level of battery energy storage with the second mix of battery chemistries (Table 14.2), the main difference in impact comes from the extraction of Li, which shows up high in both assessment methods. Therefore, a small set of transition metals including Li might constrain the implementation of a global renewable energy system. While the impacts of the main metals Fe, Al, and Cu are high, in many OECD countries their anthropogenic stocks have peaked, at approximately 12 tones *per capita* for Fe, when secondary steal (recycling) becomes more economic than primary sources.[71] The anthropogenic stocks necessary for secondary sources of metals to satisfy a significant share of the demand, from incumbent activities, as well as renewable energy technologies, may never be achieved at the predicted level of energy consumption and/or mix of technologies.

14.4 Discussion

Based on the results for energy scenarios and metals, the role of the technology mix cannot be underestimated. The choice of a high- *versus* low-tech energy system might make or break the emission reduction targets, as well

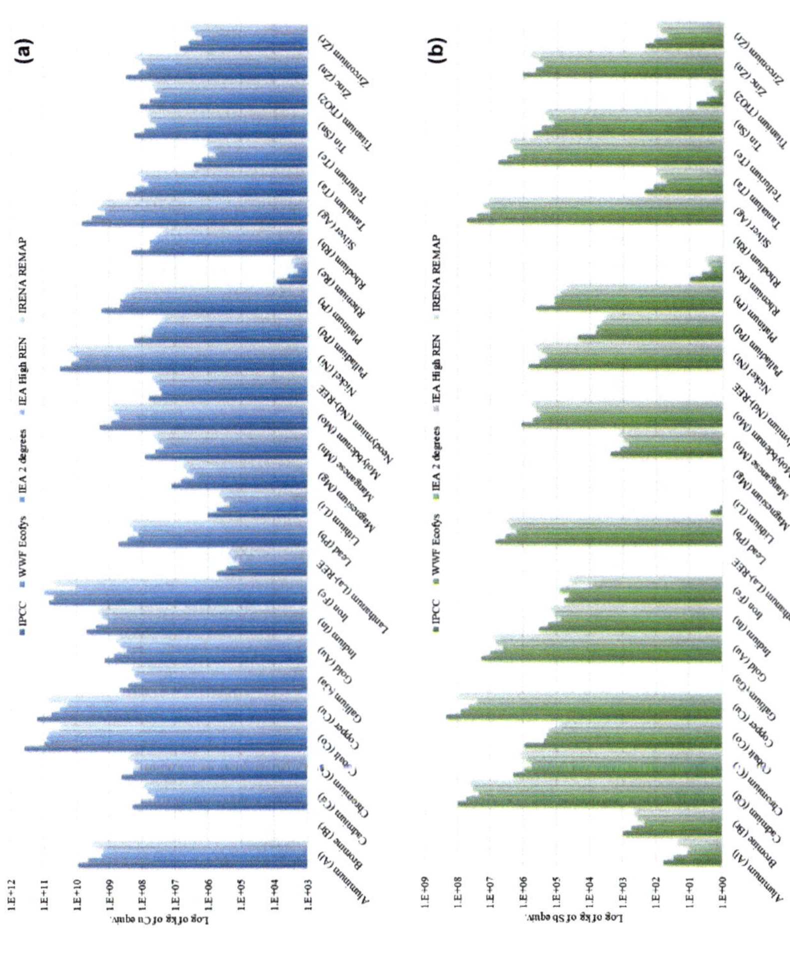

Figure 14.4 Impact scores on a log scale per metal and scenario in ReCiPe H (a) and CML (b).

as resource depletion impacts. The optimal energy system would therefore be decarbonized but also "decadmiumized" or "decobaltized" to limit the impacts on resource depletion. In absolute terms, the deployment of renewable energy technologies is unlikely to deplete metal resources up to 2050. Metal productivity gains, as well as substitutions at the technology and metal level, should become viable technically and economically before the ore grades decline and the energy costs of extraction rise.[66] However, the additional impact is non-negligible and the renewable energy industry will have to compete with the demand for specialty metals from incumbent activities, such as electronics and aerospace.

In this sense, the results are most relevant from a relative point of view, as is often the case in life cycle assessment. Indeed, informed technological choices can make a large difference, particularly when it comes to (renewable) energy technologies.[72] The identification of metals that are likely to become scarce first contributes to prioritizing research and development and industry efforts towards substitutions of technologies and/or metals *per se*. Alternative technologies can come at the expense of energy efficiency. For instance, asynchronous motors without permanent magnets in the rotors of wind turbines exist but are slightly less efficient than conventional motors. Similarly, replacing silver with copper as conductive elements in crystalline silicon solar cells would slightly reduce their conversion efficiency. Given that technological choices in energy lock us in for decades, especially in the combustion of fossil fuels, implementing a renewable energy system should take advantage of the multiple technological options now and into the future.

In the context of decarbonization to mitigate climate change, a fully renewable energy system could provide as much energy as possible, since its carbon intensity is significantly lower than the current fossil-fueled system. The results show that, even if the energy system was fully renewable, the predicted level of energy consumption in 2050 remains higher than the supply constraints that some metals can absorb. In other words, the energy transition should be towards low cadmium or low cobalt, as well as low carbon. Shifting from a fossil-based to a RE-based system does not alleviate the problem of resource depletion, it merely shifts it from fuel to metal. Nevertheless, metals are more durable in all applications than fuels in combustion, and a renewable energy system must precede a circular economy, but could also encourage reuse, remanufacture and recycling in the energy sector.

The results may therefore prove particularly relevant in steering applied research and development in line with resource constraints over the medium and long term.[73] Concentrating our efforts towards metal and/or technology substitutions is a priority, since many alternatives that do not rely on scarce metals already exist. At the top of the list of priorities, lower-tech alternatives to cadmium and cobalt intensive energy technologies should be developed. Equally important according to the results is the dependency of energy technologies on precious metals, in particular gold and silver. Stocks in use and recycling rates are high but other uses are competing for them at a comparative advantage. From a research point of view, integrating the anthropogenic stocks along with geogenic ones in the calculation of abiotic depletion potentials would most likely change the resource depletion impact of precious metals for the better.

Battery electricity storage has a significant impact on the usual suspect, lithium, as well as cobalt and nickel. While storage itself compounds the impacts of renewable energy scenarios, alternatives to batteries exist. These alternatives would certainly not make batteries obsolete, but would lower their demand to a level that might be adequate for the availability of resources. Moreover, a large and well-interconnected grid might reduce the intermittency issue on days when a region is sunny during the day and windy at night. Other forms of storage, such as power to gas and a larger share of biomass in the energy mix significantly lowers the impacts on resource depletion, as shown by the energy scenarios above, which have proportionally less electricity-generating technologies.

Finally, this resource depletion assessment focused exclusively on metals for supply-side technologies, neglecting the metal requirements to manufacture demand-side technologies such as electric vehicles, fuel cells as well as energy efficiency measures. As part of the demand side, a more accurate assessment is also needed to account for the demand for specialty metals from non-energy industries. Moreover, lifetime and maintenance of a fully renewable energy system beyond 2050 is another important research question.

14.5 Conclusions

Unlike individual scenarios, technologies or resources, this chapter takes a broad perspective on the resource depletion impacts of renewable energy systems. Out of 5 scenarios and 29 metals necessary in the lifecycle of 20 renewable energy technologies, the resources depletion impacts were both foreseeable and somewhat unexpected. The resource impacts of renewable energy scenarios are not directly proportional to their predicted level of energy supply in 2050, which leaves room for multiple technology mixes. At the metal level, the impacts are highest for a small set of transition metals, namely, cadmium, cobalt, lithium, nickel and the main metals iron, aluminum and copper. The renewable energy industry would likely mobilize a significant additional share of the demand for these four specialty metals compared to the global demand from incumbent industries.

The comparison between impact assessment methods mostly concurs. In absolute terms, there are large differences at the metal and scenario levels. Moreover, the individualist perspective in ReCiPe values more the present and yields comparatively lower impacts than the other hierarchist and egalitarian perspectives. In conclusion, the resource depletion impacts of the deployment of an energy system based exclusively on renewables are significant but offer many options per energy mixes, technologies and their metal ingredients. While past research and development efforts have focused on improving the efficiency of energy conversion technologies (*e.g.* solar cells, wind turbines), which generated a growing reliance on specialty metals, optimizing low-carbon technology mixes with respect to resource depletion impacts also lowers the risk of future resource scarcity. This also increases the potential for circular economy strategies, although the anthropogenic stocks of many metals remain far too small for secondary sources to supply a significant share of demand. The fossil fuel equivalent of our remaining carbon budget would therefore be better spent on the extraction of metals required for renewable energy technologies.

References

1. IPCC, *Global warming of 1.5°, Special report*, 2018.
2. O. Vidal, *Mineral Resources and Energy: Future Stakes in Energy Transition*, Elsevier, 1st edn., 2017.
3. D. L. P. Arrobas, K. L. Hund, M. S. Mccormick, J. Ningthoujam and J. R. Drexhage, *The Growing Role of Minerals and Metals for a Low Carbon Future*, The World Bank, 2017.
4. R. Moss, P. Willis, E. L. Tercero, E. Tzimas, J. Arendorf, P. Thompson, A. Chapman, N. Morley, E. Sims and R. Bryson, *Critical Metals in the Path Towards the Decarbonisation of the EU Energy Sector: Assessing Rare Metals as Supply-chain Bottlenecks in Low-carbon Energy Technologies*, Publications Office of the EU, 2013.
5. F. D. Meylan, V. Moreau and S. Erkman, *J. CO2 Util.*, 2015, **12**, 101–108.
6. R. U. Ayres, *Ecol. Econ.*, 1999, **29**, 473–483.
7. E. Cohen-Rosenthal, *J. Clean. Prod.*, 2004, **12**, 1111–1123.
8. V. Moreau, M. Sahakian, P. van Griethuysen and F. Vuille, *J. Ind. Ecol.*, 2017, **21**, 497–506.
9. J. A. Drielsma, R. Allington, T. Brady, J. Guinée, J. Hammarstrom, T. Hummen, A. Russell-Vaccari, L. Schneider, G. Sonnemann and P. Weihed, *Resources*, 2016, **5**, 12.
10. R. W. Scholz and F.-W. Wellmer, *Glob. Environ. Change*, 2013, **23**, 11–27.
11. V. Smil, *Energy Transitions: History, Requirements, Prospects*, Praeger, 2010.
12. A. Reller, V. Zepf and B. Achzet, in *Factor X: Re-source – Designing the Recycling Society*, ed. M. Angrick, A. Burger and H. Lehmann, Springer, Netherlands, Dordrecht, 2013, pp. 203–219.
13. O. Vidal, B. Goffe and N. Arndt, *Nat. Geosci*, 2013, **6**, 894–896.
14. L. Grandell, A. Lehtilä, M. Kivinen, T. Koljonen, S. Kihlman and L. S. Lauri, *Renew. Energy*, 2016, **95**, 53–62.
15. A. Elshkaki and T. E. Graedel, *J. Clean. Prod.*, 2013, **59**, 260–273.
16. E. Alonso, A. M. Sherman, T. J. Wallington, M. P. Everson, F. R. Field, R. Roth and R. E. Kirchain, *Environ. Sci. Technol.*, 2012, **46**, 3406–3414.
17. L. T. Peiró, G. V. Méndez and R. U. Ayres, *Environ. Sci. Technol.*, 2013, **47**, 2939–2947.

18. T. E. Graedel, *Annu. Rev. Mater. Res.*, 2011, **41**, 323–335.
19. S. Hoenderdaal, L. Tercero Espinoza, F. Marscheider-Weidemann and W. Graus, *Energy*, 2013, **49**, 344–355.
20. R. Kleijn, E. van der Voet, G. J. Kramer, L. van Oers and C. van der Giesen, *Energy*, 2011, **36**, 5640–5648.
21. O. Vidal, H. Le Boulzec and C. François, *EPJ. Web Conf.*, 2018, DOI: 10.1051/epjconf/201818900018.
22. B. A. Andersson, C. Azar, J. Holmberg and S. Karlsson, *Energy*, 1998, **23**, 407–411.
23. A. Feltrin and A. Freundlich, *Renew. Energy*, 2008, **33**, 180–185.
24. V. Fthenakis, *Renew. Sustain. Energy Rev.*, 2009, **13**, 2746–2750.
25. C. Wadia, A. P. Alivisatos and D. M. Kammen, *Environ. Sci. Technol.*, 2009, **43**, 2072–2077.
26. M. Frenzel, R. Tolosana-Delgado and J. Gutzmer, *Resour. Policy*, 2015, **46**, 45–58.
27. N. T. Nassar, T. E. Graedel and E. M. Harper, *Sci. Adv.*, 2015, DOI: 10.1126/sciadv.1400180.
28. M. Frenzel, C. Mikolajczak, M. A. Reuter and J. Gutzmer, *Resour. Policy*, 2017, **52**, 327–335.
29. L. van Oers and J. Guinée, *Resources*, 2016, **5**, 16.
30. J. Dewulf, M. E. Bösch, B. D. Meester, G. V. der Vorst, H. V. Langenhove, S. Hellweg and M. A. J. Huijbregts, *Environ. Sci. Technol.*, 2007, **41**, 8477–8483.
31. T. Sonderegger, J. Dewulf, P. Fantke, D. M. de Souza, S. Pfister, F. Stoessel, F. Verones, M. Vieira, B. Weidema and S. Hellweg, *Int. J. Life Cycle Assess.*, 2017, **22**, 1912–1927.
32. N. Adibi, Z. Lafhaj, M. Yehya and J. Payet, *J. Clean. Prod.*, 2017, **165**, 1517–1528.
33. V. Moreau, P. C. Dos Reis and F. Vuille, *Resources*, 2019, **8**, 29.
34. T. E. Graedel and B. K. Reck, *J. Ind. Ecol.*, 2016, **20**, 692–699.
35. K. Roelich, D. A. Dawson, P. Purnell, C. Knoeri, R. Revell, J. Busch and J. K. Steinberger, *Appl. Energy*, 2014, **123**, 378–386.
36. E. A. Olivetti, G. Ceder, G. G. Gaustad and X. Fu, *Joule*, 2017, **1**, 229–243.
37. J. F. Peters and M. Weil, *Resources*, 2016, **5**, 46.
38. G. Kavlak, J. McNerney, R. L. Jaffe and J. E. Trancik, *Energy Environ. Sci.*, 2015, **8**, 1651–1659.
39. L. Grandell and A. Thorenz, *Renew. Energy*, 2014, **69**, 157–165.
40. A. Elshkaki, *Resour. Policy*, 2013, **38**, 241–251.
41. J. H. M. Harmsen, A. L. Roes and M. K. Patel, *Energy*, 2013, **50**, 62–73.
42. K. Habib and H. Wenzel, *J. Clean. Prod.*, 2014, **84**, 348–359.
43. J. Lieberei and S. H. Gheewala, *Int. J. Life Cycle Assess.*, 2017, **22**, 185–198.
44. H. Lund and B. V. Mathiesen, *Energy*, 2009, **34**, 524–531.
45. D. Arvizu, T. Bruckner, H. Chum, O. Edenhofer, S. Estefen, A. Faaij, M. Fischedick, G. Hansen, G. Hiriart, O. Hohmeyer, K. G. T. Hollands, J. Huckerby, S. Kadner, Ã. Killingtveit, A. Kumar, A. Lewis, O. Lucon, P. Matschoss, L. Maurice, M. Mirza, C. Mitchell, W. Moomaw, J. Moreira,

L. J. Nilsson, J. Nyboer, R. Pichs-Madruga, J. Sathaye, J. L. Sawin, R. Schaeffer, T. A. Schei, S. Schlömer, K. Seyboth, R. Sims, G. Sinden, Y. Sokona, C. von Stechow, J. Steckel, A. Verbruggen, R. Wiser, F. Yamba and T. Zwickel, in *IPCC Special Report on Renewable Energy Sources and Climate Change Mitigation*, ed. O. Edenhofer, R. Pichs-Madruga, Y. Sokona, K. Seyboth, P. Matschoss, S. Kadner, T. Zwickel, P. Eickemeier, G. Hansen, S. Schlamer and C. von Stechow, Cambridge University Press, Cambridge, United Kingdom and New York, NY, USA, 2011.
46. Ecofys, *The Ecofys Energy Scenario*, WWF Ecofys OMA, Switzerland, 2011.
47. IEA, *Energy Technology Perspectives*, International Energy Agency, Paris, 2016.
48. IRENA, *Global Energy Transformation – A Roadmap to 2050*, International Renewable Energy Agency, Abu Dhabi, 2018.
49. Fraunhofer ISE, *Photovoltaics Report*, Fraunhofer Institute for Solar Energy, 2018.
50. S. Weckend, A. Wade and G. Heath, *End-of-life Management: Solar Photovoltaic Panels*, IRENA, 2016.
51. G. Wernet, C. Bauer, B. Steubing, J. Reinhard, E. Moreno-Ruiz and B. Weidema, *Int. J. Life Cycle Assess.*, 2016, **21**, 1218–1230.
52. A. Bauen, G. Berndes, M. Junginger, M. Londo, F. Vuille, R. Ball, T. Bole, C. Chudziak, A. Faaij and H. Mozaffarian, *Bioenergy: A Sustainable and Reliable Energy Source. A Review of Status and Prospects*, 2009, p. 33.
53. F. D. Meylan, V. Moreau and S. Erkman, *Energy Policy*, 2016, **94**, 366–376.
54. ENTSOE, *Hourly RE Generation Data*, European Network of Transmission System Operators for Electricity, Brussels, 2018.
55. US EIA, *Hourly Load Data*, US Energy Information Administration, Washington, DC, 2018.
56. M. Ram, D. Bogdanov, A. Aghahosseini, A. S. Oyewo, A. Gulag, M. Child, H.-J. Fell and C. Breyer, *Global Energy System based on 100% Renewable Energy – Power Sector*, Lappeenranta University of Technology and Energy Watch Group, Lapeeranta and Berlin, 2017.
57. G. Pleßmann, M. Erdmann, M. Hlusiak and C. Breyer, *Energy Procedia*, 2014, **46**, 22–31.
58. M. Hiremath, K. Derendorf and T. Vogt, *Environ. Sci. Technol.*, 2015, **49**, 4825–4833.
59. G. Majeau-Bettez, T. R. Hawkins and A. H. Strømman, *Environ. Sci. Technol.*, 2011, **45**, 4548–4554.
60. M. Weil, S. Ziemann and J. Peters, in *Behaviour of Lithium-Ion Batteries in Electric Vehicles: Battery Health, Performance, Safety, and Cost*, ed. G. Pistoia and B. Liaw, Springer International Publishing, Cham, 2018, pp. 59–74.
61. IEA, *Global EV Outlook 2018 – Towards Cross-modal Electrification*, International Energy Agency, Paris, 2018.
62. T. E. Graedel, J. Allwood, J.-P. Birat, M. Buchert, C. Hagelüken, B. K. Reck, S. F. Sibley and G. Sonnemann, *J. Ind. Ecol*, 2011, **15**, 355–366.

63. K. Binnemans, P. T. Jones, B. Blanpain, T. Van Gerven, Y. Yang, A. Walton and M. Buchert, *J. Clean. Prod.*, 2013, **51**, 1–22.
64. M. Reuter, C. Hudson, A. van Schaik, K. Heiskanen, C. Meskers and C. Hagelüken, *Metal Recycling Opportunities, Limits, Infrastructure. A Report of the Working Group on the Global Metal Flows to the International Resource Panel*, UNEP, 2013.
65. A. M. Bradshaw and T. Hamacher, *ChemSusChem*, 2012, **5**, 550–562.
66. F. Fizaine and V. Court, *Ecol. Econ.*, 2015, **110**, 106–118.
67. *ILCD Handbook: General Guide for Life Cycle Assessment – Provisions and Action Steps*, 2010.
68. USGS, *Mineral Commodity Summaries*, 2018.
69. V. C. Valero and A. V. Delgado, *Thanatia: The Destiny of the Earth's Mineral Resources: a Thermodynamic Cradle-to-cradle Assessment*, World Scientific, 2015.
70. M. A. J. Huijbregts, Z. J. N. Steinmann, P. M. F. Elshout, G. Stam, F. Verones, M. Vieira, M. Zijp, A. Hollander and R. van Zelm, *Int. J. Life Cycle Assess.*, 2017, **22**, 138–147.
71. J. Allwood and J. Cullen, *Sustainable Materials with Both Eyes Open*, UIT, Cambridge, UK, 2012.
72. A. Månberger and B. Stenqvist, *Energy Policy*, 2018, **119**, 226–241.
73. A. Zuser and H. Rechberger, *Resour. Conserv. Recycl.*, 2011, **56**, 56–65.

CHAPTER 15

Linking Life Cycle Assessment to the Wider Circular Economy

S. J. MCLAREN

Massey University, School of Agriculture and Environment, Tennent Drive, Palmerston North 4474, New Zealand
Email: s.mclaren@massey.ac.nz

15.1 Introduction

The question of how to integrate consideration of environmental aspects into decision-making received increasing attention from the 1970s onwards, as people became concerned about resource depletion and pollution of the natural environment. In the 1980s, the publication of "Our Common Future"[1] led to widespread interest in sustainable development, including the need for a greater focus on environmental sustainability. This led to a proliferation of environmental assessment tools over the next couple of decades, as scientists investigated different ways of addressing the needs of decision-makers for environmental information in situations with (often) limited data availability. These assessment tools were supported by the development of increasingly sophisticated models to predict the environmental impacts associated with different proposed human activities; examples include: the European Union System for the Evaluation of Substances (EUSES) for toxicity assessment; the Regional Acidification Information and Simulation (RAINS) model for assessing acid rain; and environmentally extended multi-regional input-output analysis (EE-MRIO) databases for modelling the environmental impacts of international trade. At the same time, other scientists and environmental activists were working on how to conceptualise more sustainable societies, and they advanced a number of concepts that gained traction

Life Cycle Assessment: A Metric for the Circular Economy
Edited by Aiduan Borrion, Mairi J. Black and Onesmus Mwabonje

Published by the Royal Society of Chemistry, www.rsc.org

amongst the wider population. These included industrial ecology, with its focus on "closing the loop" for flows of materials and energy in industrial ecosystems; producer responsibility and product stewardship, where the manufacturer retains responsibility for its products throughout their life cycle from cradle to grave; the food miles concept, which asserts that more local production and consumption food systems are associated with reduced environmental and social impacts; and the precautionary principle, which states that if there is insufficient information about possible harm to people and/or the environment from a proposed action, then steps should be taken to avoid or reduce that possible harm.

The distinction between concepts and environmental assessment tools is an important one. A concept provides us with guideposts for how to think about a problem situation. It is closely linked with the social sciences idea of framing, which Rein and Schön[2] describe as, "a perspective from which an amorphous, ill-defined, problematic situation can be made sense of and acted on". Many environmental issues can be described as arising from amorphous and ill-defined problem situations, and hence sustainability concepts are particularly important for helping us to make sense of these situations. Rein and Schön[2] go on to explain, "Framing is a way of selecting, organising, interpreting, and making sense of a complex reality to provide guideposts for knowing, analysing, persuading, and acting." Thus, sustainability concepts provide the guideposts within which we both develop, and then choose to use, one or more environmental assessment tools to provide quantitative information to support decision-making. This means that there are usually concepts and associated assessment tools that go hand-in-hand, because they are based on the same framing of decision situations. For example, the life cycle thinking concept is associated with Life Cycle Assessment (LCA); the precautionary principle is associated with risk assessment; and industrial ecology is associated with Material Flow Analysis (MFA).

This chapter, then, is concerned with exploring the use of concepts in environmental sustainability with particular reference to one concept: the circular economy, which has gained significant attention and traction amongst both government and industry decision-makers in recent years. As there are many interpretations of the circular economy, this chapter draws particularly on the circular economy concept and its application, as expounded by McDonough and Braungart[3,4] in their cradle-to-cradle approach, and the Ellen MacArthur Foundation.[5,6]

The use of the circular economy concept is compared and contrasted with the use of LCA, an environmental assessment tool developed at the end of the 1980s, which is now widely used for assessing the environmental impacts of products, processes, activities, organisations, and policies.[7] As Chapter 3 provides an overview of LCA methods and their application, this chapter does not introduce LCA methodology, but assumes prior knowledge of the approach. Section 15.2 of this chapter discusses the role of concepts and tools in supporting decision-making.

The rest of the chapter is organised around a number of common fundamental characteristics that have shaped the development of environmental sustainability concepts and assessment tools. These characteristics are discussed in Section 15.3, focusing on how each one has been interpreted in the circular economy concept and in LCA. Based upon this analysis, recommendations are made about the use of the circular economy concept and LCA in Section 15.4.

15.2 Concepts and Tools in Environmental Sustainability

Regarding the distinction between concepts and tools, as introduced in Section 15.1, a concept can be regarded as "an idea about how to achieve sustainability", whilst a tool, in the context of environmental sustainability, is defined as "an approach that typically consists of a systematic step-by-step procedure and a mathematical model".[8]

The term circular economy has become popular recently, but actually is based on an idea of "closing the loop" for materials in the economy that has been in use for many years,[6,9,10] as noted in Chapter 2. Concepts that have made use of this idea in the past include: industrial ecology, industrial ecosystems, cleaner production, product-service systems, eco-efficiency, environmental economics, the performance economy, biomimicry, cradle-to-cradle design, and regenerative design.[10,11] All these concepts, including the circular economy concept, posit the potential environmental (and economic in some cases) benefits of increased reuse, refurbishment, remanufacture, repair, and recycling of materials and/or products. The more recent, and explicit, descriptions of the circular economy also include a requirement for renewable energy sources to be integral to closed-loop systems. Thus, for example, the Ellen MacArthur Foundation states that one of the principles of the circular economy is that it runs on renewable energy.[5]

The idea of having closed loops for materials in the economy, that are powered by renewable energies, is easily understood. However, what is less well understood is whether these closed-loop systems do actually deliver environmental benefits in all circumstances, and the nature of these environmental benefits. In the case of the circular economy, researchers have also questioned whether these systems can deliver the economic benefits that are asserted by their proponents.[12]

The critical analysis of the circular economy has centred on the following observations:[10]

1. Following the second law of thermodynamics, all economic processes lead to some waste and byproducts, and so it is not possible to have a 100% closed-loop system (after Georgescu-Roegen);[13] note that his work was later updated by others to distinguish between energy and materials, recognising that it is not possible to close the loop for useful

energy (exergy) but (arguably) it could theoretically be closed for materials given an infinite supply of solar energy.[14]
2. The world economy is expanding and so more materials will inevitably need to be brought into the economy, to account for the increase in stocks, such as buildings and other infrastructure. These stocks act as long-term sinks for many materials, and so these materials cannot be recycled/reused within short timeframes.[15]
3. Current economic systems are based on a need for continuous economic growth, and this is delivered through increased consumption of goods and services. Therefore, additional materials will have to be brought into the economy to account for the increased consumption.[15]
4. Secondary use of materials leads to cheaper products, and this causes a rebound effect, whereby people have more money and so they buy more goods, leading to increased overall consumption, and hence higher demand for materials in the economy.[10]
5. A focus on developing a circular economy distracts attention from other approaches that may be more relevant in pursuit of a sustainable future for humanity. In particular, priority should be given to approaches that focus on stabilising, or even reducing, the volume and material composition of products in the economy[16] and/or the inflow of materials into the economy.[17,18]

A counterargument to observations (1) and (2) is that the circular economy concept is not meant to be literally interpreted as having a goal of 100% closed-loop systems for materials, but instead its value lies in providing a general goal and direction for restructuring economies. The implication is that more work needs to be done to identify optimal residence times of different materials in the economy. However, observations (3), (4) and (5) are less easy to refute. Indeed, they can be generalised to a concern about whether current economic systems are actually capable of supporting more sustainable societies, or whether we need to be investigating alternative economic systems that are aligned with sustainability goals, including facilitation of closed loops for materials. The discipline of ecological economics is concerned with this research agenda, and this aspect has received more widespread attention recently with the publication of *Doughnut Economics*.[19]

Korhonen *et al.*,[11] drawing on the work of Gallie,[20] suggest that the circular economy is therefore an "essentially contested concept" (ECC), where an ECC means that "there is agreement on the means and goals of a concept but disagreements on how to define it, which units of analyses to use to capture the dynamism, what the conceptual cornerstones are and what methodology of enquiry is appropriate". Of course, this "fuzziness" in the circular economy concept has catalysed ongoing research on its legitimacy and interpretation in different contexts. As such, it may be argued that its fuzziness is actually its strength and that, in stimulating us to consider different ways of framing and managing our economic systems, it is doing a great job!

LCA, in contrast to the circular economy concept, is an environmental assessment tool with a procedure and accompanying quantitative model that has been (fairly) precisely defined over the last thirty years. It is closely linked with the concept of life cycle thinking, whereby economic activities are considered along their full life cycles from initial extraction of raw materials, through refining, manufacture, distribution, use and on to end-of-life management. Its role in supporting environmental sustainability is clear: it provides quantitative environmental information on the environmental impacts of alternative systems using a life cycle perspective. Of course, there are ongoing debates about aspects of the methodology, such as attributional *versus* consequential modelling, allocation approaches, and so on, but these debates occur within the context of an agreed overarching methodology and purpose for this analytical tool.

The historical development and use of LCA has perhaps been more conventional than the circular economy concept, in that it has occurred hand-in-hand with a closely related concept *i.e.* life cycle thinking. The circular economy concept, on the other hand, emerged without an accompanying analytical tool to assess progress towards its environmental goals. Thus there is ongoing research into which analytical tools are best used to support its implementation[21] (Chapter 3). This book contributes to the discussion by exploring the use of LCA as an analytical tool in conjunction with the circular economy concept.

15.3 Some Fundamental Characteristics in Framing of Environmental Problems

A number of fundamental characteristics of environmental sustainability approaches are significant in determining the types of decision support that can be provided by these approaches. As well as the distinction between concepts and tools discussed in Section 15.2, they include: a focus on just the environmental, or additionally the social and/or economic, dimensions of sustainability; strong *versus* weak sustainability; relative *versus* absolute sustainability; consideration of humans as part of, or separate from, the natural environment; and assessment of the positive or negative aspects of sustainability. These characteristics are discussed below, with particular reference to the circular economy and LCA.

15.3.1 Environmental, Social and Economic Dimensions of Sustainability

Sustainability approaches are often categorised as addressing one or more of the three dimensions of sustainability: the environmental, economic and social dimensions.[22] The assumption is that sustainable development cannot occur without consideration of all three dimensions in society. In some countries, such as New Zealand, a cultural dimension is additionally identified as separate and distinct from the social dimension.[23]

Studies on the circular economy to date have been largely focused on the environmental and economic dimensions and Millar *et al.*[10] note that these studies have only made unsubstantiated suggestions, at best, about the social benefits of the circular economy. Regarding the economic dimension, on the one hand, some ambitious claims have been made about the potential of the circular economy to deliver economic benefits to society; for example, the Ellen MacArthur Foundation[6] estimated a primary resource benefit of up to €0.6 trillion per year by 2030 to Europe's economies, and an additional €1.2 trillion in non-resource and externality benefits. At the same time, researchers have suggested that closing the loop for materials will struggle to be economically successful because:

1. People prefer the convenience of buying new products and the higher perceived value of new products;[24,25]
2. An expanding world population and increasingly affluent lifestyles mean that new materials will always be in demand;[12,15]
3. There is a physical limit to the fraction of a material that can be recycled due to inevitable contamination.[26]

Raworth[19] goes further, and makes the argument that it will be impossible for us to move to more sustainable societies unless we fundamentally rethink how our economic systems operate; for example, current economic systems do not (and may be unable to) account for the environmental and social externalities of current production systems, and are dependent upon continuous economic growth with its associated material consumption. Thus, the economic viability of closed-loop systems for materials that is asserted by some circular economy proponents remains a contested assertion.

LCA, on the other hand, is solely focused on the environmental dimension of sustainability. The endpoints in its analysis are identified in the ISO standards as ecosystem health, human health and resource depletion[27] (ISO, 2006, Section 3.36). However, there is ongoing interest and research into expanding its remit to address economic and social aspects, and hence undertaking Life Cycle Sustainability Assessment (LCSA).

For both the circular economy and LCA, then, the main goal is concerned with environmental sustainability. Other analytical tools exist to assess the economic and social dimensions of sustainability, and can be used alongside these approaches. For example, Life Cycle Costing (LCC) can be, and is used alongside LCA, to assess the economic impacts of a product system.

15.3.2 Strong and Weak Conceptions of Sustainability

A fundamental debate regarding sustainable development concerns whether we should subscribe to a strong or a weak conception of sustainability. This distinction is illustrated in Figure 15.1. According to a strong sustainability perspective (Figure 15.1a), society is recognised as operating within the constraints imposed by the environment, and the economy is recognised as operating within the constraints imposed by society. In practical terms, this means that society sets rules and regulations, so that human activities operate in such a way as to conserve ecosystems and natural resources, and protect biogeochemical cycles, and that economic activities contribute to the flourishing of society.

The alternative weak sustainability perspective is represented in Figure 15.1b. When utilising a weak sustainability perspective, it is recognised that there are trade-offs between the different dimensions of sustainability. Effectively, in any decision situation, it may be preferable to accept some environmental degradation or social impacts in order to achieve better economic outcomes.

This distinction gained attention at the beginning of the 1990s through the work of Daly,[28] and Costanza and Daly,[29] who discussed it in terms of the substitutability (or not) of natural, economic and social capital. Today, it can hardly be maintained that a strong sustainability perspective is followed widely, given that almost half the world's economic wealth is owned by 1% of the global population,[30] and that we are in the midst of the sixth mass extinction of the world's species.[31]

Environmental sustainability approaches may or may not be aligned with a strong or weak sustainability perspective. Proponents of the circular economy concept, such as the Ellen MacArthur Foundation, are agnostic with respect to these perspectives because it is (generally) assumed that closed loops for materials can be achieved at the same time as growing the economy, and so no trade-offs are required between environmental and economic outcomes. Indeed, they stated that, "The goal is … to generate cyclical, cradle-to-cradle 'metabolisms' that enable materials to maintain their status as resources and accumulate intelligence over time (upcycling). This inherently generates a synergistic relationship between ecological and

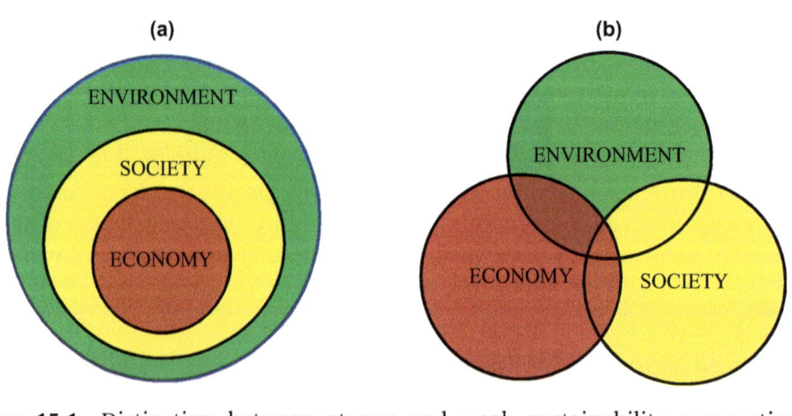

Figure 15.1 Distinction between strong and weak sustainability perspectives.[56] (a) Strong Sustainability Perspective. (b) Weak Sustainability Perspective.

economic systems, a positive recoupling of the relationship between economy and ecology."[32] McDonough and Braungart state, "The key is not to make human industries and systems smaller ... but to design them to get bigger and better in a way that replenishes, restores, and nourishes the rest of the world."[3]

LCA is concerned only with assessment of environmental impacts and so trade-offs with economic and social outcomes are not considered. However, the conversion of environmental impacts into financial values in some LCA Impact Assessment methods (using techniques such as contingent valuation) can be viewed as implementation of a weak sustainability perspective, since representing these diverse impacts using a common unit of measurement (dollars, yen, *etc.*) promotes the perception (either intentionally or unintentionally) that they can be traded off against economic costs and benefits.[33]

Given the critical state of many of the Earth's natural ecosystems, it may be argued that more attention should be given to aligning environmental sustainability approaches with a strong sustainability perspective. Indeed, there has been increasing interest in development of new types of Earth system governance that prioritise actions to conserve the Earth's ecosystems.[34,35,54,55] Aligned with this global sustainability perspective, environmental assessment approaches such as the planetary boundaries have emerged in the last ten years. The planetary boundaries approach measures the extent to which human activities are aligned with keeping the Earth system in the stable state that has characterised the Holocene epoch.[36,37] Essentially, the planetary boundaries approach is concerned with setting carrying capacities for different critical processes such as climate change, biogeochemical flows of nitrogen and phosphorus, ocean acidification and freshwater use. The use of the term "carrying capacity" in this context is somewhat different from its historical use in ecological studies, where it refers specifically to the maximum number of individuals of a species that can be supported in an ecosystem. In this newer context, it can be defined as "the maximum sustained environmental intervention a natural system can withstand without experiencing negative changes in structure or functioning that are difficult or impossible to revert".[38] The Ellen MacArthur Foundation[39] explains its mission as being about "creating a new system that can meet our needs within planetary boundaries" and realising "an economy that is regenerative and restorative by design". And LCA researchers have begun to explore use of planetary boundaries in Impact Assessment approaches,[40–42] thus aligning LCA with an absolute sustainability perspective since the Impact Assessment results of LCA studies are interpreted relative to their allocated shares of the "safe operating space" for human activities.

15.3.3 Relative and Absolute Perspectives in Environmental Sustainability

Most environmental sustainability approaches can be categorised as *relative* rather than *absolute*. This means that they either implicitly or explicitly rank a particular system in relation to a reference system.[43,44] For example, an LCA study will typically address questions such as: "Is product system A better than product system B?" or "Which activity in the examined life cycle is responsible for the most of the environmental impacts?" In contrast, an absolute environmental assessment approach is concerned with the performance of a system relative to the overall carrying capacity of a geographical area which may be defined on a scale ranging from a local ecosystem through to the entire Earth.[38] The distinction between relative and absolute approaches has become more important as we approach, and in some cases exceed, the planetary boundaries for critical Earth system processes such as climate change and stratospheric ozone depletion, and local ecosystem boundaries for biogeochemical flows of nitrogen and phosphorus, genetic diversity of species, and freshwater use.[36,37] It means that a focus on just improving the environmental profile of an economic system (be it a product, industry sector or country economy) is not sufficient; instead, we must assess whether these economic systems are compatible with a sustainable future in an absolute sense.

The circular economy concept is based on an absolute sustainability perspective, to the extent that it recognises the finite nature of natural resources, and promulgates an overarching solution that involves retaining these resources within the economy through reuse, refurbishment, recycling, *etc.* (*i.e.* closing the loop for these resources). Both McDonough and Braungart[3] and the Ellen MacArthur Foundation[6] make a distinction between biological and technical nutrients when considering this closed-loop management of resources in the circular economy. McDonough and Braungart[3] define biological nutrients as materials or products that can be broken down into simpler organic compounds through natural degradation processes mediated by micro-organisms, and technical nutrients as materials or products that are used in industrial systems and which can (at least potentially) be recycled within industrial systems.

Wood, paper and food are examples of biological nutrients; steel, plastic and glass are examples of technical nutrients. After use in the economy, biological nutrients can be returned to the biosphere, where they will safely degrade, and their constituent chemicals become available again for future use, whilst the aim with technical nutrients is to retain them within the technosphere through reuse, refurbishment, recycling, *etc.* However, in the context of environmental sustainability, such an approach must be shown to reduce net environmental impacts compared with linear systems of production and consumption. Critics have suggested that there may be trade-offs against: (1) the environmental impacts associated with the additional energy required to recycle biological and technical nutrients (*e.g.* additional materials for new renewable energy infrastructure, and land conversion for cultivating more biofuels); (2) the potential environmental impacts associated with releasing biological nutrients into naturally nutrient-poor ecosystems; and (3) allowing, and even advocating for, continued economic growth within a circular economy.[45,46] Thus, the question of whether a circular economy approach, as envisaged by McDonough and Braungart[3] and the Ellen MacArthur Foundation,[6] can deliver absolute sustainability remains open due to potential trade-offs with these other environmental impacts.

LCA, on the other hand, is recognised as a relative environmental assessment approach. As such, most LCA studies are concerned with comparing alternative systems delivering the same service (defined through a "functional unit") or with identifying environmental hotspots in the life cycle of a studied system (*i.e.* processes in the life cycle with the highest environmental impacts relative to other processes in that life cycle). It is thus concerned with evaluating the relative eco-efficiency of alternatives, but provides no insights into whether preferred alternatives do actually contribute to absolute sustainability. However, as noted in Section 15.3.2, researchers are currently developing absolute sustainability-based impact assessment methods based on the planetary boundaries.

15.3.4 Humans as Part of, or Separate from, the Environment

A fundamental characteristic concern whether human activities are regarded as part of, or separate from, the natural environment (aka the "ecosphere"). People's views on this topic are shaped by social values and religious beliefs, and it is also related to the well-rehearsed debate in environmental management about whether the environment should be protected for the benefit of humans (*i.e.* it has instrumental value for humans) or for its own sake (invoking the intrinsic value of the environment).[47,48] A third perspective is that the instrumental *versus* intrinsic values dichotomy is too narrow, and that relational values are a more promising basis for environmental management. According to this perspective, people are in a relationship with nature, and identifying both individual and collective notions of a "good life" with respect to this ongoing relationship are of primary importance.[47]

In the circular economy concept, both McDonough and Braungart,[3] and the Ellen MacArthur Foundation[6] address the question of whether human activities are part of, or separate from, the natural environment by distinguishing between two separate management systems for biological and technical nutrients (as discussed in Section 15.3.3). For biological nutrients, it is assumed that human activities contribute to a wider "natural" recycling system for these nutrients, and that using more of these nutrients and then returning them to the environment is beneficial. For technological nutrients, these are regarded as separate from the ecosphere and the objective is to keep them within the technosphere and, where possible, shift over to use of more renewable resources (aka biological nutrients).[6] The circular economy thus communicates a message of humans coexisting with ecosystems in regenerative systems that benefit both humans and other species. As Braungart *et al.*[49] state, "cradle-to-cradle design provides a practical design framework for creating products and industrial systems in a positive relationship with ecological health and abundance, and long-term economic growth".

LCA methodology, on the other hand, is built upon a modelling approach where human activity takes place within a "technosphere" that is separate and distinct from the ecosphere, and the environmental impacts of flows of materials and energy crossing the boundary from the technosphere into the ecosphere are assessed at Impact Assessment. For example, waste that is landfilled remains part of the technosphere because the landfill is an object that has been created, and is actively managed, by humans; however, emissions from the landfill, such as methane and metals in leachate, are assessed for their environmental impacts when they are emitted from the landfill into the ecosphere. In LCA terminology, these emissions cross the "system boundary" into the environment, where the system boundary serves to delineate the activities constituting the system under analysis (located in the technosphere) from the ecosphere. This distinction does become somewhat blurred at times. For example, consider agricultural crops: should rainwater transpired through the plants be assessed as part of the water scarcity impact category at Impact Assessment? Is it meaningful to assess land occupation relative to a notional natural vegetation state even if that state is desert or under the sea (for reclaimed land)? Also, a logical interpretation of this modelling approach is that the preferred state of the ecosphere is one without human interference and therefore a world without *Homo sapiens.* Yet, at the same time, human health impacts are assessed at Impact Assessment in LCA, and human health is one of the three Areas of Protection (alongside ecosystem quality and natural resources) recognised as being the endpoints of concern in environmental assessment using LCA.[50]

15.3.5 Assessment of Positive or Negative Aspects of Sustainability

A common criticism of environmentalists and the environmental movement more generally, is its focus on negative messages. The dominant discourse is around the negative environmental impacts of human activities, and narratives tend to focus on subjects such as using fewer products, reducing energy consumption, eating less meat, and so on. The same is true for environmental assessment approaches which generally assess the negative environmental impacts of human activities.

It is not surprising, then, that the circular economy, with its vision of a regenerative and restorative economy is immediately appealing. As McDonough and Braungart explain, "The key is not to make human industries and systems smaller, as efficiency advocates propound, but to design them to get bigger and better in a way that replenishes, restores and nourishes the rest of the world."[3]

LCA methodology does actually allow for representation of some positive aspects which, perversely, are usually quantified as negative values! For example, carbon sequestration is assessed as contributing to mitigating climate change at Impact Assessment in LCA; a change in land use may be measured as positive if it involves restoration of natural ecosystems; and release of treated water from an industrial facility with a higher quality than the incoming water may be regarded as contributing to restoration of ecosystem quality. However, the idea of focusing more generally on positive rather than negative aspects is at an early stage of development in the LCA research community, and is being progressed through elaboration of approaches for environmental handprinting.[51–53]

15.4 Implications for Use of the Circular Economy and LCA

The state of the environment both now and in future projections is a subject of concern. Therefore, it is important to ensure that the most appropriate concepts and tools are used to support decision-making. In particular, it is increasingly clear that utilisation of a weak sustainability perspective is threatening the continued existence of the Earth system in a state that is hospitable to the human species. However, there is a dearth of environmental sustainability approaches that are aligned with a strong sustainability perspective. This presents an opportunity for those interested in implementing the circular economy, and LCA researchers: can these approaches be adapted and/or extended to more explicitly make the links between economic activities and the limits to economic growth with respect to ecological carrying capacities? There is promising work already underway in the LCA research community, and there is potential to integrate ecological carrying capacities into development of both conceptual and practical models of the circular economy.

The circular economy concept has achieved widespread attention, at least partly because it provides a positive message about economic growth being potentially compatible with environmental sustainability. Thus, it speaks to two of the three dimensions of sustainability, and its positive message is naturally attractive to audiences. However, the assertion that a circular economy can deliver economic as well as environmental benefits is a contested assertion, and further research is needed to show whether closed-loop systems for materials can actually deliver economic benefits within current economic frameworks. If not, then what new business models and economic systems are capable of supporting these closed-loop systems? Research on product-service systems (PSSs), reverse logistics, the sharing economy, and other suggested business models is ongoing in order to identify an "upgraded economic governance system" for facilitating a widespread transition to a circular economy.[6] Others (such as Raworth[19]) argue that a much deeper re-evaluation of our economic systems is needed if we are to develop appropriate mechanisms support this transition.

Furthermore, in order to validate the circular economy concept in the context of environmental sustainability, more case studies are required to investigate the environmental trade-offs associated with development of closed-loop systems for materials. Is it possible to continue to expand the economy without consuming more natural resources, supported by increasing supplies of renewably-sourced energy? If not, then what are the acceptable rates of consumption of natural resources and are there limits to the amount of renewable energy that can be provided to drive this circular economy? Addressing these questions requires the use of environmental assessment tools, and LCA is an appropriate candidate for such studies, with its focus on the cradle-to-grave analysis of environmental impacts associated with economic systems.

LCA, used by itself to support decision-making, conventionally provides information about whether System A is better or worse than System B, and provides no information about the absolute sustainability of either of these systems. Indeed, it has been criticised for its ability to support the implementation of "highly optimised wrong systems".[45,46] This suggests that there is merit in considering the complementary use of these two approaches: the circular economy because it "draws a picture" of future sustainable economic systems, and LCA because it provides information on whether these future economic systems are more or less sustainable than each other and the present situation. For both approaches though, the idea of evaluating environmental performance relative to carrying capacities for both resources and emissions of pollutants (as developed in the planetary boundaries approach) seems a promising way forward in evaluating the absolute sustainability of alternative future economic systems.

Finally, a framing of environmental assessment approaches that sets humans as either an integral part of, or apart from, the environment is significant. It is outside the scope of this chapter to discuss this characteristic of environmental sustainability approaches in detail. However, as Lakoff notes, "We are an inseparable part of Nature. Yet we separate self from other, and conceptualize Nature as other. This separation is so deep in our conceptual system that we cannot simply wipe it from our brains. It is a terribly false frame that will not go away."[54] In the circular economy concept as promulgated by McDonough and Braungart,[3] and the Ellen MacArthur Foundation,[6] the idea of humans being separate from the environment has been intentionally upended, and human activities are seen as part of a flourishing biosphere that is in balance with human activities. Furthermore, the bigger the economy, the more the positive effects on the biosphere: "The key is not to make human industries and systems smaller... but to design them to get bigger and better in a way that replenishes, restores, and nourishes the rest of the world."[3] This is appealing and stands in contrast to LCA, where the focus is predominantly upon the negative environmental impacts of human activities.

Baumann and Cowell[8] suggest that five basic relationships can be defined between different environmental management approaches:

- Consecutive: where the outputs from one approach become the input data for another approach (*e.g.* use of toxicity assessment results in LCA).
- Complementary: where different approaches use the same basis for comparison but given different results because they investigate different dimensions (*e.g.* use of Life Cycle Costing and LCA).
- Encompassing: where one or more approaches are an integral part of another approach (*e.g.* Material Flow Analysis is integral to the practice of Industrial Ecology).
- Overlapping: where approaches are essentially the same (*e.g.* carbon footprinting is equivalent to assessment of climate change in LCA).
- Competing/incompatible: where approaches have the same unit of analysis and investigate the same issue but give different results because their scopes are different (*e.g.* Environmental Impact Assessments usually focuses on direct impacts related to an activity ("project") in a specific geographical location whilst LCA assesses both direct and indirect impacts arising from a series of activities usually occurring in different locations but linked *via* a supply chain).

The ideal relationship between the circular economy and LCA may be best described as an encompassing one. The circular economy provides us with a positive vision of how to proceed towards sustainable societies. However, its detailed elaboration in practice should be tested through LCA studies that provide insights about the environmental trade-offs that may exist in closing the loop for materials. In the end, as illustrated through the case studies in this book, this synergy may provide us with the insights upon which to re-invent our economic systems so that they work for both societies and the environment.

References

1. WCED (World Commission on Environment and Development), *Our Common Future*, Oxford University Press, Oxford, 1987.
2. M. Rein and D. Schön, in *The Argumentative Turn in Policy Analysis and Planning*, ed. F. Fischer and J. Forester, Duke University Press, Durham, 1993, p. 145.
3. W. McDonough and M. Braungart, *Cradle to Cradle—Remaking the Way We Make Things*, North Point Press, New York, 2002.
4. W. McDonough and M. Braungart, *The Upcycle: Beyond Sustainability-Designing for Abundance*, North Point Press, New York, 2013.
5. EMAF (Ellen MacArthur Foundation), *Towards the Circular Economy. An Economic and Business Rationale for An Accelerated Transition*, Ellen MacArthur Foundation, Cowes, 2013.

6. EMAF (Ellen MacArthur Foundation), Sun and McKinsey Centre for Business and Environment, *Growth within: A circular economy vision for a competitive Europe*, Available at: https://www.ellenmacarthurfoundation.org/assets/downloads/publications/EllenMacArthurFoundation_Growth-Within_July15.pdf, (Accessed June 2019).
7. S. Hellweg and L. M. I. Canals, *Science*, 2014, **344**(6188), 1109.
8. H. Baumann and S. J. Cowell, *Greener Management International*, 1999, **Summer 1999**, 109.
9. Y. Kalmykova, M. Sadagopan and L. Rosado, *Resour., Conserv. Recycl.*, 2018, **135**, 190.
10. N. Millar, E. McLaughlin and T. Börger, *Ecol. Econ.*, 2019, **158, 11.**
11. J. Korhonen, C. Nuur, A. Feldmann and S. E. Birkie, *J. Cleaner Prod.*, 2018, **175**, 544.
12. T. Zink and R. Geyer, *J. Ind. Ecol.*, 2017, **21**(3), 593.
13. N. Georgescu-Roegen, *The Entropy Law and the Economic Problem*, Harvard University Press, Harvard, 1971.
14. J. Korhonen, A. Honkasalo and J. Seppala, *Ecol. Econ.*, 2018, **143**, 37.
15. S. Pauliuk, *Resour., Conserv. Recycl.*, 2018, **129**, 81.
16. J. M. Allwood, in *Handbook of Recycling: State-of-the-Art for Practitioners, Analysts, and Scientists*, ed. E. Worrell and M. A. Reuter, Elsevier, Waltham, vol. 30, 2014, p. 445.
17. R. Geyer, B. Kuczenski, T. Zink and A. Henderson, *J. Ind. Ecol.*, 2015, **20**(5), 1010.
18. T. Zink and R. Geyer, *J. Ind. Ecol.*, 2019, **23**(3), 541.
19. K. Raworth, *Doughnut Economics: Seven Ways to Think Like a 21st Century Economist*, Penguin Random House, London, 2017.
20. W. B. Gallie, *Proceedings of the Aristotelian society*, vol. 55, 1956, p. 167.
21. A. Parchomenko, D. Nelen, J. Gillabel and H. Rechberger, *J. Cleaner Prod.*, 2019, **210**, 200.
22. B. Purvis, Y. Mao and D. Robinson, *Sustainability Sci.*, 2019, **14**, 681.
23. S. Pizzirani, S. J. McLaren and J. K. Seadon, *Int. J. LCA*, 2014, **19**(6), 1316.
24. A. Tukker, *J. Cleaner Prod.*, 2015, **97**, 76.
25. P. Hopkinson, M. Zils, P. Hawkins and S. Roper, *Calif. Manag. Rev.*, 2018, **60**(3), 71.
26. M. K. Eriksen, A. Damgaard, A. Boldrin and T. F. Astrup, *J. Ind. Ecol.*, 2019, **23**(1), 156.
27. ISO, *ISO 14040, Environmental Management – Life Cycle Assessment – Principles and Framework*, International Organisation for Standardisation, Geneva, 2006.
28. H. E. Daly, *Ecological Economics: The Science and Management of Sustainability*, ed. R. Costanza, Columbia University Press, 1991.
29. R. Costanza and H. E. Daly, *Conserv. Biol.*, 1992, **6**, 37.
30. Global Wealth Report, Available at: https://www.credit-suisse.com/corporate/en/articles/news-and-expertise/global-wealth-report-2018-us-and-china-in-the-lead-201810.html, (Accessed July 2019).

31. G. Ceballos, P. R. Ehrlich and R. Diroz, *Proc. Natl. Acad. Sci. U. S. A.*, 2017, **30**, E6089.
32. EMAF (Ellen MacArthur Foundation), *Efficiency vs Effectiveness.* Available at https://www.ellenmacarthurfoundation.org/news/efficiency-vs-effectiveness, (Accessed July 2019).
33. M. Pizzol, B. Weidema, M. Brandão and P. Osset, *J. Cleaner Prod.*, 2015, **86**, 170.
34. F. Biermann, K. Abbott, S. Andresen, K. Bäckstrand, S. Bernstein, M. M. Betsill, H. Bulkeley, B. Cashore, J. Clapp and C. Folke, *Science*, 2012, **335**(6074), 1306.
35. F. Biermann, K. Abbott, S. Andresen, K. Bäckstrand, S. Bernstein, M. M. Betsill, H. Bulkeley, B. Cashore, J. Clapp, C. Folke, A. Gupta, J. Gupta, P. M. Haas, A. Jordan, N. Kanie, T. Kluvánková-Oravská, L. Lebel, D. Liverman, J. Meadowcroft, R. B. Mitchell, P. Newell, S. Oberthür, L. Olsson, P. Pattberg, R. Sánchez-Rodríguez, H. Schroeder, A. Underdal, S. Camargo Vieira, C. Vogel, O. R. Young, A. Brock and R. Zondervan, *Science*, 2012, **335**(6074), 1306.
36. J. Rockström, W. Steffen, K. Noone, Å. Persson, F. S. Chapin III, E. Lambin, T. Lenton, M. Scheffer, C. Folke, H. J. Schellnhuber, B. Nykvist, C. A. De Wit, T. Hughes, S. van der Leeuw, H. Rodhe, S. Sörlin, P. K. Snyder, R. Costanza, U. Svedin, M. Falkenmark, L. Karlberg, R. W. Corell, V. J. Fabry, J. Hansen, B. Walk, D. Liverman, K. Richardson, P. Crutzen and J. Foley, *Ecol. Soc.*, 2009, **14**(2), 32.
37. W. Steffen, K. Richardson, J. Rockstrom, S. E. Cornell, I. Fetzer, E. M. Bennett, R. Biggs, S. R. Carpenter, W. de Vries, C. A. de Wit, C. Folke, D. Gerten, J. Heinke, G. M. Mace, L. M. Persson, V. Ramanathan, B. Reyers and S. Sorlin, *Science*, 2015, **347**(6223), 1259855.
38. A. Bjørn and M. Z. Hauschild, *Int. J. Life Cycle Assess.*, 2015, **20**(7), 1005.
39. EMAF (Ellen MacArthur Foundation), *What is the Circular Economy?* Available at https://www.ellenmacarthurfoundation.org/circular-economy/what-is-the-circular-economy, (Accessed June 2019).
40. A. Bjørn, M. Margni, P.-O. Roy, C. Bulle and M. Z. Hauschild, *Ecol. Indic.*, 2016, **63**, 1.
41. C. Chandrakumar, S. J. McLaren, N. P. Jayamaha and T. Ramilan, *J. Ind. Ecol.*, 2019, **23**(4), 906.
42. M. W. Ryberg, M. Owsianiak, K. Richardson and M. Z. Hauschild, *Ecol. Indic.*, 2018, **88**, 250.
43. C. Chandrakumar and S. J. McLaren, *Ecol. Indic.*, 2018, **90**, 477.
44. M. Z. Hauschild, *Procedia CIRP*, 2015, **29**, 107.
45. A. Bjørn and M. Z. Hauschild, *J. Ind. Ecol.*, 2013, **17**(2), 321.
46. A. Bjørn and M. Z. Hauschild, *Life Cycle Assessment: Theory and Practice*, ed. M. Z. Hauschild, R. K. Rosenbaum and S. I. Olsen, Springer Nature, Switzerland, 2018, pp. 605–632.
47. K. M. A. Chan, P. Balvanera, K. Benessaiah, M. Chapman, S. Díaz, E. Gómez-Baggethun, R. Gould, N. Hannahs, K. Jax, S. Klain, G. W. Luck,

B. Martín-López, B. Muraca, B. Norton, K. Ott, U. Pascual, T. Satterfield, M. Tadaki, J. Taggart and N. Turner, *Proc. Natl. Acad. Sci. U. S. A.*, 2018, **113**, 1462.
48. C. Groves, *Sustainability Sci.*, 2019, **14**(4), 915–924.
49. M. Braungart, W. McDonough and A. Bollinger, *J. Cleaner Prod.*, 2007, **15**(13–14), 1337.
50. R. Frischknecht, P. Fantke, L. Tschümperlin, M. Niero, A. Antón, J. Bare, A. M. Boulay, F. Cherubini, M. Z. Hauschild, A. Henderson, A. Levasseur, T. E. McKone, O. Michelsen, L. M. Canals, S. Pfister, B. Ridoutt, R. K. Rosenbaum, F. Verones, B. Vigon and O. Jolliet, *Int. J. Life Cycle Assess.*, 2016, **21**, 429.
51. D. Goleman, *Handprints not footprints*, Time article dated 12th March 2012, http://content.time.com/time/subscriber/article/0,33009,2108015,00.html, (Accessed May 2019).
52. G. Norris, *Handprint-Based Net Positive Assessment*, Harvard TH Chan School of Public Health, Boston, Available at: http://hwpi.harvard.edu/files/chge/files/handprint-based_netpositive_assessment.pdf, (Accessed June 2019).
53. K. Grönman, T. Pajula, J. Sillman, M. Leino, S. Vatanen, H. Kasurinen, A. Soininen and R. Soukka, *J. Cleaner Prod.*, 2019, **206**, 1059.
54. G. Lakoff, *Environ. Commun.*, 2010, **4**(1), 70.
55. F. Biermann, X. Bai, N. Bondre, W. Broadgate, C. T. Arthur Chen, O. P. Dube, J. W. Erisman, M. Glaser, S. van der Hel, M. C. Lemos, S. Seitzinger and K. C. Seto, *Glob. Environ. Change*, 2016, **39**, 341.
56. B. Giddings, B. Hopwood and G. O'Brien, *Sustainable Dev.*, 2002, **10**, 187–196.

Subject Index